FDA IN THE TWENTY-FIRST CENTURY

FDA
in the
TWENTY-FIRST CENTURY

*The Challenges of Regulating Drugs
and New Technologies*

EDITED BY
Holly Fernandez Lynch
and I. Glenn Cohen

Columbia University Press
New York

Columbia University Press
Publishers Since 1893
New York   Chichester, West Sussex
cup.columbia.edu

Copyright © 2015 Columbia University Press
All rights reserved

Library of Congress Cataloging-in-Publication Data
FDA in the twenty-first century : the challenges of regulating drugs
and new technologies / Edited by Holly Fernandez Lynch
and I. Glenn Cohen.
  pages cm
Includes bibliographical references and index.
ISBN 978-0-231-17118-2 (cloth : alk. paper) —
ISBN 978-0-231-54007-0 (e-book)
 1. United States. Food and Drug Administration.
 2. Drugs–Law and legislation–United States.
  I. Lynch, Holly Fernandez, editor.
   II. Cohen, I. Glenn, editor.
  III. Title: Food and Drug Administration
   in the twenty-first century.
 KF3871.F34 2015
 363.19'20973–dc23
  2015001340

Columbia University Press books are printed on permanent
and durable acid-free paper.

This book is printed on paper with recycled content.
Printed in the United States of America

c 10 9 8 7 6 5 4 3 2 1

Cover design: Noah Arlow

References to websites (URLs) were accurate at the time of writing.
Neither the author nor Columbia University Press is responsible for URLs
that may have expired or changed since the manuscript was prepared.

To Philip Katz,
for nurturing a young associate's interest in FDA,
even if she didn't know anything about anything.
—HFL

To Auntie Choco and Lily,
redheads in Heaven whom I will never forget.
—IGC

Contents

Acknowledgments
xiii

Introduction
Holly Fernandez Lynch and I. Glenn Cohen
1

CHAPTER ONE
Historical Themes and Developments at
FDA Over the Past Fifty Years
Peter Barton Hutt
17

**PART ONE
FDA in a Changing World**

Introduction
Holly Fernandez Lynch
35

CHAPTER TWO
A Global and Innovative Regulatory Environment
for the U.S. FDA
Howard Sklamberg and Jennifer Devine
39

CHAPTER THREE
FDA and the Rise of the Empowered Patient
Lewis A. Grossman
59

CHAPTER FOUR
After the FDA: A Twentieth-Century Agency
in a Postmodern World
Theodore W. Ruger
76

CHAPTER FIVE
The Future of Prospective Medicine Under the Food
and Drug Administration Amendments Act of 2007
Barbara J. Evans
92

PART TWO
Preserving Public Trust
and Demanding Accountability

Introduction
Christopher Robertson
109

CHAPTER SIX
Global Trends Toward Transparency in
Participant-Level Clinical Trials Data
*Alla Digilova, Rebecca Li, Mark Barnes,
and Barbara Bierer*
115

CHAPTER SEVEN
Conflicts of Interest in FDA Advisory Committees:
The Paradox of Multiple Financial Ties
Genevieve Pham-Kanter
134

CHAPTER EIGHT
The Crime of Being in Charge: Executive Culpability
and Collateral Consequences
Katrice Bridges Copeland
146

CHAPTER NINE
Recalibrating Enforcement in the Biomedical Industry:
Deterrence and the Primacy of Protecting the Public Health
Patrick O'Leary
162

PART THREE
Protecting the Public Within Constitutional Limits

Introduction
I. Glenn Cohen
179

CHAPTER TEN
Prospects for Regulation of Off-Label Drug Promotion
in an Era of Expanding Commercial Speech Protection
Aaron S. Kesselheim and Michelle M. Mello
184

CHAPTER ELEVEN
The FDCA as the Test for Truth
of Promotional Claims
Christopher Robertson
204

CHAPTER TWELVE
Why FDA's Ban on Off-Label Promotion Violates
the First Amendment: A Study in the Values of
Commercial Speech Protection
Coleen Klasmeier and Martin H. Redish
219

PART FOUR
**Timing Is Everything: Balancing Access
and Uncertainty**

Introduction
W. Nicholson Price II
247

CHAPTER THIRTEEN
Speed Versus Safety in Drug Development
R. Alta Charo
251

CHAPTER FOURTEEN
Overcoming "Premarket Syndrome":
Promoting Better Postmarket Surveillance
in an Evolving Drug-Development Context
Shannon Gibson and Trudo Lemmens
268

CHAPTER FIFTEEN
FDA's Public Health Imperative: An Increased Role
for Active Postmarket Analysis
Efthimios Parasidis
286

PART FIVE
Old and New Issues in Drug Regulation

Introduction
R. Alta Charo
303

CHAPTER SIXTEEN
The Drug Efficacy Study and Its Manifold Legacies
Daniel Carpenter, Jeremy Greene, and Susan Moffitt
306

CHAPTER SEVENTEEN
Drug Safety Communication: The Evolving Environment
Geoffrey Levitt
328

CHAPTER EIGHTEEN
Innovation Policy Failures in the Manufacturing of Drugs
W. Nicholson Price II
343

PART SIX
Regulatory Exclusivities and the Regulation of Generic Drugs and Biosimilars

Introduction
Benjamin N. Roin
361

CHAPTER NINETEEN
From "Recycled Molecule" to Orphan Drug: Lessons from Makena
Kate Greenwood
366

CHAPTER TWENTY
FDA, Negotiated Rulemaking, and Generics: A Proposal
Marie Boyd
382

CHAPTER TWENTY-ONE
The "Follow-On" Challenge: Statutory Exclusivities and Patent Dances
Arti Rai
402

CHAPTER TWENTY-TWO
FDA Regulation of Biosimilars
Henry Grabowski and Erika Lietzan
414

PART SEVEN
New Wine in Old Bottles: FDA's Role in Regulating New Technologies

Introduction
Frances H. Miller
433

CHAPTER TWENTY-THREE
Analog Agency in a Digital World
Nathan Cortez
438

CHAPTER TWENTY-FOUR
Twenty-First-Century Technology with Twentieth-Century Baggage:
FDA Regulation of Regenerative Medicine
Margaret Foster Riley
455

CHAPTER TWENTY-FIVE
Device-ive Maneuvers: FDA's Risk Assessment
of Bifurcated Direct-to-Consumer Genetic Testing
Elizabeth R. Pike and Kayte Spector-Bagdady
470

CHAPTER TWENTY-SIX
A New Regulatory Function for E-Prescriptions:
Linking FDA to Physicians and Patient Records
Andrew English, David Rosenberg, and Huaou Yan
486

CHAPTER TWENTY-SEVEN
Race and the FDA
Jonathan Kahn
501

Contributors
517

Index
519

Acknowledgments

A BOOK like this could not have happened without the support and assistance of a number of people. We thank Cristine Hutchison-Jones for helping us mount the original conference that gave rise to this volume, as well as Nicholson Price and Jeffrey Skopek for their assistance with the conference. Louis Fisher and Thomas Blackmon helped us line edit and format the entire book, acting really as "deputy editors." We are always grateful to Harvard Law School and the Petrie-Flom Center for Health Law Policy, Biotechnology, and Bioethics for their support of our activities. I. Glenn Cohen also thanks the Greenwall Foundation Faculty Scholar in Bioethics program for its financial and intellectual support during the gestation of this book. And last, but certainly not least, we thank all of our contributors for their tremendous efforts.

FDA IN THE TWENTY-FIRST CENTURY

Introduction

HOLLY FERNANDEZ LYNCH
AND I. GLENN COHEN

Don't bother about being modern.
Unfortunately it is the one thing that,
whatever you do, you cannot avoid.
—SALVADOR DALÍ

HOW SHOULD FDA regulate stem-cell therapies? Genetic testing? How can it discourage off-label promotion without violating the First Amendment? How should the interface between patent law and market exclusivity differ for biologics rather than small molecules? These are but a smattering of the new and enduring challenges facing the U.S. Food and Drug Administration (FDA) as the agency endeavors to successfully regulate drugs and new technologies in today's world, and a quick sample of the issues tackled in this book. The book's contributing authors grapple with a wide range of challenges, from globalization to corporate integrity to personalized medicine to mobile health and beyond. Often they offer competing visions of what success for the agency would look like and the best ways to achieve it.

The basic metrics of success are clear from FDA's authorizing statute, which focuses on ensuring that drugs and devices are both safe and effective. However, a truly successful agency must balance getting promising products to market quickly against rigorously protecting patients across a product's entire life cycle, as well as encouraging innovation, ensuring patients have adequate information to make their own decisions, and remaining nimble enough to confront a world that could not even be imagined at the time the agency first ventured down the path of drug regulation.

To understand the magnitude of the challenge before today's FDA, consider what the world looked like in 1906, the year that Congress passed the original Federal Food and Drugs Act. The Wright Brothers had just invented the first successful airplane, and the globalization of medical technologies was so far off as to be even beyond a pipe dream, although interstate commerce posed many important challenges on its own. Snake oil flourished, and the most fundamental problem was simply making sure that drugs were neither misbranded nor adulterated.

By 1938, when the Federal Food, Drug, and Cosmetic Act was passed requiring that new drug products be demonstrated safe before marketing, the world had already changed dramatically. The United States had been through one world war and stood on the precipice of its second, and the country was reeling from almost a decade of Great Depression. A range of medical advances, particularly the development of new vaccines and contraceptive drugs, had occurred, but we were still fifteen years from Watson and Crick's description of the structure of the DNA molecule. Patient autonomy was nonexistent, and the concept of patient advocates likely alien. Acute diseases were beginning to be replaced with chronic diseases, and Blue Cross–Blue Shield private, voluntary health insurance plans had just been born. Perhaps most importantly, the idea of blinded, randomized, controlled clinical trials did not really take hold until World War II.

When the next major development in FDA legislation occurred with the Kefauver-Harris Drug Amendments in 1962, requiring drug manufacturers to prove effectiveness before marketing, we were still more than a decade away from the launch of Microsoft and Apple, and even further away from the popular uptake of personal computers. The HIV epidemic, which in many ways shaped the modern FDA, was two decades away. When the Hatch-Waxman Act paved the way for generic drugs in 1984, there was still no World Wide Web, no WebMD, and no social media, although patient activism and advocacy had begun to take root. Even by the 1997 Food and Drug Administration Modernization Act, we were still six years from publication of the complete human genome, seven years from Facebook, and ten years from the iPhone, all massive recent changes that have fundamentally reshaped the landscape for how drugs are developed, marketed, and used—enter personalized medicine; patient-to-patient dialogue and a plethora of inexpert information available about drugs and their benefits, uses, and side effects; viral campaigns for compassionate use; and more.

Today's FDA is, of course, still concerned with ensuring that our drug supply is not adulterated or misbranded (among its many other portfolios we do not touch on in this book, including food regulation), but its mission is now far more complex, as is the world in which it regulates. We now inhabit a world in which genetic information is available with increasing ease, globalization is a fact of life, and clinical trials have become an industry unto themselves. Mountains of health information are available through electronic health records and other data sources, and patients are active partners in the health care endeavor, with reams of resources available at the swipe of a finger over their smartphones.

Over the past eleven decades, FDA has done a tremendous job of keeping up with the times. Indeed, in a 2010 survey conducted by PricewaterhouseCoopers, 93 percent of respondents were confident about the safety and effectiveness of drugs and devices approved for use in the United States, and two-thirds agreed that the United States has the highest standards in the world for drug safety and effectiveness (PharmaList 2010). FDA and Congress have consistently worked to ensure that the agency does not stagnate, with progressive movement to regulate biological products, avoid drug abuse, offer information directly to patients, respond to the HIV epidemic, encourage the development of data and products for pediatric and orphan populations, promote clinical trial transparency, prepare for bioterrorism, encourage both innovation and competition, and generally offer flexible paths to approval. On the other hand, the same PricewaterhouseCoopers survey found that only 8 percent of drug and device makers believe that FDA is doing enough to advance personalized medicine (PharmaList 2010). And more than half of American consumers would be willing to use drugs and devices approved outside the United States before they were approved by FDA (PharmaList 2010), suggesting that the agency's standards may be—or at least are perceived to be—overprotective. In many ways, FDA is (understandably) reactionary and cautious, walking the tightrope between speed and safety, with its hands tied behind its back in terms of statutory authority and adequate resources, all while juggling the regulation not only of drugs but also of devices, foods, cosmetics, tobacco, animal products, and more. Can

FDA really do it all, and do it well, and how might it best manage the challenges it faces in today's world? Those are the questions at the heart of this book.

In 2013, we brought together a range of leading academics, government officials, practitioners, and representatives from industry at Harvard Law School for a conference to discuss how FDA is faring in the twenty-first century and, in particular, to assess its greatest challenges, most enduring lessons, and best approaches to modern drug regulation. Ultimately, we asked these experts to evaluate the agency and begin to chart a course for its future. The result of that discussion is memorialized in the twenty-seven chapters that follow, most of which were generated in the context of that 2013 meeting, while a few others were brought into the fray later as we identified important gaps and additional topics. This volume makes no attempt to be comprehensive, however, and there are certainly some gaps remaining. For example, we do not cover pricing, quality system challenges, new approaches to clinical trials, regulatory science, the interface between FDA and other regulatory agencies at the federal and state level, a host of relevant jurisdictional issues, and a variety of other critical topics. Exclusion is no suggestion of unimportance, but simply reflective of reality: there is only so much we can do with one book. The topics we have chosen to cover are themselves critical, and offer important—if selective—insight into FDA's regulation of drugs and new technologies. Even with this focus, however, several authors have used FDA's regulation of food to illustrate key concepts.

As a final editorial note, we wish to emphasize that each of these chapters represents a unique perspective, and they sometimes conflict with one another. As editors, we thought it essential to retain that conflict in order to provide a full sense of the range of open issues and diversity of opinion as to how FDA ought to proceed—and diversity of opinion abounds, on everything from whether FDA remains a relevant agency to its approach to corporate bad actors and drug marketing to how best to handle biosimilars and innovative technologies.

Following this introduction, the book begins with a chapter by Peter Barton Hutt, a founding father of modern FDA law without whom no volume like ours could possibly be complete. Hutt's chapter, "Historical Themes and Developments at FDA Over the Past Fifty Years," provides an overview of FDA's history as a regulatory institution that now touches products comprising twenty-five cents out of every dollar spent by U.S. consumers, outlining agency management, rulemaking, use of guidance documents, enforcement, the birth and impact of user fees, and more.

Against this history, the book moves into the present day and the multitude of new challenges facing the agency. The chapters in part 1, "FDA in a Changing World," provide a high-level review of major developments in the background context against which FDA must now regulate, which have put the agency on the precipice of sweeping change.

The first chapter in this part, "A Global and Innovative Regulatory Environment for the U.S. FDA," offers an insider's view from within FDA. Authored by Howard Sklamberg, FDA deputy commissioner for global regulatory operations and policy, and Jennifer Devine, then-FDA associate commissioner for global regulatory operations and policy, the chapter focuses on the way in which the supply chain for therapeutics has become globalized, thrusting FDA into a new world of regulatory challenges. As the authors put it: "Decades ago, it is likely the drugs in your medicine cabinet would have come from a domestic source, but now, the reality is that they could just as easily have been made abroad or contain ingredients or components sourced from abroad." After describing this phenomenon, the authors turn to how FDA has tried to manage these developments, including through congressional changes to agency authority, partnerships with other national regulators, and additional steps. They also critically assess what further changes are needed.

Lewis Grossman's chapter, "FDA and the Rise of the Empowered Patient," examines a different development that has required rethinking FDA's role: the rise of consumer-directed health care and patient advocacy more generally. After contextualizing this movement in the historical episodes that gave rise to it (e.g., the clash with supporters of the experimental cancer treatment laetrile in the 1970s and AIDS activism of the late 1980s), Grossman explains how the Food and Drug Administration Safety and Innovation Act of 2012 (FDASIA) represents a further advance in the rise of a patient-centered perspective in the FDA regulatory process. He then examines the challenges and opportunities this may pose going forward.

Next, in "After the FDA: A Twentieth-Century Agency in a Postmodern World," Theodore Ruger argues that FDA's historical success in narrow regulatory categories is no longer sufficient—the agency must readjust its regulatory priorities. As he puts it, "the deeper threat to FDA's role in the regulatory firmament of the twenty-first century is not that it is inadequately performing its core functions in assessing the safety of food and therapeutic products, but rather that this focus on product 'safety'—at least in the relatively acontextual manner FDA has tended to assess that variable—is increasingly unimportant in a new world of focus on product cost, consumer behavior, and consumption patterns." Like the other authors in this part, Ruger looks both back—to the successes of what he calls the "FDA century"—and forward—to how FDA could use its existing authority to address the major public health issues of the twenty-first century.

Barbara Evans's chapter, "The Future of Prospective Medicine Under the Food and Drug Administration Amendments Act of 2007," rounds out this part, taking a look at a different challenge and possible future for FDA: prospective medicine. Defining prospective medicine's key features to be "personalized, predictive, preventive, and participatory," including technologies such as testing gene variants or other biomarkers to identify future health, Evans shows that FDA's product-focused legal and institutional approach is not well suited for prospective interventions that act on healthy patients over years and decades to prevent disease. She then analyzes how the new powers given to FDA under the Food and Drug Administration Amendments Act of 2007 (FDAAA) might help the agency cope with these new product types and examines how willing FDA will be to make the necessary legal interpretations and implementing steps.

Part 2 of the book, "Preserving Public Trust and Demanding Accountability," highlights in particular FDA's role in encouraging transparency, as well as its enforcement approach when wrongdoing occurs. It begins with "Global Trends Toward Transparency in Participant-Level Clinical Trials Data" by Alla Digilova, Rebecca Li, Mark Barnes, and Barbara Bierer, all of the Multi-Regional Clinical Trials Center at Harvard University. This group documents how "the complementary effects of regulatory trends in the United States and Europe, academic journals' new data disclosure requirements, and voluntary efforts by the industry have led to an unprecedented growth of data-sharing mechanisms within the clinical research enterprise." The authors then discuss the advantages and risks of this increased transparency and go on to sketch their own version of a well-designed data-sharing model, which, they argue, has more benefits and fewer risks than some of the others currently being proposed or used.

Genevieve Pham-Kanter addresses another area where there has been more emphasis on transparency in the last several years: conflicts of interest. Her chapter, "Conflicts of Interest in FDA Advisory Committees: The Paradox of Multiple Financial Ties," documents the special rules that govern FDA advisory committee members and their ability to get waivers as to financial conflicts of interest that would otherwise be disqualifying. Pham-Kanter presents data from an empirical study examining the financial interests and voting behavior of all individuals who attended and voted in meetings of advisory committees convened by the FDA Center for Drug Evaluation and Research between 1997 and 2011. The study encompasses data on 1,379 unique people who cast at least one vote during the fifteen years. She uses this data set to compare those who had no financial ties to any study sponsor, a single exclusive financial tie to a study sponsor, and multiple nonexclusive ties to study sponsors to produce some intriguing findings.

The next two chapters, Katrice Bridges Copeland's "The Crime of Being in Charge: Executive Culpability and Collateral Consequences" and Patrick O'Leary's "Recalibrating Enforcement in the Biomedical Industry: Deterrence and the Primacy of Protecting the Public Health," focus on the intersection of white-collar criminal liability and FDA regulation. Copeland begins by detailing the history of FDA's strategy of cooperative enforcement with pharmaceutical companies to avoid the collateral consequences of criminal conviction. Traditionally, FDA has allowed pharmaceutical companies to avoid the "death sentence" of exclusion from federally funded health care programs by entering into "Corporate Integrity Agreements" requiring the companies to pay large fines and enact compliance measures designed to prevent illegal promotional activity from recurring. Copeland then details FDA's shift in policy in March 2010 to make use of individual misdemeanor prosecutions "to hold responsible corporate officials accountable." She argues that with this new policy FDA may have overreached, and, "absent a showing of moral blameworthiness on the part of health-care executives who were in charge at the time the misconduct occurred, a period of exclusion longer than the three-year base level for permissive exclusions is a grossly disproportionate remedy." On this point, O'Leary argues that "prosecutors' emphasis on ever-larger monetary settlements and Corporate Integrity Agreements (CIAs) has imposed significant costs on industry without achieving real deterrence, and threats of increased reliance on administrative and criminal sanctions against corporate officers at offending firms have raised practical and philosophical concerns." But O'Leary does not think FDA needs more enforcement powers to get the job done; instead, he argues for better coordination and alignment of enforcement policies at FDA and other agencies and refocusing enforcement decisions on "the objective of promoting and protecting the public health."

Part 3, "Protecting the Public Within Constitutional Limits," offers a debate, of sorts, on the interplay between off-label promotion and the First Amendment. On one side of the debate, Aaron Kesselheim and Michelle Mello's "Prospects for Regulation of Off-Label Drug Promotion in an Era of Expanding Commercial Speech Protection" begins by reviewing FDA's traditional attempts to limit the promotion of unapproved (off-label) uses for approved drugs. They then examine the Supreme Court's commercial speech jurisprudence. The two are tied together in the December 2012 decision by the U.S. Court of Appeals for the Second Circuit in *United States v. Caronia*, which held that the First Amendment to the U.S. Constitution prohibited the prosecution of a pharmaceutical sales representative caught on tape promoting a drug for conditions not approved by FDA. The authors then recommend several potential pathways that remain open to FDA in prohibiting off-label promotion notwithstanding the court's decision.

On the same side of the debate, Christopher Robertson's chapter, "The FDCA as the Test for Truth of Promotional Claims," argues that the First Amendment challenges identified in the previous chapter and upheld in *Caronia* have relevance even beyond FDA's authority to regulate promotional off-label speech. If taken seriously, he argues, these constitutional claims may undermine the FDCA's foundational requirement for premarket authorization of new drugs itself. This threat motivates Robertson to explore whether FDA regulation of promotion could be reconstructed in a constitutionally sufficient way. Among the ways he proposes doing so is to examine how courts and commentators seem to be presuming the truth about off-label promotional claims and presuming that FDA is motivated by a paternalistic desire to keep that truth out of the hands of consumers, both of which he thinks may be unwarranted presumptions.

The other side of the debate is set forth by Coleen Klasmeier and Martin H. Redish in "Why FDA's Ban on Off-Label Promotion Violates the First Amendment." Klasmeier and Redish argue that FDA's current approach—leaving off-label uses essentially unregulated but making a manufacturer's "promotion" of an off-label use categorically illegal—amounts to the agency "swimming half way across a river, thereby pleasing no one and avoiding none of [the] quite legitimate concerns." The authors seek to "establish the unambiguous unconstitutionality of FDA's current prohibition on off-label promotion as measured by controlling Supreme Court decisions concerning the First Amendment's protection of commercial speech." They then seek to buttress this conclusion by formulating "four basic postulates of American constitutional and political theory that, [they] assert, underlie the social contract between citizen and government implicit in a societal commitment to liberal democracy." These postulates, they claim, make "the ban on manufacturer promotion of off-label use . . . even more bizarre."

Part 4 of the book, "Timing Is Everything: Balancing Access and Uncertainty," examines FDA's various categories of premarket approval schemes and postmarket surveillance and also puts them in context of how other national and supranational regulators behave. In "Speed Versus Safety in Drug Development," R. Alta Charo sketches what an ideal pharmaceutical development and regulatory system would look like: it "delivers new drugs or new indications for old drugs in a timely manner, with an assurance of both safety and effectiveness"; "incentivizes innovation, particularly for those medical products that address conditions for which we need options that are more effective, less risky, and more affordable than existing offerings"; and "is capable of self-correction when errors are made" since it can be "more important to build in the capacity to detect and correct inevitable errors than to build a system so chock full of protections that error minimization comes with overwhelming rigidity and stifling regulation." Charo then examines how FDA historically has managed and currently manages these trade-offs between speed and safety, among other things contrasting FDA with the European Medicines Agency.

Shannon Gibson and Trudo Lemmens's chapter, "Overcoming 'Premarket Syndrome': Promoting Better Postmarket Surveillance in an Evolving Drug-Development Context," focuses on what they perceive as a regulatory fixation on premarket activities, which they dub "premarket syndrome." Among the negative consequences of this fixation, they argue, is the conduct of premarket trials under controlled conditions that generally do not reflect how a drug will be used in the real world, the failure to asses whether a drug is actually more effective than existing therapies, the short duration of trials that hide rare or longer-term side effects until after the drug is already widely prescribed, and the frequent prescription of drugs to patients and disease groups never assessed in clinical trials. They examine this "syndrome" in the wake of two more recent developments: the increased interest in developing drugs for niche markets and the recent trend of "drug regulatory systems in various jurisdictions [that] have proposed reforms that move away from the 'artificial dichotomy' of the pre- versus postmarket stages," what they instead refer to as a "life-cycle approach."

The implications of this life-cycle approach are also front and center in Efthimios Parasidis's chapter, "FDA's Public Health Imperative: An Increased Role for Active Postmarket Analysis." Parasidis argues that in those circumstances when it is within FDA's authority to do so, the agency should *require* that sponsors conduct active postmarket analysis for the life of their products. By postmarket analysis he means thorough, timely, and continuous monitoring of risks and benefits in real-world patients. Parasidis argues that this can be accomplished through a combination of health information technology, observational studies, biomedical informatics, and, where appropriate, postmarket clinical trials. While he envisions a role for FDA in "framing postmarket obligations and reviewing the results," he places primary responsibility on sponsors to conduct and complete these studies.

Part 5 of the book, "Old and New Issues in Drug Regulation," starts with a historical perspective and then moves on to consider evolving issues in drug-safety communication and the oft-overlooked area of drug manufacture. Daniel Carpenter, Jeremy Greene, and Susan Moffitt's chapter, "The Drug Efficacy Study and Its Manifold Legacies," examines FDA's Drug Efficacy Study Initiative (DESI) and what it can tell us about the future of regulation in this area. DESI occurred as a result of the Kefauver-Harris Amendments passed by Congress in 1962, which (in reaction to the thalidomide tragedy) converted the previous FDA preclearance requirement to an affirmative approval standard and added to the 1938 requirements of safety a criterion of effectiveness. What to do about the more than 4,000 already approved drugs? The National Academy of Sciences and the National Research Council were tasked with conducting an efficacy review of products on the market and convened a Policy Advisory Committee of twenty-seven members and twenty-one separate panels of therapeutically specific expertise. The authors tell this remarkable story, discuss the data on its successes and failures, and examine its implications for current FDA policy.

Next, Geoffrey Levitt, in "Drug Safety Communication: The Evolving Environment," argues that three overlapping trends have caused upheaval in the realm of communicating about the safety of approved drugs: (1) increasing doubts in some quarters about the ability and willingness of regulators and industry to communicate promptly and accurately about drug safety; (2) movements toward greater disclosure of certain kinds of health-related information, in particular the results of clinical research (discussed in greater depth earlier in this book by Digilova et al.), coupled with the ever-increasing ease of access to such information through electronic channels; and (3) the "open-source" or "open-science" movement, which is a broader trend toward increased transparency of scientific and technological information. Levitt sees these as converging to decenter regulators and drug companies from controllers of information and instead requiring them to "contend with a constant flow of safety analyses and communications from a wide variety of independent third parties, often putting them in a reactive or even defensive position." In this morass he offers a call for new "rules of the road" that will "ensure the quality and integrity of both drug safety information and of the manner in which it is communicated to the various stakeholders who rely on it."

Next, in "Innovation Policy Failures in the Manufacturing of Drugs," W. Nicholson Price II examines an area of drug innovation that has been sorely neglected by both academics and policy makers: innovation in manufacturing. He shows that improvements in manufacturing efficiency would not only save the industry between $15 and $90 billion annually but would also improve quality and reduce the number of drug shortages and recalls. He traces the lack of innovation in this space to "the current combination of regulatory barriers to manufacturing innovation and poorly aligned intellectual property incentives," and he considers several proposals to improve the situation.

Part 6, "Regulatory Exclusivities and the Regulation of Generic Drugs and Biosimilars," considers some of the ways in which FDA encourages both advancement and competition, as well as some of the pitfalls and implications of the current approach. Kate Greenwood's chapter, "From 'Recycled Molecule' to Orphan Drug: Lessons from Makena," examines the disputes that have arisen over the Orphan Drug Act's provision of market exclusivity to so-called recycled molecules—older drugs that may have already been available to patients in compounded or generic form before their designation as "orphan drugs" and approval for marketing as such. Greenwood gives in-depth examination of the controversy surrounding FDA's action for one such drug, Makena, a recycled molecule that FDA approved for orphan status in 2011 to treat pregnant women at high risk of giving birth prematurely. She argues against amending the Orphan Drug Act to modify the exclusivity period or to allow FDA to take a case-by-case approach to awards of exclusivity and instead argues for providing a formal mechanism through which "patients or others could challenge as inadequate a company's patient-assistance program or other efforts to ensure that individual patients who cannot afford a drug are nonetheless able to access it."

In "FDA, Negotiated Rulemaking, and Generics: A Proposal," Marie Boyd focuses on the implications of the U.S. Supreme Court's holding that state failure-to-warn claims against generic drug manufacturers are preempted. This outcome had the dual effects of eliminating the protections that state tort law can provide consumers of generic drugs through the law's compensation and information disclosure functions, as well as exposing "a gap in the federal regulation of generic drug labeling in which no manufacturer is responsible for updating the labeling of generic drugs if the corresponding brand-name drug is no longer marketed." FDA has, in response, published a notice of proposed rulemaking that would permit generic drug manufacturers to update their product labeling in certain circumstances. Boyd argues for a different approach, focused on negotiated rulemaking.

Arti Rai's chapter, "The 'Follow-On' Challenge: Statutory Exclusivities and Patent Dances," looks at a different aspect of the generic industry, the way in which the Biologics Price Competition and Innovation Act of 2010 enables or fails to enable what is essentially a generic biologics industry (often called "biosimilars"). Rai contrasts this Act's provisions with the more familiar Hatch-Waxman Act provisions for small molecules, focusing not only on their differing exclusivity periods but also on "the regimes' different mechanisms for addressing questions regarding patent validity and infringement." In the case of biologics, the Act provides what she characterizes as a complex "patent dance" procedure "through which originator and follow-on firms exchange information regarding patents and commercial marketing." Rai critically examines both of these policy decisions and charts some of the rough seas ahead.

Henry Grabowski and Erika Lietzan give their own take on biologics in their chapter, "FDA Regulation of Biosimilars." They summarize the experience of industry over the first four years of this regime and its near future, concluding that getting approval for biosimilars requires "a significantly greater investment of time and resources than are required of generic firms," which—when combined with the low likelihood that biosimilars will be rated as interchangeable with their reference products—will lead to "fewer entrants and a pattern of biosimilar competition that is more likely to resemble branded competition than generic competition for the foreseeable future." They express some hope that scientific advances could reduce the regulatory costs of "developing biosimilars and allow demonstration of interchangeability through analytical characterization (e.g., 'fingerprinting' structural analysis)." The authors also identify a series of "thorny" legal and regulatory issues that still need to be resolved in this space, "including permissible extrapolation of safety and effectiveness from one indication (demonstrated in clinical trials) to other indications (not tested), naming requirements, and the use of bridging studies to justify use of global marketing dossiers."

Finally, the last part of this book, "FDA's Role in Regulating New Technologies," highlights some cutting-edge issues and ideas that are testing the agency's limits. In "Analog Agency in a Digital World," Nathan Cortez takes the long and wide view on FDA's difficulties with emerging technologies, focusing on its regulation of computer software that has culminated recently in its struggles with mobile health devices (the use of iPhones and the like for health purposes). What he finds is the tale "of an old agency applying an old regulatory framework to very new technologies" and unwillingness or inability by Congress or FDA to update the framework. While recognizing that other agencies, such as the Office of the National Coordinator for Health Information Technology, the Federal Trade Commission, and the Federal Communications Commission, have a role to play, this chapter stands as a passionate call for the federal government to stop addressing these questions "under a very old statutory framework" and instead for "Congress to consider a twenty-first-century framework for software devices."

Striking the same key in a different chord, Margaret Foster Riley's "Twenty-First-Century Technology with Twentieth-Century Baggage: FDA Regulation of Regenerative Medicine" focuses on FDA's regulation of therapeutics involving pluripotent stem cells. She first examines how FDA has regulated and proposed to regulate this area and then turns to legal challenges that have been raised to FDA's authority in this area, including its authority over autologous stem cell treatments, its potential to regulate the "practice of medicine," and problems involving the Commerce Clause authority for FDA's actions when the relevant activities are primarily intrastate. Riley also discusses potential future issues and many of the ethical challenges relating to enhancement technologies.

Elizabeth R. Pike and Kayte Spector-Bagdady turn to a different frontier for FDA regulation, genetic testing, in their chapter, "Device-ive Maneuvers: FDA's Risk Assessment of Bifurcated Direct-to-Consumer Genetic Testing." Using FDA's confrontation with the company 23andMe as the backdrop, their chapter examines FDA's challenges in regulating genomic information as a medical device: classifying these products according to levels of risk; implementing standard risk-mitigation strategies; ensuring the safety and effectiveness of genomic information, which requires ensuring analytic and clinical validity; and minimizing risks of disclosure. This chapter proposes a path forward for satisfying these requirements. The authors argue that FDA should require that labs doing this work ensure that genomic information is based on analytically valid genomic data, focus on validating tests for the riskiest types of information, and minimize the risks of disclosure through labeling requirements; moreover, the fact that the most concerning therapeutic actions are taken in response to genomic information provides an additional safety mechanism.

The next chapter offers a new idea, rather than a critique of the agency's approach to new technologies. In "A New Regulatory Function for E-Prescriptions: Linking FDA to Physicians and Patient Records," Andrew English, David Rosenberg, and Huaou Yan propose deploying e-prescription functionality to address a different regulatory challenge for FDA: overseeing the postmarket safety and efficacy of medical products. After discussing the postmarket surveillance challenges faced by FDA, covered in greater depth earlier in this book, the authors move to examining how e-prescriptions might help: "by leveraging e-prescriptions, invaluable data about drug and device usage and effects not fully captured by existing systems could be comprehensively obtained and delivered directly to FDA or other regulators. Simultaneously, the same platform could deliver critical warnings regarding drug and device use to practitioners at the very moment they need that information the most."

Finally, Jonathan Kahn's chapter, "Race and the FDA," examines the agency's "evolving practices with respect to the use of racial and ethnic categories in pharmaceutical research and development" against a backdrop of a richer historical and institutional context of the place of race in medicine. He discusses the way these tendencies are fermented by "a wide array of federal mandates that dictate the characterization and application of genetically based biomedical interventions, such as pharmaceuticals and diagnostic tests, in relation to socially defined categories of race." That is, in conditioning the awarding of grants and approvals on the collection of data according to the Office of Management and Budget's race categories, the federal government provides powerful incentives to "introduce race into biomedical contexts, regardless of its relevance." In this story he sees a troubling through line in which the drive to "regularize" racial categories produces a tendency to geneticize race, and he warns that "in the drive to harmonize international drug development, we must be careful to avoid adopting a harmonized conception of race as genetic."

Each of the chapters in this book describes, analyzes, and evaluates the myriad new and enduring challenges faced by FDA and offers suggestions—some incremental and some massive—for change. This is not a mere academic exercise but rather one at the heart of our health and health care: How much risk are we willing to accept, how much delay? How will we demand accountability from corporate executives when things go wrong, how much will we allow sales representatives to say? How much will we jam square pegs into round holes, and when will we say, "it's time to start from scratch"? This isn't your grandmother's FDA, but what should the agency's next generation look like? Read on to find out what some of the leading experts have to say, and then decide for yourself. But one thing is for sure—FDA is constantly moving, constantly changing, constantly working to meet its challenges head-on. Tomorrow's FDA will not look like today's, and that is a good thing.

"To improve is to change; to be perfect is to change often."
—WINSTON CHURCHILL

CHAPTER ONE

Historical Themes and Developments at FDA Over the Past Fifty Years

PETER BARTON HUTT

I. INTRODUCTION

This chapter provides a historical overview of seven important policies that the Food and Drug Administration (FDA) has addressed over the past fifty years. Each of these policies has had antecedents initially under the Federal Food and Drugs Act of 1906 (34 Stat. 768) and then under the Federal Food, Drug, and Cosmetic Act (FDCA) of 1938 (52 Stat. 1040) that replaced the 1906 act. Each of them has evolved over time and will continue to evolve in the future. FDA has always been a dynamic organization and must continually change to reflect new insights into both risk assessment regarding the products it regulates and risk management to provide a reasonable balance between fostering innovation and protecting the public health.

The origin of FDA extends back to 1839, when Congress first appropriated funds to the Patent Office for "the collection of agricultural statistics, and for other agricultural purposes." An Agricultural Division was subsequently created in the Patent Office and a Chemical Laboratory was established within the Agriculture Division.[1] When the

United States Department of Agriculture (USDA) was created by Congress in 1862, the Agriculture Division of the Patent Office formed the nucleus of the new department, and its Chemical Laboratory became the organizational antecedent of what is now FDA. It was officially named the Chemical Division in 1862, the Division of Chemistry in 1890, the Bureau of Chemistry in 1901, the Food, Drug, and Insecticide Administration in 1927, and the Food and Drug Administration in 1930. It resided in the USDA until it was transferred to the Federal Security Administration in 1940, the Department of Health, Education, and Welfare in 1953, and finally the Department of Health and Human Services in 1979, where it resides today. Throughout this chapter, "FDA" will be used to refer to all of these antecedent organizations.

Today FDA regulates food, drugs, cosmetics, medical devices, and tobacco. Together, these consumer products comprise twenty-five cents out of every dollar spent by consumers in the United States. The scope and power of FDA is therefore enormously important for both economic and health reasons. There is a growing academic literature that explores the history and work of FDA, to which this volume makes an important contribution.

II. FDA MANAGEMENT

From its inception, even before it was given the regulatory authority conveyed by the 1906 act, FDA was administered by a single individual, who earlier was called the chemist or chief chemist but since 1940 has been called the commissioner. Throughout this time, the agency has been divided into two components: (1) the Headquarters staff located in Washington, D.C., and the vicinity, and (2) the Field Force, located throughout the entire United States. Although management of the Field Force remains largely unchanged throughout the history of FDA, the organization and management of FDA Headquarters has evolved over time. In the early days of FDA, the Headquarters was organized along functional responsibilities.[2] The major functions were scientific research, regulation, and the field. During that time, the vast majority of work related to food, and thus there was no need for FDA to be organized by product categories. With enactment of the Drug Amendments of 1962 (76 Stat. 780) and the emergence of the modern medical device industry,[3] the agency was completely reorganized along product lines in 1970. There are now

TABLE 1.1

Year	Congressional Appropriation (Dollars)	Employees
1900	17,100	—
1910	930,560	467
1920	1,391,571	374
1930	1,849,140	530
1940	2,741,000	719
1950	4,802,500	955
1960	13,800,000	1,678
1970	72,352,000	4,252
1980	312,796,000	7,517
1990	560,271,000	7,815
2000	1,048,000,000	7,728
2010	2,370,000,000	9,368
2014	2,557,963,000	9,940

Compiled from FDA reports. This table does not include user fees or employees paid with user fees.

centers within FDA for food, drugs, biological products, animal feed and drugs, medical devices, and tobacco.

Although there is no reliable historical statistical series reflecting the resources of FDA throughout its history, table 1.1 presents the best available data from 1900 to the present. The modest resources available to the agency through 1960 were reflected in a highly personal and close-knit management approach. The commissioner's team was small and comprised individuals who had spent their entire careers at the agency. They knew most of the people in the Field Force and could reach out for information quickly and efficiently. Following enactment of the FDCA in 1938, there were only two congressional hearings before 1960 (Hutt 1983), and thus there was very little congressional distraction. FDA had not yet become the subject of the intense press interest that it is today. In short, the agency was a small, tightly run organization that attracted little public attention. By 1970, all that had changed. The thalidomide disaster (Mintz 1962), followed by the Drug Amendments of 1962, changed the agency forever.

From 1862 through 1965, there were fourteen FDA commissioners. Since the last career commissioner retired from FDA in 1965, there have also been fourteen FDA commissioners. Each has reflected his or her own personal management style. For some, their personal management philosophy has been clearly articulated and has had a major impact on the agency. For others, there has been relatively little impact and thus only minor change when a new commissioner has arrived. The following brief summary illustrates the differences that have occurred.

Charles C. Edwards, MD, was FDA commissioner from 1969 through 1973. Although he was trained as a surgeon, he spent the years before he came to FDA as a medical management consultant. He recruited business managers to the agency and initiated a project management system (PMS) to organize the increasing work at Headquarters. Every morning, Dr. Edwards and his immediate staff met with a director of one of the product bureaus (now called centers) and the bureau staff to review important programs and issues. Everyone knowledgeable with respect to an issue was included in the discussion, but the commissioner made the ultimate decision. No issue went unresolved for longer than a week or two.

As the issues proliferated and became more technical and complex and the resources of the agency increased dramatically, the close supervision and decisional authority of the commissioner eroded. By 1990, the weekly meetings of the commissioner with the center directors was abandoned. Since then, the commissioner has become involved in only the most important issues in the agency, as determined largely by Congress and the media. It is rare that a truly important scientific or medical issue will be reviewed by the commissioner, and even then it is usually not the commissioner who makes the final decision.

The personal interests of a commissioner nonetheless remain an important force within FDA. One of the more significant issues that is discussed within both FDA and the regulated industry is the extent to which relations between these two interested parties should be on a collaborative and cooperative level or at arm's length. Some commissioners have unequivocally established their own policy on this issue (Hutt, Merrill, and Grossman 2007). Others have simply declined to enter the debate. Under the commissioners who have not established clear policy, the FDA employees who have daily contact with the regulated industry make their own personal determinations of what is an appropriate relationship.

Commissioners also leave their imprint on the agency by the type of personnel they recruit as part of the Headquarters team. As noted already, Commissioner Edwards recruited business managers to help him run the agency. In contrast, Commissioner McClellan recruited Ph.D. economists. The differences that occur from these approaches have had significant effects on the work of the agency.

As the number of FDA employees increases, FDA managers at every level within the agency know less and less about what is actually happening when applications of various types are reviewed and regulatory decisions are made. FDA has repeatedly expressed a commitment to preventing and resolving inconsistencies within the agency as a whole, the centers that review similar products, the offices within each center that have overlapping issues, and even down to the lowest agency employees who have direct regulatory responsibilities. But the potential for the lack of consistent policy and regulatory decisions only increases with the ever-larger size of the agency. Together with recent decisions that higher officials cannot overrule lower officials on regulatory decisions (Compare Food and Drug Administration 2009 with *Ivy Sports Medicine, Inc. v. Sebelius* 2013:938 and *Sports Medicine, Inc. v. Burwell* 2014:767), the increasing size of the agency has presented difficult and as yet unresolved management issues.

III. FDA RULEMAKING

FDA rulemaking began shortly after enactment of the 1906 act. In one year, the agency promulgated four regulations governing its regulatory authority that were virtually unamended for the next thirty years (Hutt 1981).

From 1906 to 1970, FDA used rulemaking for the technical decisions required by the FDCA for establishing food standards (21 CFR Parts 130–169) and food additive regulations (21 CFR Parts 170–186), but most new policy was developed through litigation and informal guidance. By the 1970s, however, it was clear to the agency that it needed to engage in substantial rulemaking for all of the products for which it was responsible in order to develop new policies and procedures.

The 1970s were the decade of FDA rulemaking. The agency reinterpreted the informal rulemaking provision in Section 701(a) of the FDCA as authorizing substantive rulemaking, not just general policy statements (e.g., *National Nutritional Foods Ass'n v. Weinberger* 1975:688). Major new programs were established, such as a transformation of all food labeling (Hutt 1989a), the requirement of nutrition labeling for half of the food supply (Hutt 1995), the development of the OTC Drug Review (21 CFR Part 330), the requirement for ingredient labeling on cosmetics (21 CFR § 701.3), and the initial regulations governing medical devices even before enactment of the Medical Device Amendments of 1976 (Hutt 1989b). From a procedural standpoint, FDA promulgated regulations governing implementation of the Freedom of Information Act (21 CFR Part 20) and the government in the Sunshine Act (21 CFR Part 14), as well as the broad procedural regulations that established requirements for all aspects of agency activities (21 CFR Parts 10, 12, 13, 15, 16). For the first time, moreover, it imposed the requirement of lengthy and detailed preambles to all regulations so that agency employees, the regulated industry, and the public would better understand agency requirements (Hutt 1972).

By establishing its policies as enforceable regulations, FDA sought to discourage unnecessary litigation. The agency invented the Regulatory Letter (later renamed the Warning Letter) in order to provide strong incentives for industry to comply with FDA regulations in an efficient and effective way (Hutt et al. 2007).

The era of FDA reliance on rulemaking lasted for roughly twenty years. In a series of statutes and executive orders, Congress and the president imposed requirements on rulemaking that crippled this form of agency policy making (Hutt 2008). The requirements for review of agency regulations both within the Department of HHS and in the Office of Management and Budget have dramatically slowed down even the smaller regulations that have been issued in the last twenty years. Thus, rulemaking is used by FDA today only where Congress explicitly requires it.

IV. FDA USE OF GUIDANCE

From its inception, FDA was a pioneer in the development of informal guidance (Lewis 2011; Martini 2009). FDA issued bulletins beginning in 1890 and circulars beginning in 1894 that presented its regulatory position on product issues. From 1902 to 1927, FDA issued 212 "Food Inspection Decisions," and from 1938 to 1946 it issued 439 "Trade Correspondence" circulars for the same purpose. None of these had the force and effect of law. They were, however, invaluable in providing FDA views on regulatory matters for the regulated industry. Following enactment of the Administrative Procedure Act in 1946 (60 Stat. 237), FDA issued 131 policy statements on regulatory matters. In 1968, these were converted to Compliance Policy Guides, and they survive in this form to this day.

As part of its procedural regulations that were proposed in 1975 (40 Fed. Reg. 22950; 40 Fed. Reg. 40682) and promulgated in 1977 (42 Fed. Reg. 4680), FDA defined what it then called "Guidelines" and now calls "Guidance" (21 CFR § 10.90(b)). Following enactment of the statutory authority for the development of Guidance as part of the Food and Drug Administration Modernization Act of 1997 (Section 701(h) of the FDCA, 21 USC § 371(h)), FDA replaced its Guidelines regulation with its broader regulation governing all of Good Guidance Practices (21 CFR § 10.115). To date, the agency has issued more than 3,000 draft and final Guidances. Some, but not all, have been the subject of a brief notice published in the *Federal Register*. It is difficult, perhaps even impossible, to keep track of all these documents.

FDA has supplemented its informal Guidance in three ways. First, beginning in the early 1970s it has included extensive preambles explaining its proposed and final regulations (Hutt 1972). These are invaluable in understanding FDA policy. For example, the regulation defining the difference between a permitted structure/function claim for a food or dietary supplement and an illegal disease prevention claim is only half of one column, but the preamble takes up fifty pages in the *Federal Register* (65 Fed. Reg. 1000). Second, the FOI Act regulations promulgated by FDA in 1974 (21 CFR Part 20) are among the most liberal in the federal government. Extensive information on FDA policy issues is available from this source as well. Finally, it is widely acknowledged that FDA's website is a model for the federal government. In order to avoid having thousands of FOI Act requests, FDA finds it easier simply to put significant documents on its website.

Nonetheless, one fundamental legal question that has received widespread discussion remains unresolved. In many instances, FDA has issued Guidance in one form or another that the regulated industry regards as an attempt illegally to engage in rulemaking without complying with the Administrative Procedure Act. It can be anticipated that as FDA pushes forward with additional Guidance, this issue will ultimately be addressed in the courts.

V. FDA ENFORCEMENT

Nothing stirs more passion than the perennial debate whether FDA enforcement is increasing or decreasing. The raw enforcement data easily show that formal court enforcement actions have decreased (table 1.2). But that does not translate to a reduction in total FDA enforcement or in industry compliance. Formal court enforcement action has been replaced by three administrative compliance programs: (1) informal enforcement techniques, such as Warning Letters and Recall Requests; (2) the premarket approval programs for new drugs, OTC drugs, medical devices, and biological products, which involve FDA premarket scrutiny and approval of matters that once required enforcement surveillance; and (3) regulations, preambles, and guidance that most often provide a clear regulatory pathway for the regulated industry and thus reduce uncertainty and confusion that formerly resulted in court enforcement action.

On balance, these administrative programs unquestionably provide more efficient and effective compliance than formal court action. The only troubling informal enforcement technique is FDA's unfettered use of publicity. Examples include the finding of aminotriazole in cranberries (1959), salmonella in Borden's Starlac nonfat dry milk (1966), phenylpropanolamine (2000), and ephedra (2003). Each was an action taken by FDA by press release without any opportunity for public participation. For each one, publicity destroyed the ingredient, product, or industry involved, without an opportunity for the injured parties to participate. Publicity has become FDA's most powerful weapon in today's culture of instant communication and should be used sparingly and only when absolutely necessary.

VI. THE IMPACT OF USER FEES

User fees were imposed for certification of coal-tar colors (Section 706 of the 1938 FDCA, 52 Stat. 1049) beginning with enactment of the FDCA, for certification of insulin in 1941 (55 Stat. 851), and for all antibiotics in 1962 (76 Stat. 780). When user fees were proposed to finance the new drug approval system in the early 1970s, however, FDA vigorously opposed such an approach, primarily on two grounds (Kuhlik 1992). First, FDA regulation of new drugs benefits everyone,

TABLE 1.2

Year	Congressional Appropriations	Employees	Seizure Actions	Injunction Actions	Criminal Actions (Field Force)	Warning Letters	Recalls
1940	2,288,000	719	1,697	0	337	—	—
1950	4,884,000	955	1,460	15	378	—	17
1960	13,800,000	1,678	1,002	24	248	—	39
1970	72,352,000	4,252	600	24	42	—	1,427
1980	312,796,000	7,517	428	49	24	311	819
1990	600,979,000	7,629	144	9	19	498	2,352
2000	1,048,149,000	7,728	36	9	2	1,154	3,716
2010	2,370,000,000	9,368	10	17	0[a]	673	3,799

[a]All criminal actions have been transferred from the Field Force to the Office of Criminal Investigations.
Compiled from FDA reports. This table does not include user fees or employees paid with user fees.

not just the industry. Second, FDA did not want to be financed by the industry because of the appearance of conflict of interest. Industry also opposed these fees. As a result, the issue was tabled for twenty years.

By the 1990s, however, it was clear that FDA could not conduct its new drug review and approval work by congressional appropriations alone. In contrast to the situation in the early 1970s, the FDA commissioner in the early 1990s strongly supported user fees. The turning point came when industry realized that it would actually save money by paying user fees that would get new drugs to the market faster. The cost of an approved new drug was so great that the amount of money paid as a user fee was less than the amount of money that would be paid just in carrying costs for three months. Thus, the Prescription Drug User Fee Act was passed by Congress in 1992 and renewed every five years thereafter (106 Stat. 449; 111 Stat. 2298; 116 Stat. 687; 121 Stat. 825; 126 Stat. 996). Since then, user fee statutes have been enacted for medical devices (116 Stat. 1588; 118 Stat. 572; 121 Stat. 842; 126 Stat. 1002), animal drugs (117 Stat. 1361; 122 Stat. 3509; 127 Stat. 451), animal generic drugs (122 Stat. 3515; 127 Stat. 464), human generic drugs (126 Stat. 1008), and biosimilar biological products (126 Stat. 1026). An attempt to impose user fees for the food industry, however, failed when the FDA Food Safety Modernization Act was enacted (124 Stat. 3885).

The impact of these user fees on the two industries that do not have them, i.e., food and cosmetics, is not widely appreciated. All user fee statutes require that Congress must increase appropriations proportionately as the industry user fees increase. Thus, unless congressional appropriations increase as user fees do, there will be less and less money for those programs that are not supported by user fees (Hutt 2008). In short, when there is a flat budget, the programs supported by user fees automatically take money away from the programs not supported by user fees, by operation of law.

For example, the new drug review and approval program is supported by user fees. The OTC Drug Review is not. As a result, the OTC Drug Review, which was started in 1972 (21 CFR Part 330), has completely stagnated. Forty years later, it is still only partly completed (79 Fed. Reg. 10,168). Faced with a lack of funds, the affirmation of generally recognized as safe (GRAS) status for food ingredients that was begun in 1973 (21 CFR § 170.35) was abandoned by FDA in 1997 and replaced by a simple notification process (62 Fed. Reg. 18,938). These are but two examples where user fees have devastated programs that do not have such funding.

When this was clearly pointed out in a 2007 report (Hutt 2008) prepared for the FDA Science Board, Congress held a hearing where both political parties pledged a substantial increase in FDA appropriations in order to repair the damage that had been done (U.S. Congress 2008). As a result of that report, FDA appropriations doubled between 2008 and 2013. In the current fiscal era, however, that type of extraordinary appropriation is not likely to be repeated. It remains to be seen whether more stable funding for FDA can be obtained without user fees.

VII. LEGACY PRODUCT PROGRAMS

Congress and FDA are very adept at creating new statutory and regulatory programs to review categories of ingredients or products. These are always well intentioned, but when the commissioner or the members of Congress who created the programs retire, there is no one to step into the void and make certain that the necessary funding is maintained. The six uncompleted programs of this nature shown in table 1.3 serve to illustrate the problem. Other programs have suffered as well. For example, FDA began the review of the food ingredient GRAS list at the direct request of President Nixon ("Consumer Protection" 1969), in 1969, forty-five years ago. When it became clear that FDA could not

TABLE 1.3

The Program	Year Begun	Years Since Program Began
Review and Approval of all Color Additives (74 Stat. 397)	1960	54 years
Review of New Drugs with Effective NDAs During 1938–1962 (76 Stat. 780, 788)	1962	52 years
OTC Drug Review of All Nonprescription Drugs Marketed Prior to 1972 (21 CFR part 330)	1972	42 years
Review of Antibiotics Used in Animal Feed for Growth Promotion (21 CFR § 510.110)	1972	42 years
Review of All Biological Products Licensed During 1902–1972 (21 CFR §§ 601.25, 26)	1972	42 years
Review of All Pre-Amendments Class III Medical Devices (70 Stat. 539, 552–553)	1976	37 years

allocate sufficient funds to the project, FDA quietly just gave up and abandoned the program without completing it. For each one of these programs there is no line appropriation by Congress and no user fee, and each new commissioner has shown no interest in completing the work begun by a previous commissioner.

VIII. SCIENCE, POLICY, AND LAW

As long as FDA has existed, there have been heated disagreements about whether FDA decisions are scientific, policy, or legal in nature. Commissioner Wiley's autobiography shows his continuous struggle with the USDA lawyer assigned as his chief counsel about implementation of the 1906 act (Wiley 1929). Commissioner Edwards held a famous debate with the chairman of a House subcommittee over whether FDA is a scientific or a regulatory agency (U.S. Congress 1970).

Both the 1906 and 1938 acts were written in an era when Congress gave broad authority to a government agency and left it to the discretion of the agency to determine how best to implement that authority. The basic provisions of the FDCA are sufficiently broad, ambiguous, and undefined that there is a wide degree of discretion and judgment in determining how implementation should be undertaken. This discretion is explicitly given by statute to the secretary of HHS. The secretary, in turn, has delegated it to the commissioner (Food and Drug Administration 2005). Some matters are delegated further to other officials within the agency.

It is difficult to find any significant issue faced by FDA that is not ultimately a matter of policy, informed by both scientific and legal considerations. But recently, FDA scientists have attempted to assert the primacy of their contributions to issues. Two examples will serve to illustrate this matter.

First, when Plan B was proposed to be switched from prescription to nonprescription status, the physicians and scientists who initially reviewed the matter concluded that it should be switched for any female sixteen years or older. The higher level managers who reviewed this recommendation, who were also scientists and physicians, concluded that the cutoff age should be eighteen, not sixteen. The lower level scientists claimed that their recommendation was "scientific" and therefore could not be changed by the higher level managers. At issue, from a matter of statutory law, was whether the additional two years is one of the statutory "collateral measures necessary to" use of the drug (Section 503(b)(1) of the FDCA, 21 USC § 353(b)(1)). By any analysis, this is not a scientific issue. It is a matter of public policy. Yet a court determined otherwise (*Tummino v. Hamburg* 2013:162).

A second example of public policy adopted in the guise of science is the convention of the confidence level of 0.05 for statistical significance with respect to a finding in a clinical trial. This is not set by statute or even by an FDA regulation. Whether to use a confidence level of 0.1, 0.01, or 0.05, or indeed any other level, is pure public policy. It just signifies the acceptable level of confidence in the result. It may be that some people would be quite willing to accept a drug with a smaller level of confidence than other people, and some people might even wish to have a higher level of confidence. That kind of issue represents policy, not science.

Thus, while these types of debates will not soon (if ever) be resolved, it appears that few, if any, FDA decisions involve solely a scientific determination. The statute and public policy always have a role in such decisions.

IX. CONCLUSION

The matters described and discussed here represent only a small sample of intractable FDA questions on which there is wide latitude for disagreement. The following chapters of this book provide a wealth of other issues that also inspire continuing debate. It is hoped that those who read these chapters will encounter and debate them with an open mind and thus pursue the spirit with which the coeditors have conceived this book.

NOTES

1. For the organizational history of FDA within the federal government, see Hutt 1990.

2. For the organizational history of FDA within the agency, see Brannon 1984.

3. The Medical Device Amendments were enacted in 90 Stat. 539 (1976).

REFERENCES

Brannon, Michael. 1984. "Organizing and Reorganizing FDA." *Food and Drug Law* 75:135–173.

"Consumer Protection." 1969. *Weekly Compilation of Presidential Documents* 5 (November): 1516.

Food and Drug Administration. 2005. "Section 1410.10: Regulatory Delegations of Authority to the Commissioner Food and Drugs." *FDA Staff Manual Guides*. http://www.fda.gov/AboutFDA/ReportsManualsForms/StaffManualGuides/ucm080711.htm.

Food and Drug Administration. 2009. "Review of the ReGen Menaflex: Departures from Processes, Procedures, and Practices Leave the Basis for a Review Decision in Question." *Preliminary Report* (September).

Hutt, Peter Barton. 1972. "Public Information and Public Participation in the Food and Drug Administration." *Quarterly Bulletin of the Association of Food and Drug Officials of the United States* 36:216–217.

Hutt, Peter Barton. 1981. "About Fairness in Applying the Law." *FDA Consumer* 15(5):23.

Hutt, Peter Barton. 1983. "Investigations and Reports Respecting FDA Regulation of New Drugs (Part 1)." *Clinical Pharmacology and Therapeutics* 33:537–539.

Hutt, Peter Barton. 1989a. "Regulating the Misbranding of Food." *Food Technology* 43:288.

Hutt, Peter Barton. 1989b. "A History of Government Regulation of Adulteration and Misbranding of Medical Devices." *Food Drug Cosmetic L.J.* 44 (99):110–112.

Hutt, Peter Barton. 1990. "A Historical Introduction." *Food Drug Cosmetic L.J.* 45:17.

Hutt, Peter Barton. 1995. "A Brief History of FDA Regulation Relating to the Nutrient Content of Food." In *Nutrition Labeling Handbook*, ed. Ralph Shapiro, 1–29 New York, NY: CRC Press.

Hutt, Peter Barton. 2008. "The State of Science at the Food and Drug Administration." *Administrative Law Review* 60:431, Tables 2 and 3.

Hutt, Peter Barton, Richard A. Merrill, and Lewis A. Grossman. 2007. *Food and Drug Law: Cases and Materials*. St. Paul, MN: Foundation Press. 1529–1531.

Ivy Sports Medicine, Inc. v. Sebelius, 938 F. Supp. 2d 47 (D.D.C. 2013), *rev'd Ivy Sports Medicine, Inc. v. Burwell*, 767 F.3d 81 (D.C. Cir. 2014).

Kuhlik, Bruce N. 1992. "Industry Funding of Improvements in the FDA's New Drug Approval Process: The Prescription Drug User Fee Act of 1992." *Food and Drug L.J.* 47:483.

Lewis, Kevin M. 2011. "Informal Guidance and the FDA." *Food and Drug L.J.* 66:507.

Martini, Kasey L. 2009. "A Historical Look at FDA's Approach to Regulation and Policymaking." In *An Electronic Book of Harvard Law School Student Papers on Food and Drug Law*, ed. Peter Barton Hutt. http://www.law.harvard.edu/faculty/hutt/book_index.html

Mintz, Morton. 1962. "'Heroine' of FDA Keeps Bad Drug Off Market." *Washington Post* (July 15). http://www.washingtonpost.com/wp-srv/washtech/longterm/thalidomide/keystories/071598drug.htm.

National Nutritional Foods Association v. Weinberger, 512 F.2d 688 (2d. Cir. 1975).

Tummino v. Hamburg, 936 F. Supp. 2d 162 (E.D.N.Y. 2013).

U.S. Congress, House of Representatives, Subcommittee of the Committee on Government Relations. 1970. *Regulatory Policies of the Food and Drug Administration*. 91st Cong., 2nd sess., 137–138.

U.S. Congress, Subcommittee on Oversight and Investigations of the Committee on Energy and Commerce. 2008. *Science and Mission at Risk: FDA's Self-Assessment*. 110th Cong., 2nd sess.

Wiley, Harvey W. 1929. *The History of a Crime Against the Food Law*. New York: Ayer Company Publishers. 156–160.

PART ONE

FDA in a Changing World

Introduction

HOLLY FERNANDEZ LYNCH

A COLLEAGUE of mine is fond of reminding his students to challenge the status quo because the world as it exists is not necessarily the way it has to be. Indeed, the status quo may better reflect a range of historical accidents that led down a particular path than it does a well-thought-out, strategic set of choices. And even the best-laid plans may require revision as the world changes around them. Politics and bureaucracy aside, nearly everything is changeable.

With that principle in mind, we might ask, "Is the FDA we have today the FDA we really need, and if not, how should it be changed?" The agency as it currently exists is the product of a long history of reaction to circumstances as they presented themselves—reaction to sulfanilamide, thalidomide, HIV/AIDS, Vioxx, and on and on—with a steady vacillation between emphasis on speed and emphasis on safety. Moreover, the world around FDA has changed dramatically since the agency's birth. The chapters in this section evaluate how well a reactionary FDA has been able to keep up with some of those changes, including globalization, patient advocacy, rising health care costs, and prospective medicine. In some ways, it seems, FDA has risen to the meet these challenges, and in others it has not, as a result of either jurisdictional limitations or cabined readings of available authority.

In this part's first chapter, Howard Sklamberg and Jennifer Devine, deputy and then-associate commissioner for global regulatory operations and policy at FDA, respectively, tout the agency's expansive responses to the globalization of its regulated products. Sklamberg and Devine offer numerous examples of FDA's innovative approaches to partnering with foreign authorities, placing staff on the ground abroad, expanding facilities inspections, and sharing scientific knowledge, all of which are important steps toward keeping American medicine cabinets safe. But do they go far enough? If we think about the agency we might build from scratch, perhaps not. Rather than focusing on what globalization means for the manufacture and trade of approved drugs, we might think much more broadly about developing a global marketing approval system that goes far beyond harmonization of requirements to actual reliance on decisions made by other authorities. Especially in the information age and with increasing efforts to harness the power of big data, consider how much more efficient and accurate approval decisions and subsequent safety monitoring could be on a global scale. If we challenge the status quo, we might start to see globalization as less of a challenge and more of an opportunity.

In chapter 3, Lewis Grossman describes the myriad ways in which patient advocacy and empowered consumers have influenced how FDA regulates. Of course, patients have an important role in determining whether any given product's benefits outweigh its risks since that is a normative rather than scientific evaluation. Grossman's chapter demonstrates FDA's efforts, increasing over several decades, to give patients a seat at the table. But at what cost? Do we want an agency focused on speed, on safety, on both, or on something else entirely? Perhaps the empowered consumer would prefer an agency whose mission is not to serve as a gatekeeper to products reaching the market but rather as an independent and expert evaluator of information about those products that patients can then use to make informed decisions themselves. This need not bring us back to the bad old days of snake oil salesmen—to the contrary, patients would have what they need to distinguish snake oil from gold—that is, assuming that patients are in fact empowered, which is not always the case, as Grossman reminds us.

And even empowered patients are likely to be interested in more than a product's bare risks and benefits, for example, information about how that product stacked up against the competition.

It is precisely FDA's failure to provide this sort of comparative information that Theodore Ruger highlights in his chapter. Ruger notes that although FDA is jurisdictionally limited in some important ways, it has exhibited "willful ignorance to comparative efficacy and cost," focusing instead on safety and effectiveness as demonstrated against the gold standard of a placebo control. Ruger suggests that this failure to innovate despite authority to do so may severely marginalize the agency, as the decision by payers whether to include a product on their reimbursable formulary begins to overshadow the regulatory decision of whether to allow a product on the market in the first place. After all, what good is a drug that no one can afford? However, it may be that even on a blank slate, we would choose to separate questions about whether a drug works and is safe from questions about how much a drug should cost or whether it should be covered, such that our current bifurcated system gets things right in at least one regard. Nonetheless, it would be important to generate reliable information about a product as compared to available alternatives, and that is sorely missing from the current equation.

In the final chapter of this part, Barbara Evans hones in on FDA's failure to fully embrace the era of prospective medicine, or interventions to help healthy people stay well. Evans emphasizes that prospective medicine poses a number of unique challenges to FDA's status quo, the most important of which is that a new product intended to predict or prevent disease might not be demonstrated to be definitively effective for years or decades—a three-year trial, or even ten years, would not be anywhere near sufficient to evaluate the direct effectiveness of an Alzheimer's prevention drug, for example. A lot of the work of protecting patients will have to happen in the post-marketing time period, but FDA has not adequately advanced its authority to do so. Against this background, we might prefer a world in which drug approvals were time limited and had to be renewed, just like IRB-approved research protocols, so that there are predefined and automatic points in time to reassess based on new information and context. We might also develop a much more robust system for tapping into electronic health records to gather data outside of the research setting, which might entail bypassing patient consent or articulating some social obligation to contribute data in general or as a condition of receiving a drug. FDA has already begun to head down some of these paths, but it could go substantially further.

These chapters offer an important perspective and opportunity to assess whether we are happy with the agency we have or whether dramatic changes are in order. The topics covered reflect several leading twenty-first-century challenges, but there are many others. Is FDA as it currently stands optimally situated to regulate gene therapies, for example? What about the fact that the agency largely refuses to engage in substantial questions of research ethics that are of paramount importance in a world of multinational clinical trials? And its counterparts abroad are taking huge strides to improve transparency and efforts to share data while FDA has been much more reserved.

FDA is innovating in many important ways, and it deserves substantial recognition for its efforts to manage a massive regulatory portfolio in a changing world. But we should never get complacent. Remember, the way it is is not the way it has to be. Of course, mustering sufficient political will to implement substantial change is another story.

CHAPTER TWO

A Global and Innovative Regulatory Environment for the U.S. FDA

HOWARD SKLAMBERG AND JENNIFER DEVINE

THE U.S. Food and Drug Administration (FDA) has traditionally been a domestically focused agency, for which Congress kept the regulatory lens narrow in scope with oversight of industry being concentrated on those companies within U.S. boundaries. Over the past decade, FDA's responsibilities have changed rapidly because of globalization of the supply chain, the exponential growth in imports that has followed, and the increasing breadth and complexity of FDA-regulated products resulting, in part, from scientific innovation. Decades ago, it was likely the drugs in your medicine cabinet would have come from a domestic source, but now, the reality is that they could just as easily have been made abroad or contain ingredients or components sourced from abroad. While this chapter is primarily focused on pharmaceuticals, it is important to note that this example holds true across the health and nutrition spectrum, from food to medical devices to drugs, and has set forth a new reality—a global regulatory environment for FDA. This ever evolving environment requires FDA to deepen collaborations and partnerships globally with other regulatory and public health partners—nationally and across borders—to ensure it meets its public health mission.

I. GLOBAL AND INNOVATIVE REGULATORY ENVIRONMENT

Today, the percentage of imported products consumed continues to increase, and the distinction between foreign and domestic products has become increasingly blurred—to the point of irrelevance (Food and Drug Administration 2011). FDA regulates products that account for approximately 20 percent of every consumer dollar spent in the United States (Food and Drug Administration 2015). FDA-regulated products originate in 300,000 foreign facilities in more than 150 countries, and the volume of imports to the United States has increased nearly fivefold, from 6 million product lines in 2001 to an estimated 30 million in 2014. Eighty percent of seafood consumed in the United States comes from abroad, as does 50 percent of fresh fruits and 20 percent of vegetables (ibid). For medications, approximately 40 percent of listed finished drugs come from overseas, and 80 percent of the manufacturers of active ingredients are located outside the United States (ibid.). In addition to the increasing volume of products sourced from abroad, the supply chain from manufacturer to consumer has become more and more complex, involving a web of manufacturers, suppliers, packagers, and distributors (Hamburg 2010). Risks associated with this complex supply chain are compounded by advances in technology and manufacturing, as well as language barriers, time differences, and distances. At every step in the process—from raw materials and other ingredients to manufacture, storage, sale, and distribution—a product can be contaminated, diverted, counterfeited, or adulterated.

This has made FDA oversight increasingly complicated because there are numerous individuals and firms in the supply chain that are geographically dispersed with varying levels of rigor in their regulatory systems. In the following example, a finished drug was produced with an active pharmaceutical ingredient (API) made in China and inactive ingredients (excipients) made in Europe, Japan, and the United States. These components were then shipped to India, where the finished drug was manufactured and then imported into the United States for distribution. The quality of both drug components and finished forms

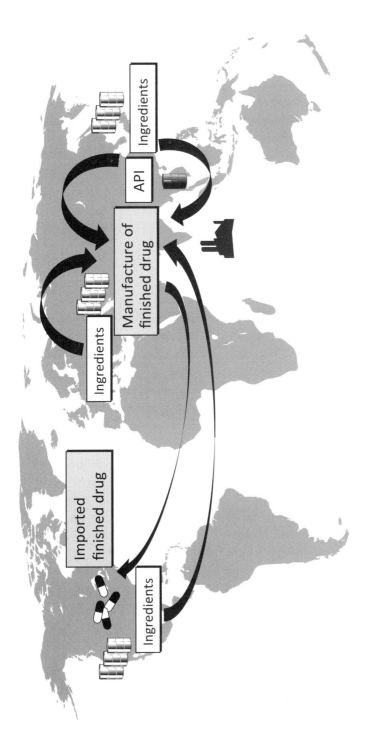

Figure 2.1 Global Drug Manufacturing Supply Chain Example

This illustration gives an example of a drug manufacturing supply chain. A finished U.S. drug may be produced using an active pharmaceutical ingredient (API) made in China and excipients made in Europe, Japan, or the United States. These components may be shipped to India, where the finished drug is manufactured and then imported into the United States for distribution.

can be compromised at various points in the supply chain. Any individual or firm involved might knowingly introduce adulterated components or unknowingly use substandard ingredients. The Internet—and the anonymity it confers—presents an additional layer of complexity by introducing more players into the system (Sklamberg 2014).

Unfortunately, there have been numerous examples of what can happen in these highly complex supply chains. Fever medicine, cough syrup, and teething products have been adulterated at one time or another with diethylene glycol, a toxic solvent found in antifreeze. Over the last twenty years, such adulterated products are thought to have caused an estimated 570 deaths worldwide, including children and adults in Haiti, Panama, and Nigeria (Autor 2011). In 2007, Chinese suppliers of wheat gluten substituted melamine, an ingredient used in making plastic, which proved toxic when it was used in U.S. pet food and dairy products. The contaminated food sickened and killed pets across the United States and put many people at risk (Hamburg 2013). Also in 2007, Chinese suppliers of heparin, a drug critical to the prevention of blood clots, substituted a lower cost, adulterated raw ingredient in their shipments to U.S. drug makers, causing deaths and severe allergic reactions (ibid.).

While it may not eliminate all problem products from the supply chain, a relatively comprehensive system of laws, regulations, and enforcement has kept supply chain incidents comparatively rare in the United States. Despite the infrequency of such incidents and FDA's continuous modernization and evolution, the heparin situation of 2007 and 2008 was a wake-up call forcing not only FDA but also other stakeholders to think more critically and foundationally about the vulnerabilities of the current regulatory environment. The incident pointed to the need to deepen collaborations and partnerships globally—internally, nationally, and across borders—to help regulate in the complex and evolving global environment. It also highlighted and galvanized stakeholders to work with Congress to pass legislation to help stakeholders navigate the realities of innovation, modernization, and globalization.

II. LEGISLATIVE AUTHORITIES

Congress has modified and expanded FDA's regulatory authorities in recognition of the globalization of the supply chain and the subsequent growth in imports, scientific innovation, and the increasing breadth and complexity of FDA-regulated products. FDA is implementing these new mandates and leveraging additional authorities in an effort to continue to meet its public health mission. Some of the major legislation Congress has recently passed included the FDA Food Safety and Modernization Act, the Food and Drug Administration Safety and Innovation Act, and the Drug Quality and Security Act.

A. The FDA Food Safety and Modernization Act (FSMA)

Every year, approximately one in six Americans suffers from a foodborne illness (Centers for Disease Control and Prevention 2014). The U.S. Centers for Disease Control and Prevention estimates an annual total of 48 million illnesses, 128,000 hospitalizations, and 3,000 deaths (ibid.). Under FSMA (Pub. L. No. 111-353), FDA is creating a modernized food safety system that, when fully implemented, should reduce the number of food-related illnesses. This law requires that FDA establish prevention-oriented standards that cover each stage of the food supply chain and grants FDA new administrative tools to hold all parties in the supply chain responsible for the production and distribution of safe food for humans and animals (including pet food and nonmedicated animal feed).

FDA has proposed FSMA-mandated rules that, when final, will establish the comprehensive framework of modern, prevention-oriented standards mandated by FSMA, from the grower of fresh produce, to food and feed manufacturers and processors, through the transportation to retail of foods and feeds produced in the United States or overseas and consumed in the United States. FSMA allows for the creation of an imported food safety system that enables FDA to hold imported foods to the same standards as domestic foods. Importers now have a responsibility to demonstrate the safety of the food they bring into the United States (Food and Drug Administration 2014a). In addition, FSMA mandates that FDA collaborate with and leverage other government agencies, both domestic and international. The statute explicitly recognizes that all food safety agencies need to work together in an integrated way to achieve shared public health goals (ibid.). Under these provisions, FDA is working to leverage the efforts of local, state, and foreign government counterparts (ibid.). FDA continues to work on implementing FSMA and its new authorities.

B. The Food and Drug Administration Safety and Innovation Act (FDASIA)

Complementing its new legislative authorities for foods, FDA is implementing several new authorities to help further ensure the safety and integrity of drugs imported into and sold in the United States. In 2012, Congress passed the Food and Drug Administration Safety and Innovation Act (FDASIA), expanding FDA's authorities and strengthening the agency's ability to safeguard and advance public health by giving it the authority to collect user fees from industry to fund reviews of innovator drugs, medical devices, generic drugs, and biosimilar biological products; promoting innovation to speed patient access to safe and effective products; increasing stakeholder involvement in FDA processes; and enhancing the safety of the drug supply chain (Pub. L. No. 112-144). FDASIA built on previous foundational legislation—like the FDA Modernization Act (FDAMA), which enabled FDA to begin to work internationally to strengthen standards when setting forth specific drug supply chain provisions.[1]

FDASIA Title VII strengthens drug safety by giving FDA new authority to protect the integrity of an increasingly global drug supply chain, for example, by increasing FDA's ability to collect and analyze data to enable risk-informed decision making, advancing risk-based approaches to facility inspections, partnering with foreign regulatory authorities, and driving safety and quality throughout the supply chain through strengthened enforcement tools.

The law provides FDA with the authority to administratively detain drugs believed to be adulterated or misbranded (Section 709) and the authority to destroy certain adulterated, misbranded, or counterfeit drugs offered for import (Section 708). The law also requires foreign and domestic companies to provide additional firm registration and drug listing information to ensure that FDA has accurate and up-to-date information about foreign and domestic manufacturers (Sections 701–703, 714). In addition, the law gives FDA the ability to work with foreign governments in a number of different capacities.

Since enactment of FDASIA, FDA has been working diligently to implement the Title VII supply chain authorities in a meaningful way that strives to maximize its public health impact. To date, FDA issued a proposed and final rule to extend the agency's administrative detention authority to include drugs intended for human or animal use, in addition to the authority that is already in place for foods, tobacco, and devices; issued a proposed rule regarding administrative destruction of imported drugs refused admission into the United States; issued draft and final guidance defining conduct that the agency considers delaying, denying, limiting, or refusing inspection, resulting in a drug being deemed adulterated; issued draft and final guidance addressing specification of the unique facility identifier system for drug establishment registration; and successfully worked with the U.S. Sentencing Commission on higher penalties relating to adulterated and counterfeit drugs (Food and Drug Administration 2014c). The agency had already taken steps toward development of a risk-based inspection schedule, before FDASIA. However, the enhancements provided by FDASIA will further assist the agency in responding to the complexities of an increasingly globalized supply chain.

In addition, FDA hosted a public meeting in 2013 and also requested written comments (78 Fed. Reg. 36,711 (Jun. 19, 2013)) to solicit feedback from the public about implementation of Title VII generally and to specifically address the provisions related to standards for admission of imported drugs and commercial drug importers, including registration requirements and good importer practices (Food and Drug Administration 2013). FDA continues to work on implementing FDASIA Title VII, prioritizing its efforts to achieve the greatest public health impact and deploy its limited resources most effectively.

C. The Drug Quality and Security Act (DQSA)

The Drug Quality and Security Act was signed into law in 2013 and provided a significant step toward having new and stronger drug quality and safety laws. The law enhances FDA's oversight of certain entities that prepare compounded drugs and enables certain prescription drugs to be traced as they move through the U.S. drug supply chain.

Specifically, Title II of DQSA, the Drug Supply Chain Security Act (DSCSA), outlines critical steps to build an electronic, interoperable system to identify and trace certain prescription drugs as they are distributed in the United States by 2023. DSCSA aims to facilitate the exchange of information to verify product legitimacy, enhance detection and notification of an illegitimate product, and facilitate product recalls. Drug manufacturers, wholesale drug distributors, repackagers, and many dispensers (primarily pharmacies) will be called on to work in cooperation with the FDA to develop the new system over the next nine years.

The law requires the FDA to establish standards, issue guidance documents, and develop a pilot program(s), in addition to other efforts, to support effective implementation and compliance. In 2014, the FDA issued a draft guidance on identifying suspect drug products in the supply chain and notification, and a draft guidance on establishing standards for the interoperable exchange of transaction information for trading partners (Food and Drug Administration 2014b).

III. GOING FORWARD: POSITIONING FDA TO OPERATE IN A GLOBAL AND INNOVATIVE REGULATORY ENVIRONMENT

Scientific innovation, globalization, and the increasing complexity of regulated products as well as new legal authorities provide FDA a unique opportunity to improve oversight of regulated products. Building on its long history of adaptation to fulfill its mission, FDA is moving forward and taking the necessary measures to ensure it is fully prepared to operate in this new dynamic environment.

A. Global Regulatory Operations and Policy Directorate

In June 2011, the commissioner of the FDA established the Global Regulatory Operations and Policy Directorate (GO) to provide oversight, strategic leadership, and policy direction to FDA's domestic and international product quality and safety efforts. The GO directorate includes the Office of Regulatory Affairs (ORA) and the Office of International Programs (OIP). ORA and OIP are responsible for conducting domestic and foreign inspections and advancing FDA's global engagement work, including deepening collaborations with local, state, federal, and foreign regulatory and public health partners.

Since its inception, GO has been working steadily to further develop a global product "safety net" that protects public health. GO is engaged in a variety of activities, internally and externally, to support its mission of advancing FDA's role as a modern public health regulatory agency.

B. FDA's Program Alignment

FDA is also working internally to achieve greater operational and programmatic alignment. Recognizing that the agency must change in fundamental ways to adapt to this new globalized climate, in 2013, Commissioner Hamburg instituted a Program Alignment Group to identify and develop plans to modify agency compliance and regulatory functions and processes in order to address the challenges associated with globalization and achieve mission-critical agency objectives (Food and Drug Administration 2014e). The overarching goal of program alignment is for the agency to modernize and strengthen the way it fulfills its public health role by changing how its inspectional workforce is organized, from alignment by domestic regions to distinct commodity-based and vertically integrated regulatory programs with a focus on specific product areas, such as pharmaceuticals, food, animal feed, medical devices, biologics, or tobacco.

The FDA Directorates, Centers for Disease Control and Prevention, and the ORA have been collaborating closely to define the changes that need to be made in order to align FDA more strategically and operationally to meet the greater demands placed on the agency. As a result, each regulatory program has established detailed action plans that identify critical actions needed to jointly fulfill FDA's mission in the key areas of specialization, training, work planning, compliance policy and enforcement strategy, imports, laboratory optimization, and information technology (ibid.).

C. Developing a Proactive Approach on Trade

Spurred by globalization and innovation, we have reached a turning point where trade, economic development, and global health intersect as never before. For FDA-regulated products, many goods and services are no longer sourced from one supplier; product development and distribution come from multiple sites in multiple countries. FDA understands that, in order to be effective in a dynamic, globalized world, the agency must take an active approach toward its engagement with the global public health and trade policy arenas. Thus, FDA has established an Office of Public Health and Trade (OPHT) in the OIP and has embarked on a new approach of active engagement that uses FDA's scientific, technical, and policy expertise to champion science-based regulation as a critical enabler of trade globally. This approach seeks to develop more cohesive strategies and policies relevant across FDA-regulated commodities and assure product safety and quality and stronger supply chains while promoting increased market access for U.S. exporters, increased trade, and enhanced economic development in low- and middle-income countries. The establishment of OPHT/OIP consolidates responsibility for policy coordination for FDA inputs into trade negotiations in one office and helps establish clear, coordinated, and consistent policy positions on significant issues related to trade. By ensuring a central, unified approach, FDA has a heightened opportunity to shape consistent, meaningful obligations under free trade agreements to support FDA's regulatory approaches and public health mission.

D. FDA's International Presence

In 2008, FDA received a special U.S. congressional appropriation to establish offices overseas and opened the first office in China the same year. Today, FDA's nine posts in seven countries on four continents provide the agency with the ability to work more closely with foreign regulators and industry and to educate foreign stakeholders about FDA standards, regulations, and procedures. FDA's strategically placed locations enhance its ability to build mutually beneficial relationships with foreign counterparts; gain a greater understanding of the host country's regulatory, public health, cultural, security, economic, and geopolitical landscape; and identify any developments in these arenas that may affect the quality, safety, or availability of FDA-regulated products bound for the United States.

FDA also has a foreign inspectorate consisting of a dedicated cadre of staff members whose sole job is to conduct inspections internationally. Currently, foreign inspections are being conducted by FDA staff

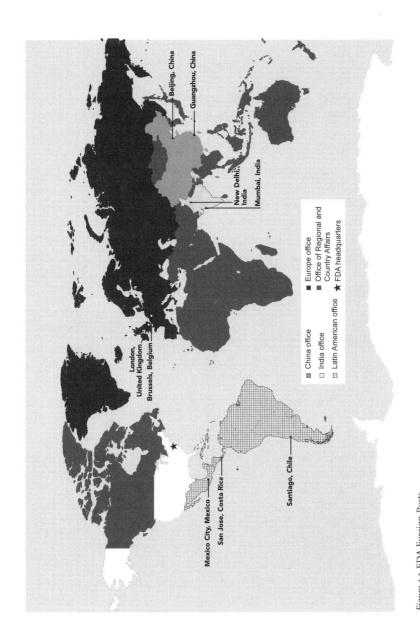

Figure 2.2 FDA Foreign Posts

Source: Food and Drug Administration 2015.

in the United States and in Latin American, Chinese, and Indian foreign posts. The presence of FDA staff in foreign offices throughout the world and the increased emphasis on international inspections have led to greater oversight of foreign regulated commodities. In addition, the establishment of foreign posts, combined with a variety of other activities led by FDA staff in the United States, has enhanced FDA's ability to deepen its collaborations and partnerships around the globe.

E. Deepening Our Global Collaborations and Partnerships

Hundreds of thousands of entities worldwide are processing, manufacturing, producing, and distributing FDA-regulated products for the U.S. market. These entities use increasingly diverse and complicated processes and manage complex and extended supply chains. The burgeoning scale and complexity of the innovative global regulatory system make it difficult for FDA to meet its public health mission on its own or by employing historic approaches and tools that were often limited to a domestic focus, such as checking shipments at the U.S. border or seizing product within the United States. FDA's success depends on working with other federal agencies and state, local, and international stakeholders.

FDA has worked collaboratively throughout its history with local, state, and federal agencies in a number of different areas including establishing national standards; developing and delivering training and certification programs to ensure a highly skilled workforce; developing mechanisms to share information; creating and executing laboratory proficiency programs; and responding quickly to problems related to regulated products. To complement these national efforts, FDA has worked internationally in formal and informal dialogues and through developing collaborative relationships among regulatory colleagues and international organizations. These global engagements have been helpful in building relationships with counterparts and exchanging or sharing information. FDA intends to continue to build on these relationships and focus on meaningful collaborations that leverage knowledge and resources. Following are some examples of FDA's many collaborations and partnerships from a medical product perspective; similar efforts are ongoing within the food commodity area at FDA. These collaborations and partnerships usually relate to information sharing, harmonization/convergence, and mutual reliance/oversight efforts.

F. Information Sharing

As collaborations and partnerships increase, so does the need for timely exchange of scientific and regulatory data and approaches. FDA continues to explore areas for alignment with global stakeholders in an effort to increase knowledge and understanding and allow regulatory agencies and other stakeholders to leverage their ability to detect, prevent, and respond to public health concerns.

FDA has long promoted engagements with regulatory counterparts related to information sharing through binding and nonbinding agreements. FDA has entered into more than 135 international arrangements with counterparts around the world that enable sharing of scientific knowledge, expertise, standards, regulations, and, in some cases, inspection or other nonpublic information (Food and Drug Administration 2014d).

In addition to these agreements, FDA has been supporting and working on numerous multilateral information-sharing platforms. For example, the Pan American Health Organization launched the Regional Platform on Access and Innovation for Health Technologies (PRAIS), an information hub to facilitate medical product data sharing in the Americas, serving as a regional model of a data-sharing system and network that can be expanded globally. The platform can enable an environment for regulatory exchange. FDA also supported a World Health Organization (WHO) multiregional platform for monitoring substandard, spurious, falsely labeled, falsified, and counterfeit (SSFFC) medical products to help improve data sources and collection methodologies globally. This global monitoring system is a means to share information on a global scale regarding counterfeit medical products. These information-sharing collaborations are an important aspect of FDA's efforts to leverage scientific and regulatory expertise and encourage others to do so as well.

G. Mutual Reliance/Recognition

Launched in May 2014, the Mutual Reliance Initiative is a strategic collaboration between the FDA and European Union (EU) to evaluate whether reliance on each other's inspection reports of human drug manufacturing facilities is possible. Strengthening reliance upon each other's expertise and resources will result in greater efficiencies for both regulatory systems and provide a more practical means to oversee the large number of drug manufacturing sites outside of the U.S. and EU. Both the FDA and EU have dedicated teams to assess the risks and benefits of entering into a mutual reliance agreement.

In this context, the term "mutual reliance" means that the EU and member state authorities and FDA would rely upon each other's factual findings from manufacturing quality/Good Manufacturing Practice (GMP) inspections to the greatest extent possible and generally not re-inspect a facility already inspected by the other party unless special circumstances exist. However, because of the differing legal systems and other factors, from time to time, the U.S. and EU might come to different conclusions about an application or enforcement action, and in such situations each would exercise the right to make its own decisions.

A mutual reliance agreement ultimately would allow both the FDA and EU authorities to reallocate and leverage limited resources to increase oversight of manufacturing operations with potentially higher risk to quality and safety and thus to public health, not only within our respective regions, but also in other parts of the world that manufacture pharmaceuticals for the U.S. and EU markets. This collaboration and redeployment of resources would benefit patients, as well, by refocusing efforts to better address problems before they result in adverse public health outcomes.

H. Harmonization and Convergence

FDA has a long history of supporting and participating in harmonization efforts of scientists and regulators. FDA is working through multilateral harmonization efforts such as the International Medical Device Regulators Forum (IMDRF) and the International Conference on Harmonisation of Technical Requirements for Registration of Pharmaceuticals for Human Use (ICH) to ensure the broadest possible reach of science-based standards and guidelines. FDA's participation in these multilateral harmonization efforts is significant and ongoing. For example, medical device regulators built upon the strong foundational work of the Global Harmonization Task Force on Medical Devices (GHTF)[2] and established the IMDRF (International

Medical Device Regulators Forum 2014). This new forum has an expanded membership of regulatory bodies, observer organizations, and related communities and is moving toward greater international uniformity and convergence in regulatory practices for medical devices. The ICH originally brought together the regulatory authorities and pharmaceutical industries of Europe, Japan, and the United States to increase harmonization in the interpretation and application of technical requirements for drug marketing approvals. Over time, the group has gradually evolved to become more global in nature, inviting observer organizations and interested parties into the discussion. Altogether, ICH has developed more than sixty harmonized guidelines aimed at eliminating duplication in the development and registration process. FDA is now involved in harmonization efforts in most regulated product areas. These harmonization efforts involve aligning different countries on science-based regulatory standards for the quality, safety, and efficacy of imported and exported products. Harmonization helps governments realize efficiencies in developing, implementing, and enforcing standards.

I. International Coalition

In addition to these ongoing initiatives related to information sharing, mutual reliance, and harmonization, an effort under development is the newly established International Coalition of Medicines Regulatory Authorities (ICMRA), which is a voluntary, senior-level, strategic coordinating, advocacy, and leadership entity that provides direction for a range of areas and activities that are common to many regulatory authorities' missions and goals. The ICMRA is anchored in the recognition by regulators of the need to address current and emerging human medicine regulatory and safety challenges globally, strategically, and transparently in ways where collective approaches and collaboration expand the regulatory reach of any one regulator in effective and sustainable ways. Over time, ICMRA will enable a global framework in which ICMRA participants will work on selected joint initiatives to stimulate collective thinking and action; pilot new ways of working to build mutual reliance; and facilitate early and timely identification of emerging public health crises that intersect with medical regulatory authorities, resulting in a coordinated, multilateral response to such crises.

F. Strengthening Regulatory Systems

The increase in global trade highlights the need to strengthen regulatory systems. FDA has had a century to develop its regulatory infrastructure, but many countries are still in the early stages of building theirs. Stronger regulatory systems will be increasingly important to FDA's ability to fulfill its mission to ensure the safety of the increasing volume of food, feed, medical products, and cosmetics that enter the United States from around the world. According to an Institute of Medicine study commissioned by FDA, *Ensuring Safe Foods and Medical Products Through Stronger Regulatory Systems Abroad*, the fundamental elements of regulatory systems are that they are independent, predictable, and recognize, collect, and transmit evidence when breaches of law occur (Institute of Medicine of the National Academies 2012). The report highlights these core elements that every system, rich or poor, should ideally have and makes recommendations for increased attention to them by the international community.

The international community increasingly recognizes the importance of strong regulatory systems. The Pan American Health Organization's (PAHO) Directing Council passed a resolution on strengthening regulatory systems in 2010 (Pan American Health Organization 2010), and at the 2012 World Health Assembly, member states passed a resolution that will facilitate international collaboration on SSFFC (World Health Organization 2012). Most recently, the United States cosponsored a resolution on strengthening regulatory systems that went to the World Health Assembly in May 2014. This resolution was based on collaborative work among FDA, WHO, and other member states, including Australia, Nigeria, South Africa, Mexico, and Switzerland. The resolution called for WHO to continue to support countries in the area of strengthening regulatory systems; to integrate regulatory systems into health systems's strengthening efforts; and to support subregional and regional networks to build regulatory capacity and cooperation.

Partnerships are also emerging. For example, FDA works in partnership with the Bill and Melinda Gates Foundation, the World Bank, the African Union, and others in support of the African Medicines Regulatory Harmonization (AMRH) initiative, which seeks to speed product registrations and improve market access and safety surveillance for critical medicines and vaccines. FDA is also providing support to PAHO to explore the potential of developing a Caribbean-wide regulatory system. Because many countries in the Caribbean are small, they are not able to perform critical regulatory functions on their own (e.g., marketing authorization, etc.), and some do not have regulatory authority over medicines. The building blocks for a Caribbean-wide regulatory system are already in place. The Caribbean has implemented regional coordination mechanisms through, for example, the Caribbean Community and Common Market and established a regional public health agency that includes a regional drug testing laboratory. Ministers of health also approved a Caribbean Pharmaceutical Policy in which implementing a subregional regulatory system is a priority. The broader context is that such a regional approach can be a model for how to sustainably build regulatory capacity in other parts of the world where there are similar resource constraints.

The 2012 Institute of Medicine report *Ensuring Safe Foods and Medicines Through Stronger Regulatory Systems Abroad* also recommended the development of a global curriculum of fundamental regulator competencies in response to the identified lack of high quality and consistent training for food and drug regulatory staff in ensuring food and drug safety across the globe, particularly in low- and middle-income countries. Acting on this recommendation, FDA partnered with key stakeholders, including WHO, PAHO, the Bill and Melinda Gates Foundation, the Regulatory Affairs Professional Society (RAPS), the International Food Protection Training Institute (IFPTI), and the Drug Information Association (DIA), to begin a discussion about the task of developing a global curriculum that governments, multilateral organizations, educational institutions, and others could use to train food and drug regulatory staff, particularly in low- and middle-income countries. The group subsequently wrote and published a stakeholder discussion paper in June 2013 (Preston 2013) and then began an effort to produce basic competencies that could lead to the development of a general global curriculum. This collaboration, led by RAPS and IFPTI, is scheduled for completion at the end of 2015. Its goal is to strengthen the skills of critically important regulatory professionals. By defining essential competencies and then developing curriculum modules, this project can meet the needs of different countries and regions, whether low resourced or not, and can adapt to changing science and technology.

IV. CONCLUSION

The vast and complex public health and regulatory panorama outlined in this chapter may appear daunting. However, we believe FDA is ready; indeed, for some time, the agency has moved in to address global regulatory challenges. Commissioner Hamburg noted, "Our job requires greater coordination of regulatory standards and practices across nations to ensure safety and quality, regardless of where a product is produced. We are working through bilateral and multilateral agreements, as well as through international organizations, specialized partnerships and various coordinating bodies. And at the highest levels, we have embarked on creating a new model or framework for global governance" (Hamburg 2014). This ongoing work complements the significant legislative authority received by the agency in the past two years.

Moreover, for some time, FDA has been reorganizing its structure to transform from a domestic agency operating in a globalized world to a truly global agency fully prepared for a complex regulatory environment that takes into account the risks across a product's life. We know this requires information and workload sharing with the help of regulatory partners, data-driven analytics, and allocation of resources achieved through public and private partnerships. Global approaches to public health problems will become standard.

The rapid and profound rise of global commerce and trade requires that FDA continue to evolve and meet new demands—the demands of global approaches to public health problems have become inherent in FDA's daily work and long-term strategy. FDA has significantly strengthened its ability to respond to these challenges and know where and how it must do more in its mission of protecting and promoting the health of the American public.

NOTES

1. The FDA Modernization Act (FDAMA) of 1997 granted FDA authority to "participate through appropriate processes with representatives of other countries to reduce the burden of regulation, harmonize regulatory requirements, and achieve appropriate reciprocal arrangements." (21 USC § 393(b)(3),

added by Pub. L. No. 105-115, 111 Stat. 2369, § 406). This language recognized and mandated FDA's participation in harmonization efforts, such as the International Conference on Harmonisation of Technical Requirements for Registration of Pharmaceuticals for Human Use (ICH).

2. GHTF was formed in 1992 in an effort to develop greater uniformity among national medical device regulatory systems. Historical information concerning GHTF is available at http://www.imdrf.org/ghtf/ghtf-archives.asp.

REFERENCES

Autor, Deborah M. 2011. "Securing the Pharmaceutical Supply Chain." Food and Drug Administration (September 14). http://www.fda.gov/newsevents/testimony/ucm271073.htm.

Centers for Disease Control and Prevention. 2014. "Estimates of Foodborne Illness in the United States." http://www.cdc.gov/foodborneburden/index.html.

Food and Drug Administration. 2011. "Pathway to Global Product Safety and Quality" (July 7). http://www.fda.gov/downloads/aboutfda/centersoffices/officeofglobalregulatoryoperationsandpolicy/globalproductpathway/ucm262528.pdf.

Food and Drug Administration. 2013. "Public Meeting: Implementation of Drug Supply Chan Provisions of Title VII of FDASIA" (July 12). http://www.fda.gov/regulatoryinformation/legislation/federalfooddrugandcosmeticactfdcact/significantamendmentstothefdcact/fdasia/ucm357783.htm.

Food and Drug Administration. 2014a. "Background on the FDA Food Safety Modernization Act (FSMA)." http://www.fda.gov/Food/GuidanceRegulation/FSMA/ucm239907.htm.

Food and Drug Administration. 2014b. "Drug Supply Chain Security Act (DSCSA)." http://www.fda.gov/drugs/drugsafety/drugintegrityandsupplychainsecurity/drugsupplychainsecurityact/

Food and Drug Administration. 2014c. "Food and Drug Administration Safety and Innovation Act (FDASIA)." http://www.fda.gov/regulatoryinformation/legislation/federalfooddrugandcosmeticactfdcact/significantamendmentstothefdcact/fdasia/ucm20027187.htm.

Food and Drug Administration. 2014d. "International Arrangements." http://www.fda.gov/InternationalPrograms/Agreements/default.htm.

Food and Drug Administration. 2014e. "Program Alignment Group Recommendations—Decision" (February 3). http://www.fda.gov/AboutFDA/CentersOffices/ucm392738.htm.

Food and Drug Administration. 2015. "GO 101" (January). http://www.fda.gov/downloads/AboutFDA/CentersOffices/OfficeofGlobalRegulatoryOperationsandPolicy/UCM405133.pdf.

Hamburg, Margaret A. 2010. "Remarks at the Consumer Healthcare Products Association Annual Executive Conference." Food and Drug Administration (March 12). http://www.fda.gov/newsevents/speeches/speecharchives/ucm210009.htm.

Hamburg, Margaret A. 2013. "President's Fiscal Year 2014 Budget Request for the FDA." Food and Drug Administration (April 26). http://www.fda.gov/NewsEvents/Testimony/ucm351980.htm.

Hamburg, Margaret A. 2014. "Speech by Dr. Hamburg to the Food and Drug Law Institute (FDLI) 2014 Annual Conference." Food and Drug Administration (April 23). http://www.fda.gov/NewsEvents/Speeches/ucm394646.htm.

Institute of Medicine of the National Academies. 2012. "Ensuring Safe Foods and Medical Products Through Stronger Regulatory Systems Abroad." National Academies Press. April. http://www.iom.edu/Reports/2012/Ensuring-Safe-Foods-and-Medical-Products-Through-Stronger-Regulatory-Systems-Abroad.aspx.

International Medical Device Regulators Forum. 2014. "About IMDRF." http://www.imdrf.org/about/about.asp.

Pan American Health Organization. 2010. "Strengthening National Regulatory Authorities for Medicines and Biologicals." CD50.R9, 50th Directing Council, 62nd Sess. http://new.paho.org/hq/dmdocuments/2010/CD50.R9-e.pdf.

Preston, Charles et al. 2013. "Developing a Global Curriculum for Regulators." Institute of Medicine of the National Academies (June 11).

Sklamberg, Howard. 2014. "Counterfeit Drugs: Fighting Illegal Supply Chains." Food and Drug Administration (February 27). http://www.fda.gov/NewsEvents/Testimony/ucm387449.htm.

World Health Organization. 2012. "65th World Health Assembly Closes with New Global Health Measures" (May 26). http://www.who.int/mediacentre/news/releases/2012/wha65_closes_20120526/en/index.htm.

CHAPTER THREE

FDA and the Rise of the Empowered Patient

LEWIS A. GROSSMAN

AS RECENTLY as the 1970s, patients played virtually no role in the U.S. Food and Drug Administration's (FDA's) process for approving drugs (Best 2012; Carpenter 2004). The drug development and review process was the exclusive domain of government bureaucrats, scientific experts, and the pharmaceutical industry. Not coincidentally, the agency provided desperately ill patients with access to promising investigational drugs only occasionally, on an ad hoc basis.

Today highly organized and well-funded patient advocacy groups regularly seek to influence FDA's decisions regarding pharmaceutical products. The agency holds public meetings with patient advocates regarding drug development issues, includes them as "special government employees" in private preapproval discussions with manufacturers, and consults with them during internal agency review of drugs. Patients also regularly testify before FDA advisory committees, which often include a patient representative as a voting member. Both Congress and FDA explicitly acknowledge the importance of the patient's perspective with respect to risk–benefit analyses, and they have established programs to hasten the approval of important drugs and to provide early access to desperately needed unapproved drugs. The Food and Drug Administration Safety and Innovation Act of 2012 (FDASIA) reflects and requires this patient-centered ethos in various ways discussed in this chapter.

While the extent of patients' actual impact on FDA decision making remains unclear, they irrefutably wield much more influence than they did four decades ago. This chapter explores a constellation of trends and events that underlie this dramatic shift, examines the current role of patients in drug regulation, and considers the future of patient-focused drug development.

I. THE CULTURAL FOUNDATION

Various cultural developments of the past half century underlie the transformation of the American patient from a passive, uninformed recipient of medications into an active, knowledgeable participant in drug development and regulation.

A. The Decline of Trust

One critical trend has been the citizenry's declining trust in the leaders of major institutions, including those that formerly exercised exclusive control over the drug supply. This decay of confidence has driven patients to intervene in the drug regulatory process rather than passively accept the decisions of governmental and scientific experts.

A loss of faith in large institutions and professional expertise was one of the defining characteristics of the 1970s (Berkowitz 2006; Schulman 2002; Carroll 2000), and Americans have remained skeptical about the motives and competence of "the establishment" ever since. Between the mid-1960s and 1980, the percentage of poll respondents saying "you can trust" the leaders of various institutions plummeted with respect to all institutions considered, including the federal government, the medical profession, and major companies. Although the public's confidence has waxed and waned since 1980, its trust level has never approached its mid-1960s peak (Harrisinteractive.com 2011; Website of American National Election Studies). FDA appears to have maintained the public's esteem longer than other institutions. Since the turn of the twenty-first century, however, confidence in the FDA has fallen precipitously, as well (Carpenter 2010:12, 749–50).

B. The Rights Revolution

The 1970s have also frequently been identified as the period of the "rights revolution" (Berkowitz 2006:133–57; Sandbrook 2011:249–50). Concepts like women's rights, gay rights, environmental rights, disability rights, and consumer rights dominated the national conversation. Another important aspect of the rights revolution that blossomed in the 1970s was the notion of "patients' rights." The genesis of the patients' rights movement appears to have been the drafting in 1970 of twenty-six such rights by the National Welfare Rights Organization. This action precipitated a widespread discussion that culminated in the adoption of a "Patient's Bill of Rights" in 1973 by the American Hospital Association. This document, and indeed the entire patients' rights movement, was premised on the notion that patients would no longer simply accept the ministrations of a presumptively wise and beneficent medical system but would instead become active, informed agents in their own treatment. It is important to note that the patients' rights movement overlapped in significant ways with other rights movements—for example, those for disability rights, racial civil rights, women's rights, and—most importantly for this chapter—gay rights.

The rights culture and rhetoric that coalesced in the 1970s lived beyond it (Walker 1998:33). One author argues that the rights revolution "survived the Reagan and Bush years unscathed and even enhanced" (Sandbrook 2011:250). This assertion is certainly true with respect to patients' rights.

C. The Changing Health Information Environment

Another noteworthy cultural development of the past several decades is the enormous increase in health information available to the general public. Although domestic health guides have existed throughout American history, until quite recently publishers did not seek a mass market for books containing technical information about the treatments administered and prescribed by physicians. This changed in 1979, when Bantam released the first edition of *The Pill Book*, subtitled *The Illustrated Guide to the Most Prescribed Drugs in the United States*. Seventeen printings of the first edition totaled more than one million copies. Competitors, including the American Medical Association, soon offered their own volumes with similar information. For consumers not daunted by technical language, the *Physicians' Desk Reference*, containing the full physician package insert for every approved drug, became widely available in bookstores shortly after *The Pill Book* and was an immediate bestseller (Fleischer 1981).

Today, of course, the significance of printed books about prescription drugs pales in comparison to that of the Internet, which has made it easy for anyone with a computer or smartphone to find detailed medical information. During the 1990s, thousands of health-oriented sites appeared on the new World Wide Web. Commercial websites that organized this universe of information for consumers began appearing in the late 1990s. By 2007, 40 million unique users visited WebMD alone each month (Freudenheim 2007). Although advanced search-engine technology has since reduced the importance of such websites, American patients remain voracious (and in many instances, sophisticated) consumers of medical information (Duggan and Fox 2013).

Finally, patients today receive abundant information about prescription drugs directly from the manufacturers. As explored in detail elsewhere (Grossman 2014), until the 1980s, the typical consumer never saw labeling or advertising for any prescription drug, with the possible exception of the patient package insert for oral contraceptives. FDA traditionally believed (in agreement with physicians and industry) that consumers of prescription medications should be ignorant and passive beneficiaries of their doctors' care. However, a dramatic transformation in agency policies during the past several decades has led to a deluge of patient labeling and direct-to-consumer advertising of prescription drugs.

II. THE RISE OF FOOD AND DRUG ACTIVISM

The mass mobilizations of AIDS and breast cancer activists in the 1980s are often identified as the pioneering models for modern patient activism (Carpenter 2004:58; Bix 1997). But even these movements had precedents—campaigns in the 1970s involving vitamin pills, saccharin, and Laetrile (an alternative cancer therapy). These earlier social movements, though not focused on conventional prescription drugs, demonstrated how organized citizens could successfully fight to preserve consumers' freedom to make their own risk–benefit judgments about FDA-regulated products.

In August 1973, FDA issued a rule restricting the permissible combinations of nutrients in vitamin and mineral supplements (38 Fed. Reg. 20,723). The rule also declared that the agency would regulate high-potency supplements as drugs and, in some instances, *prescription* drugs. A health libertarian organization with right-wing propensities, the National Health Federation (NHF), was at the heart of the opposition, but the dissent blossomed into a genuine popular movement. The NHF choreographed a demonstration against "nutritional tyranny" in Washington. Its alarmist (and inaccurate) warnings that "the Government is going to take our vitamins away" triggered what the *New York Times* characterized as a "massive flow of letters" to Congress (Lyons 1973). Vitamin deregulation was, along with Watergate, the energy crisis, and the economy, one of the four issues that generated the most mail to Congress in 1973 ("Of Vitamins, Mineral: Fighting the FDA" 1974).

In 1974 testimony supporting congressional intervention, the legislative counsel for a health food industry trade group voiced a regulatory philosophy that seemed to reflect the views of a broad swath of Americans: "As long as he is not dealing with dangerous or untruthfully labeled food, then risktaking [sic] should be for each man to decide for himself . . ." (Senate Subcommittee on Health 1974). The next day, the U.S. Court of Appeals for the Second Circuit partially struck down FDA's vitamin and mineral regulation (*National Nutritional Foods Ass'n v. FDA* 1974:761). Two years later, Congress—without a single dissenting vote—invalidated the remainder of the rule with the Vitamin-Mineral Amendments of 1976 (90 Stat. 401, 410).

Another mass protest against FDA arose in response to the agency's April 1977 proposal to revoke the interim food additive approval for saccharin, which was then the only artificial sweetener on the market (42 Fed. Reg. 19,996). After studies demonstrated saccharin's carcinogenicity in rats, the agency did not really have any discretion in the matter, for the Federal Food, Drug, and Cosmetic Act's (FDCA's) Delaney Clause states that "no additive shall be deemed to be safe if it is found to induce cancer when ingested by man or animal" (21 USC § 409(c)(3)(A)). After publishing the proposed rule revoking the approval, however, the agency reported that "the protest is stronger and louder than any response in recent history" (Winter 1977). Outraged citizens included not only diabetics (and their physicians), but also millions of people who drank diet soda to limit caloric intake. A Harris survey found that Americans opposed the saccharin ban by 76 percent to 15 percent (Harris 1977).

As it had in the vitamin-mineral supplement controversy, Congress stepped in. By an overwhelming margin, it enacted legislation in 1977 to suspend FDA's prohibition of saccharin while also mandating warnings of its carcinogenicity in animals (91 Stat. 1451). Senator Edward Kennedy queried: "If a substance has both benefits and risks, who should decide whether the risk should be taken—the Federal Government or the individual?" (123 Cong. Rec. S29352 (Sept. 15, 1977)). He concluded, in light of saccharin's undoubted benefits and the division of opinion regarding its safety, that "the individual is in the best position to decide for himself or herself whether they [sic] want to expose themselves or their children to saccharin use."

At about the same time, the first-ever mass protest against FDA's refusal to permit marketing of a *drug* was gaining momentum. The product at issue was Laetrile (amygdalin), an alternative cancer treatment derived from apricot pits. Although Laetrile advocacy was formerly confined largely to right-wing extremists, the 1972 arrest of a California doctor who prescribed Laetrile "launched a significant SM [social movement] . . . as people from across the political spectrum united under the libertarian banner of medical freedom" (Hess 2005:522).

In May 1977, FDA held court-ordered public administrative hearings to resolve questions regarding Laetrile's legal status. These hearings, jammed with rowdy Laetrile supporters, took on an almost riotous atmosphere (Young 1992:224; "Laetrile Backers, Foes Clash at FDA Hearing" 1977). Meanwhile, U.S. Representative Steven D. Symms, citing "grass roots support" stemming from outrage over the Laetrile situation, was promoting his "Medical Freedom of Choice Bill" (H.R. 54; "Legalize Laetrile as a Cancer Drug? Interview with Representative Steven D. Symms" 1977). This law would have repealed the power FDA acquired in 1962 to review the efficacy as well as the safety of new drugs before marketing. The Symms bill gained at least

110 cosponsors in the House of Representatives (123 Cong. Rec. 18752 (June 13, 1977)). The *Washington Post* opined that the Laetrile matter was "already out of [the] control" of the "professionals" and "bureaucrats" (Editorial 1977). Even the editor of the *New England Journal of Medicine* suggested that FDA legalize Laetrile to calm the "Laetrilomania" (Ingelfinger 1977).

In July 1977, a poll showed that 58 percent of Americans believed Laetrile should be sold legally, versus 28 percent who opined that it should remain illegal (The Roper Organization 1977). In response to boisterous state legislative hearings and torrents of mail to lawmakers, half of the states passed Laetrile legalization statutes (Young 1992:221). The public's interest in the drug faded in the early 1980s (Young 1992:232–33), and the Laetrile campaign ultimately had no direct effect on federal food and drug regulation. Nonetheless, the Laetrile supporters showed how patient advocacy movements could shake FDA to its foundations—an important lesson for their successors, the AIDS activists.

III. AIDS ADVOCACY

With the terrifying spread of AIDS in the 1980s, groups such as ACT UP, Project Inform, and the Gay Men's Health Crisis commenced an epic, ultimately successful struggle to shape FDA's decisions regarding drugs for the disease (Arno and Feiden 1992; Carpenter 2010:428–57; Epstein 1996; Hilts 2003:236–54).

In 1986, under pressure from AIDS groups, FDA made the unapproved investigational drug AZT available to patients outside of formal clinical trials on a "compassionate use" basis. The next year, FDA approved the new drug application (NDA) for AZT, despite the absence of large Phase III clinical investigations and experts' doubts regarding its safety and effectiveness. Less than two years elapsed between the submission of AZT's investigational new drug (IND) application and its NDA approval—an astonishingly brief period compared to most drugs. On the very same day in 1987 that FDA approved AZT, it further responded to AIDS activists' demands by proposing a "Treatment IND" rule (52 Fed. Reg. 8850). This regulation, finalized two months later (52 Fed. Reg. 19466), permitted seriously ill people lacking satisfactory alternatives to gain access to investigational drugs that "may be effective," although this access was strictly limited to ensure that the drugs would also be tested in controlled clinical studies.

Despite these successes, the AIDS interest groups' victory was far from complete. Later in 1987, an FDA advisory committee recommended against approving the NDA for ganciclovir, a promising treatment for a blindness-inducing viral infection common among people with AIDS. The AIDS groups, outraged by the committee's recommendation, threatened action. They became even more furious when FDA imposed extremely strict access restrictions on the treatment IND for another drug, trimetrexate. Following stormy congressional hearings, FDA surrendered and broadened the terms of the trimetrexate IND.

AIDS activists nonetheless feared that the agency would remain an obstinate barrier to early access to promising drugs. In October 1988, ACT UP conducted a highly publicized symbolic takeover of FDA headquarters in suburban Maryland, protesting the agency's approach to ganciclovir, trimetrexate, and other treatments. The handbook for the action declared: "The FDA says it exists to protect consumers. Well, people with HIV are consumers too, and they need to be protected from a deadly disease" (Eigo 1988).

After this demonstration, FDA seemed more responsive to the concerns of AIDS patients. Just eight days after the takeover, the agency promulgated an interim regulation, "Subpart E," which facilitated faster development and approval of drugs for life-threatening and severely debilitating diseases. Subpart E did so by guaranteeing drug companies early consultation on study design, authorizing NDA approvals based solely on Phase II trial results, and implementing a more flexible risk–benefit analysis that considered "the severity of the disease and the absence of satisfactory alternative therapy" (21 CFR 312 Subpart E, § 312.84(a)).

The activists' success in influencing FDA policy became further apparent in connection with ddI, a drug closely related to AZT. In response to continuing pressure from the AIDS community, FDA embraced a "parallel track" approach to ddI, allowing patients who did not qualify for the ongoing Phase II trials of the drug to take it for treatment purposes if AZT did not help them. In 1989, AIDS activists, with the assistance of FDA and National Institutes of Health (NIH) officials, persuaded ddI's manufacturer to make the drug available at no cost to these patients. FDA soon officially embraced the "parallel track" mechanism more generally (55 Fed. Reg. 20856 (May 21, 1990); 57 Fed. Reg. 13250 (April 15, 1992)). Moreover, in 1991, the agency approved the NDA for ddI before the completion of the Phase II trials, based on data showing efficacy in achieving surrogate endpoints (rather than longer survival). The next year, FDA formalized this approach, as well, when it promulgated its Accelerated Approval ("Subpart H") regulations, which permit the approval of drugs for serious or life-threatening illnesses based on their effect on unvalidated surrogate endpoints reasonably likely to predict clinical benefit (21 CFR 314 Subpart H).

Congress subsequently incorporated many of the AIDS movement's demands into the FDCA itself. The Food and Drug Administration Modernization Act of 1997 (FDAMA) (111 Stat. 2296) codified FDA's Treatment IND rule, the Subpart E regulations (under the rubric "Fast Track"), and an expanded version of the Accelerated Approval rule (21 USC §§ 356, 360bbb).

IV. BEYOND AIDS

The reform of FDA's drug approval system was only one of the AIDS movement's visible triumphs in the late 1980s and early 1990s. It also effectively lobbied NIH and Congress for dramatic increases in federal funding for research into the disease (Bix 1997:781). Other disease communities noticed the AIDS activists' achievements (along with breast cancer advocates' contemporaneous success in garnering federal research funds), and a spectacular proliferation of patient advocacy organizations occurred. Whereas some of these organizations were funded by the pharmaceutical industry, others represented only the interests of disease victims. The number of large, disease-focused nonprofits swelled from 400 at the beginning of the 1990s to more than 1,000 in 2003 (Best 2012:781).

Today, following the model forged by the AIDS community, these disease groups regularly seek to sway FDA drug approval decisions. FDA advisory committee meetings, once technical affairs attended solely by scientists, bureaucrats, lawyers, and corporate officials, are now sometimes crowded with representatives of disease advocacy organizations, some of whom offer impassioned testimony. Moreover, patient advocates now often serve on the advisory committees themselves.

Their presence is another legacy of the AIDS movement. In 1991, in response to activists' demands, FDA created a position for a patient representative on the Antiviral Drugs Advisory Committee for HIV (Food and Drug Administration 2015). Inspired by this development, cancer organizations asked for similar representation on committees reviewing cancer-related therapies, and the Clinton administration granted this request in 1996 (Clinton & Gore 1996:9).[1]

Patients are now important and visible actors in the advisory committee process. Consider, for example, a September 2012 meeting of the Oncologic Drugs Advisory Committee concerning the neoadjuvant breast cancer drug Perjeta. Participants in the open public hearing session included both individual survivors and representatives of Facing Our Risk of Cancer Empowered (FORCE) and Breastcancer.org (Food and Drug Administration 2013a). The committee itself included a voting patient representative, Debra Madden, a two-time cancer survivor. The advisory committee ultimately voted unanimously in favor of granting accelerated approval to Perjeta for patients with early stage breast cancer before surgery, and FDA followed this recommendation.

It is difficult to assess precisely how influential patient activities were to this and other NDA approvals. But in a few prominent instances, patient activism has unambiguously changed outcomes. For example, in response to protests by sufferers of irritable bowel syndrome, in 2002 FDA permitted the return to the market of Lotronex, a drug previously withdrawn because of occasional severe side effects. Moreover, statistical analyses, along with anecdotal and observational evidence, indicate that effective political and media campaigns by patient groups shorten FDA drug review times (Carpenter 2004:59–60). Furthermore, the accelerated approval and fast track procedures, which owe their very existence to patient activism, have been extremely useful in getting needed treatments to desperate patients more quickly.

Another indication of FDA's growing patient orientation in drug development is the agency's creation, in cooperation with the Critical Path Institute and the pharmaceutical industry, of a Patient-Reported Outcome (PRO) Consortium. This public-private initiative's mission—to evaluate and qualify PRO instruments for use as endpoints in clinical trials—is premised on the understanding that "[m]ore and more, it has been recognized that some assessments [of investigational medical products] might best be made by the patients themselves"

(Food and Drug Administration 2010). The consortium has held annual workshops since 2010.

The impact of patient activism on FDA regulation of drugs should not be overstated. For example, treatment INDs remain rare. Although this scarcity is due primarily to manufacturers' reluctance to participate in early access programs and not to any FDA hesitancy to authorize them (Groopman 2006),[2] one reason for the industry's lack of enthusiasm is the severe restrictions the agency imposes on charging for investigational drugs (21 CFR 312.8). In 2007, these restrictions survived a constitutional challenge brought by a patient advocacy group in the U.S. Court of Appeals for the District of Columbia Circuit (*Abigail Alliance* 2007).

Nevertheless, the patient activists of the 1980s and 1990s undeniably drove FDA to reconceive the essence of its mission. The agency embraced the task not only of *protecting* the public health by preventing the sale of dangerous products, but also of *enhancing* the public health by ensuring rapid access to useful remedies (Arno and Feiden 1992:109). In 1997, Congress amended FDA's official mission statement to reflect this shift—listing health promotion *before* health protection (21 USC § 1003(b)).

A. Food and Drug Administration Safety and Innovation Act of 2012

The Food and Drug Administration Safety and Innovation Act of 2012 (126 Stat. 993) represents a further advance in the rise of a patient-centered ethos, especially in the way the statute mandates patient involvement in earlier stages of the drug development process. FDASIA adds a new section 569C to the FDCA, titled "Patient Participation in Medical Product Discussion" (21 USC § 360bbb-8c). This section obligates FDA to "develop and implement strategies to solicit the views of patients during the medical product development process and consider the perspectives of patients during regulatory discussions." To this end, it specifically instructs FDA to encourage the participation of patient representatives in agency meetings with the sponsors of drug, device, and biologic applications. In addition, FDASIA adds a new "breakthrough therapy" designation to expedite development of drugs that may offer substantial improvement over existing therapies, expands the situations in which accelerated approval is available, and broadens the surrogate endpoints on which such approval can be based (21 USC §§ 356(a), (c)).

FDASIA has prompted FDA to embrace a broad initiative titled "Patient-Focused Drug Development." Title I of FDASIA, the Prescription Drug User Fee Amendments of 2012, binds FDA to detailed performance goals for 2013 through 2017 set forth in a separate document (126 Stat. 993, 996). The goals enunciated by FDA in this document promise to move patients ever closer to the center of federal drug regulation. Under the heading of "Enhancing Benefit–Risk Assessment in Regulatory Decision-Making," FDA commits not only to increasing its use of patient consultants in regulatory discussions about specific products but also to holding four meetings per year with patient advocates regarding various disease areas (Food and Drug Administration 2013b). In its September 2013 notice of this series of meetings, FDA explained: "Patients who live with a disease have a direct stake in the outcome of the review process and are in a unique position to contribute to weighing benefit–risk considerations that can occur throughout the medical product development process" (77 Fed. Reg. 58848, 58849).

As part of its patient-centered initiative, FDA has launched a program called the "FDA Patient Network" (Food and Drug Administration 2015). Among other functions, this program educates consumers about the development and approval of medical products, announces advisory committee meetings, provides information about clinical trials and early access programs, and recruits volunteers to serve as patient representatives on advisory committees and within the product review divisions.

V. THE FUTURE

Further integration of patients into drug regulation presents both opportunities and challenges for FDA. Such measures are likely to boost the agency's reputation and public support, as they did in the 1980s and 1990s (Carpenter 2010:398). But any respect and good faith so earned could quickly dissolve if the agency's eagerness to satisfy the demands of sympathetic patients were to result in a postapproval, Vioxx-type safety fiasco. FDA can mitigate this risk, however, through liberal and creative use of its power to approve drugs subject to a Risk Evaluation and Mitigation Strategy (REMS). FDA might also face continuing pressure from patient groups to tweak its early access programs to entice more manufacturers to participate in them—perhaps by revisiting the current limits on charging for investigational drugs (21 CFR 312.8). But if these programs become more widely used, FDA may struggle to preserve the integrity of pivotal controlled clinical trials. Overall, the agency may struggle to simultaneously satisfy patients and maintain its standing as the planet's "gold standard" gatekeeper for new drugs.

In addition, FDA must consciously strive to avoid devoting a disproportionate share of its limited resources to diseases endemic to populations with the wherewithal to support the most effective advocacy organizations. The AIDS movement was successful in part because it was dominated by a corps of middle-class, highly educated white men with experience in social movement tactics (Epstein 1996:12). Not all disease groups have these advantages. And the AIDS crisis notwithstanding, prioritizing a disease because it happens to be represented by well-funded and skillful advocates may be both unwise and unjust (Dresser 2001:79–83). Furthermore, FDA should not assume that the leaders of disease organizations necessarily represent their entire diverse constituencies (Dresser 2001). Finally, the agency should carefully assess the extent to which any particular disease group represents independent interests rather than those of industry backers.

Patient groups trying to shape drug regulation face challenges, too. First, they increasingly must reckon not only with FDA and NIH but also with the Centers for Medicare and Medicaid Services (CMS). As the number of insured Americans rises, the coverage decisions of CMS are having a growing impact on patient access to particular drugs. Accordingly, patient activists are focusing more energy on this agency. For example, in 2011, patient mobilization helped induce CMS to pledge to continue covering Avastin for breast cancer after FDA withdrew approval of the drug for that use. And in light of "stakeholder input"—including furious protest by patients—CMS in 2014 withdrew a proposal to allow formulary limitations in Medicare Part D coverage for three classes of drugs (Pear 2014).

Another challenge to patient organizations arises from the increasing specificity of illness and treatment resulting from medical advances, particularly in genetics. What will happen to today's disease-based groups if growing knowledge about disorders, together with the rise of gene-based personalized medicine, fractures the unity, and perhaps even the coherence, of current disease classifications? Consider, for example, fibromyalgia and chronic fatigue syndrome—diseases of unknown, and perhaps variegated, etiologies. Each has its own national patient advocacy association, and each was the topic of one of the first "patient-focused drug development" meetings held by FDA. But one can easily imagine a future in which these syndromes disintegrate into multiple, distinct subtypes, each with its own organization. The path from the American Cancer Society (1913) to the Leukemia & Lymphoma Society (1949) to the Lymphoma Foundation of America (1986) to the Cutaneous Lymphoma Foundation (1998) presents a model that, if expanded throughout the world of morbidity, could result in uncoordinated activism and intergroup competition. Disease advocacy groups, which have long shared certain overarching goals with respect to FDA policy, may start ferociously competing with each other for the agency's resources and attention the way they already vie for NIH funds.

Even as they confront these challenges, however, patients are unlikely ever to lose their hard-earned influential role in American drug regulation.

NOTES

This work is derived in part from an article previously published in the *Administrative Law Review*: Lewis A. Grossman. 2014. "FDA and the Rise of the Empowered Consumer." *Administrative Law Review* 66: 627–77.

1. FDA gave these representatives full voting privileges later that year (Food and Drug Administration 2015).

2. In 2014, FDA stated that it had approved 99 percent of all expanded access requests since October 2009 (Silverman 2014).

REFERENCES

Abigail Alliance v. Von Eschenbach, 495 F.3rd 695 (D.C. Cir. 2007) (en banc).
Arno, Peter S. and Karyn L. Feiden. 1992. *Against the Odds: The Story of AIDS Drug Development, Politics, and Profit*. New York: HarperCollins.

Berkowitz, Edward. 2006. *Something Happened*. New York: Columbia University Press.
Best, Rachel Kahn. 2012. "Disease Politics and Medical Research Funding: Three Ways Advocacy Shapes Policy." *American Sociological Review* 77:780–803.
Bix, Amy Sue. 1997. "Diseases Chasing Money and Power: Breast Cancer and AIDS Activism Challenging Authority." *Journal of Policy History* 9:5–32.
Carpenter, Daniel P. 2004. "The Political Economy of FDA Drug Review: Processing, Politics, and Lessons for Policy." *Health Affairs* 23:52–63.
Carpenter, Daniel. 2010. *Reputation and Power: Organizational Image and Pharmaceutical Regulation at the FDA*. Princeton: Princeton University Press.
Carroll, Peter N. 2000. *It Seemed Like Nothing Happened: The Tragedy and Promise of America in the 1970s*. 1982. Reprint. New Brunswick, N.J.: Rutgers University Press.
Clinton, Bill and Al Gore. 1996. *Reinventing the Regulation of Cancer Drugs*. National Performance Review.
Dresser, Rebecca. 2001. *When Science Offers Salvation: Patient Advocacy and Research Ethics*. New York: Oxford University Press.
Duggan, Maeve, and Susannah Fox. 2013. "Health Online 2013." *Pew Research Internet Project* (January 15).
Editorial. 1977. "Why Not a Laetrile Bill." *Washington Post* (May 22).
Eigo, Jim et al. *FDA Action Handbook*, September 12, 1988. Available on ActUp website.
Epstein, Steven. 1996. *Impure Science: AIDS, Activism, and the Politics of Knowledge*. Berkeley: University of California Press.
Fleischer, Leonore. 1981. "Letter from New York: Getting to the Top." *Washington Post* (April 19).
Food and Drug Administration. 2010. "The Patient Reported Outcomes (PRO) Consortium." (November 1). http://www.fda.gov/AboutFDA/PartnershipsCollaborations/PublicPrivatePartnershipProgram/ucm231129.htm.
Food and Drug Administration. 2013a. "Oncologic Drugs Advisory Committee Meeting Transcript." (September 12). http://www.fda.gov/downloads/AdvisoryCommittees/CommitteesMeetingMaterials/Drugs/OncologicDrugsAdvisoryCommittee/UCM377714.pdf.
Food and Drug Administration. 2013b. "PDUFA Reauthorization Performance Goals and Procedures, Fiscal Years 2013 through 2017." http://www.fda.gov/downloads/ForIndustry/UserFees/PrescriptionDrugUserFee/UCM270412.pdf.
Food and Drug Administration. 2015. "About the FDA Patient Network." http://www.fda.gov/forpatients/about/default.htm.

Freudenheim, Milt. 2007. "AOL Founder Hopes to Build New Giant Among a Bevy of Health Care Web Sites." *New York Times* (April 16).
Groopman, Jerome. 2006. "The Right to a Trial: Should Dying Patients Have Access to Experimental Drugs?" *The New Yorker* (December 18), 40–47.
Grossman, Lewis A. 2014. "FDA and the Rise of the Empowered Consumer." *Administrative Law Review* 66:627–77.
Harris, Louis. 1977. "76 Per Cent [sic] Majority Opposes Ban on Saccharin." *Chicago Tribune* (April 21).
Harrisinteractive.com. 2011. "Confidence in Congress and Supreme Court Drops to Lowest Level in Many Years." (May 18).
Hess, David J. 2005. "Technology- and Product-Oriented Movements: Approximating Social Movement Studies and Science and Technology Studies." *Science, Technology, and Human Values* 30:515–35.
Hilts, Philip J. 2003. *Protecting America's Health: The FDA, Business, and One Hundred Years of Regulation*. New York: Alfred A. Knopf.
Ingelfinger, F. J. 1977. "Laetrilomania." *New England Journal of Medicine* 296:1167–68.
"Laetrile Backers, Foes Clash at FDA Hearing." 1977. *Washington Post* (May 3), at A4.
"Legalize Laetrile as a Cancer Drug? Interview with Representative Steven D. Symms." 1977. *U.S. News and World Report* (June 13), 51.
Lyons, Richard D. 1973. "Disputed Health Lobby Is Pressing for a Bill to Overturn Any Limits on Sales of Vitamins." *New York Times* (May 14).
National Nutritional Foods Ass'n v. FDA, 504 F.2d 761 (2d Cir. 1974).
"Of Vitamins, Minerals: Fighting the FDA." 1974. *Washington Post* (January 20).
Pear, Robert. 2014. "White House Withdraws Plan Allowing Limits to Medicare Coverage for Some Drugs." *New York Times* (March 20), at A13.
Sandbrook, Dominic. 2011. *Mad as Hell: The Crisis of the 1970s and the Rise of the Populist Right*. New York: Anchor Books.
Schulman, Bruce J. 2002. *The Seventies: The Great Shift in American Culture, Society, and Politics*. Cambridge, Mass.: De Capo Press.
Senate Subcommittee on Health, Senate Committee on Labor and Public Welfare. 1974. (August 14).
Silverman, Ed. 2014. "The FDA Says It's More Compassionate Than You Think." *Corporate Intelligence Blog, Wall Street Journal Online* (May 5). http://blogs.wsj.com/corporate-intelligence/2014/05/05/the-fda-says-its-more-compassionate-than-you-think/.
The Roper Organization. 1977. "Roper Reports 77-7" (July 9–16).
Walker, Samuel. 1998. *The Rights Revolution: Rights and Community in Modern America*. New York: Oxford University Press.

Website of American National Election Studies. "Trust the Federal Government 1958–2008." http://www.electionstudies.org/nesguide/toptable/tab5a_1.htm.

Winter, Christine. 1977. "Bitter Days Ahead? Consumers Protest Life sans Saccharin." *Chicago Tribune* (March 17).

Young, James Harvey. 1992. *American Health Quackery*. Princeton: Princeton University Press.

CHAPTER FOUR

After the FDA

A Twentieth-Century Agency in a Postmodern World

THEODORE W. RUGER

IN THE past decade and a half, the U.S. Food and Drug Administration (FDA) has been buffeted by and largely weathered a number of discrete and high-profile episodes of public concern over arguable regulatory failures involving pharmaceutical and food products. These episodes have generated headlines, provoked pointed criticism of FDA, and catalyzed congressional and broader public oversight of the agency's functioning (Coglianese & Ruger 2013; Harris 2004). Yet FDA is likely to recover from any dip in its prestige caused by this round of public concern just as it has overcome past chapters of momentary public concern to maintain its primary reputation as a guardian of product safety. The deeper threat to FDA's role in the regulatory firmament of the twenty-first century is not that it is inadequately performing its core functions in assessing the safety of food and therapeutic products but rather that this focus on product "safety"—at least in the relatively acontextual manner FDA has tended to assess that variable—is increasingly unimportant in a new world of focus on product cost, consumer behavior, and consumption patterns. How well or how poorly FDA is able to realign its mission to take account of these shifting regulatory priorities will have much to say about whether its prestige in the next fifty years approximates its status over the past fifty years.

For more than half a century, FDA has been a high-status icon of the federal administrative state. Few agencies have been as successful at achieving their stated regulatory goals, and few have enjoyed the reputation for technocratic expertise that the FDA has long held among the press and the public. The regulatory regime overseen by FDA has produced sizeable public health gains over the past sixty years by virtue of ensuring safer food and therapeutic products for U.S. consumers. In recent decades FDA has done much to ameliorate a countervailing critique that it has been too careful and too slow in approving new therapeutic products—the "drug lag" in U.S. approval times relative to other nations' gatekeeping agencies has largely disappeared since the 1980s (Olson 2004; Hilts 1989). Even the rare instances of high-profile failure by FDA in the past several years, when unsafe foods or pharmaceuticals have slipped past the agency's scientific safety net, such as with Vioxx in 2004, the *Salmonella* outbreak in peanut products in 2009, or October 2012's pharmacy compounding tragedy, have been met by responsive and targeted congressional and administrative action, suggesting that relevant institutions are able to learn from and swiftly correct problematic gaps in FDA's regulatory oversight (FDA Food Safety Modernization Act, Pub. L. No. 111-353, 124 Stat. 3885; Herper 2007). FDA's long track record of regulatory achievement has translated into significant reputational gains, and FDA ranks highly in public opinion polls of the most trusted, effective, and independent federal agencies. As Daniel Carpenter has written, FDA has "consistently been named or identified as one of the most popular and well-respected agencies in government" over the past several decades (Carpenter 2010:12).

FDA's longstanding stature and reputation in the American regulatory state will likely persist through the first decades of the twenty-first century or beyond. Yet if there is a looming risk to FDA's prestige and relevance in the coming decades, it is one born not of systemic regulatory incompetence but rather of an increasing marginalization from the most pressing regulatory problems involving food and drugs in the United States. The food supply of the United States is statistically as safe as it has ever been, if safety is defined by an absence of pathogenic or chemical contamination. However, illnesses relating to the consumption of food still represent the greatest preventable health problem the nation faces, attracting increasing institutional attention outside of FDA. Likewise for pharmaceuticals and other therapeutic products, the greatest challenges going forward involve allocation, price, and cost-effectiveness, as opposed to abstracted notions of "safety" and "efficacy." Yet FDA is jurisdictionally limited and bureaucratically disinclined from addressing such issues. The end result going forward is an apparent mismatch between FDA's core regulatory competencies and the most urgent policy issues in its broader domain. This dichotomous functional dynamic suggests that FDA's unmatched ascendency in the area of food and drug regulation is likely to wane as other institutions take up an increasing share of the new century's evolving regulatory challenges in this field.

To illuminate this dynamic of increasing regulatory multiplicity, this chapter first summarizes the twentieth-century development of FDA as an agency with formidable authority over the nation's food and drug supply, but also as an agency with authority and institutional competence directed at the commodified products themselves, abstracted from behavioral concerns relating to use, consumption, and societal allocation of those products. It then introduces the shifting nature of the current policy challenges relating to foods and drugs, many of which relate to issues of end-stage pricing, allocation, and consumption as opposed to the intrinsic safety (or lack thereof) of the product abstracted from the context in which it will be used, and concludes by discussing the new shape of food and drug regulation in the twenty-first century and the relatively diminished role FDA can be expected to play.

I. THE FDA CENTURY

Federal regulation of foods and drugs in the United States took shape in incremental form but early on placed FDA in a primary role for ensuring the safety of the nation's food and therapeutic products. The Pure Food and Drugs Act (PFDA) of 1906 established a predecessor office to FDA housed within the U.S. Department of Agriculture, and then the Food, Drug, and Cosmetics Act (FDCA) of 1938 fully realized the federal regulatory framework. The latter statute is, according to one leading scholar, "one of the most important regulatory statutes in American and perhaps global history" (Carpenter 2010:73).

The original FDCA included almost all of the key elements that would shape FDA's regulatory posture and preeminence for the remainder of the century: the absolute gatekeeping authority over new drugs, the emphasis on technocratic assessment of product safety, and the regulatory focus on products rather than on firm or individual behavior.

Both the FDCA and its precursor, the PFDA, reflected a regulatory concern for products rather than people—for things and words about things rather than primary conduct by producers and consumers. The FDCA's statutory language created a set of adjectival standards to be applied to products that are abstracted out of the stream of commerce: new drugs must be proven "safe" and, after amendments in 1962, "effective" before marketing, and foods must not be "adulterated" or "misbranded" or else face the prospect of government seizure (21 USC §§ 321, 355 (2006 & Supp. III 2009)). Relatedly, FDA's ample authority bears on the regulated products most heavily at temporally distinct moments in a particular product's life cycle: at the moment before initial marketing for new drugs and devices and at the period of retail sale for foods. With some exceptions, more diffuse patterns of consumption and off-label use, by individual physicians and consumers, have generally been beyond FDA's jurisdictional ambit.

This product-specific focus of FDA's jurisdiction was essential in the constitutional culture of its creation, and the agency's early power and prestige derived in part from its ability to enforce food and drug standards directly against offending products. Before the late New Deal Supreme Court reversed its earlier restrictive Commerce Clause rulings to permit greater national regulatory power, the federal government's authority over conditions of manufacturing and other features of firm behavior was extremely limited, even as applied to the largest companies (*Hammer v. Dagenhart* 1918:251). The PFDA and the FDCA avoided this problem for the FDA and its predecessor by giving the federal government authority to proceed directly against offending products in interstate commerce through in rem[1] seizure and condemnation proceedings. Beyond generating a bevy of memorable case names, such as *United States v. Two Barrels Desiccated Eggs* (1911:302), this in rem authority lent FDA ample power to enforce that was undiminished by the Supreme Court's crabbed and categorical reading of the "commerce" power in the 1920s and 1930s. It also helped foster a regulatory dynamic that would persist long after the Commerce Clause was broadened by the Court—FDA through the twentieth century continued to regulate products more tightly than it did firm or consumer behavior. Focusing on products rather than people was originally a constitutional necessity, but it became embedded in the agency's statutory architecture and institutional culture long after such doctrinal imperatives faded away and persists to a significant extent even today.

The result, then and now, is a drug regulatory regime administered by FDA that is curiously dichotomous, with substantial gatekeeping authority over certain aspects of new therapies (safety and efficacy) and virtually no authority over other crucial variables, such as a drug's pricing or conditions of use. This dichotomy, and the resultant jurisdictional incapacity on the part of FDA, is largely driven by statute. Nevertheless, in important if more nuanced ways, the agency's own practices have extended and perpetuated the acontextual nature of its approval decisions. For instance, over a half century ago, Senator Estes Kefauver and others recognized the perverse effect on overall prices of permitting approval and full patent protection of new drugs that offered only marginal gains, if any at all, over existing therapies. Many current critics have urged FDA to do more by way of requiring comparison studies before filing of a new drug application for approval of a new entity (O'Connor 2010). By continuing to adhere to placebo-controlled trials as the gold standard of safety and efficacy research, a choice invited by past agency practice but not necessarily dictated by express statutory terminology, FDA has extended a regime where new compounds are tested abstractly against a placebo rather than comparatively against existing medicines.

This dichotomous regulatory uncoupling of safety and cost is mirrored in FDA's hands-off approach to regulating the actual conditions of use of pharmaceuticals even as it exercises up-front gatekeeping authority over the ex ante safety and efficacy of those products. FDA requires scientific proof of safety and efficacy of new drugs under specific conditions of use for treatment of a specified medical condition, but once approved, physicians are free to use marketed drugs for treatment of other ailments, in dosages, durations, combinations, and patient populations dramatically different from those of the controlled trials used for approval (*Washington Legal Foundation v. Henney* 1999:81). Early on, this agency disinterest in regulating "off-label" use was a necessary compromise against physician resistance to regulation of the "practice of medicine," and it has persisted as a central limitation on FDA control of drug safety. In recent years, FDA has taken some steps to depart from this basic model and set conditions or warnings for actual use by physicians, for instance, through its Risk Evaluation and Mitigation Strategy (REMS) programs (Baker 2010), but the basic terms of this regulatory compromise remain.

FDA's regulation of food products during the twentieth century has evinced a similar product-focused dynamic. Such a regulatory posture was exemplified by FDA's ambitious midcentury effort to develop dozens of "standards of identity" for common foods, essentially commodifying large swaths of the American food supply (Merrill and Collier 1974). This effort was based on the premise that American consumers desired and were best served by a safe supply of uniform and fungible food products that were disconnected from the particulars of their agricultural production.

In sum, in its focus on technical and scientific expertise, neutrally applied to products abstracted from the stream of commerce (and the related complexities of production, consumption, and price), FDA has been the quintessential twentieth-century agency, ideal for a world of unbounded consumerism and optimistic faith in the perfecting power of impartial science. The second half of the twentieth century witnessed the apogee of the "consumer's republic" (Cohen 2003), and FDA was the ideal bureaucratic steward for a nation experiencing sustained growth in both purchasing power and scientific expertise. It is no accident that the agency's most triumphant moment of public acclaim—the 1961 ceremony honoring FDA pharmacologist Frances Kelsey for her role in preventing thalidomide's marketing in the United States—took place at the Kennedy White House, which was also responsible for the Apollo space program. The National Aeronautics and Space Administration (NASA), like FDA, reflected an optimistic faith in the virtues of scientific achievement to meet society's most urgent goals. Like NASA, FDA's fundamental role and its agency culture developed in a time relatively unconcerned with the harsh realities of resource constraints, allocative choices, and trade-offs grounded in relative benefit assessments.

II. TODAY'S FOOD AND DRUG POLICY PROBLEMS

FDA was instrumental in creating a safe and plentiful supply of food and drugs over the past century. Yet it has had relatively little to say about allocating, constraining, or encouraging firm or individual behavior along the way. The product-specific focus that was essential to FDA's developing constitutionality and institutional prestige has become a potential hindrance in the twenty-first century, when the most pressing problems involve the end-stage use and price of the regulated products rather than their abstracted safety. In this changing world, FDA's core competencies in abstract, acontextual assessment of product safety are increasingly anachronistic.

This point is illustrated by the paradox of the food "safety" situation in the United States today. By many statistical measures, the American food supply is as safe as it has ever been in history. The risk of severe pathogenic food poisoning is extraordinarily low, accounting for only a few thousand deaths per year in a nation of 300 million (Scallan et al. 2011). Even this relatively low risk will drop further with the enhanced FDA inspection and recall authority granted through the Food Safety Modernization Act of 2011, which implements changes designed to make FDA more vigilant and responsive to food safety at the factory level.

Yet on a different metric, food in the United States is far from "safe." Food consumption patterns are the single biggest consumer product threat to population health in the nation today, exceeding even tobacco in their effect on mortality and morbidity. Obesity, type 2 diabetes, coronary disease, and other health conditions related to the overconsumption of food are the largest cause of preventable deaths, accounting for over half a million deaths each year (Heron 2011). Many millions more suffer adverse health effects that impact their productivity and daily functioning. These food-related illnesses are a major driver of ever-increasing health care costs in the United States, by some measures accounting for almost a hundred billion dollars in additional medical expenditures each year (Finkelstein et al. 2009). For all of the evident urgency of addressing this obesity epidemic, FDA's core regulatory paradigm is of only marginal utility; diseases of overconsumption arise from consuming products that are neither "adulterated" nor "misbranded" by the FDCA's standards. The FDA has recently increased its information-forcing role in addressing obesity concerns, for instance, by issuing menu-labeling rules for chain restaurants requiring calorie disclosure (FDA 2014), but aside from promulgating such labeling and disclosure rules, the agency will likely play only a limited role in addressing this major public health crisis.

A similar story of mismatch between FDA's traditional jurisdiction and core competence is apparent with respect to the most urgent problems in pharmaceutical policy today. Although occasional failures in the FDA-administered drug safety regime do periodically rise to great public salience (e.g., the recent NECC meningitis outbreak), FDA continues to perform well in assessing the abstracted "safety" and "efficacy" of new drugs before they enter the market. Yet most current policy discourse about pharmaceuticals emphasizes drug pricing and comparative effectiveness as the key questions for public and private regulation, and these considerations are outside of FDA's traditional regulatory interests. Despite arguably possessing broad statutory authority to define efficacy as including a comparative component, FDA has codified by longstanding regulation a policy of assessing efficacy against placebo controls (O'Connor 2010).

This willful ignorance to comparative efficacy and cost is at odds with current health policy consensus. Policy scholars, including key architects of the 2010 Patient Protection and Affordable Care Act (ACA), have identified variations in physician practice—including in the use of pharmaceuticals—as a major drag on health care quality and a major driver of increasing health costs (Fischer et al. 2003). FDA's historical inability (or unwillingness) to specify precise conditions of use for the drug products it approves has facilitated the regime of diffuse individualized physician practice that gives rise to such suboptimal variation. Although FDA has in the past decades tentatively done more to structure conditions of use for limited categories of products, the most important structures for public or private regulation that would smooth out such variation in the delivery of pharmaceuticals within the health care system look to institutions other than FDA, like public and private payors, to fix this problematic authority structure.

Looking ahead, even as FDA continues to perform its traditional function in assessing the abstract safety and efficacy of marketed products, it is operating in a world where such concerns, when divorced from the realities of price, comparative effectiveness, and physician usage, are progressively less important. As some scholars of this phenomenon have recently described, "[t]o gain market access, most prescription drugs will need to obtain a positive reimbursement or coverage decision from payers in addition to satisfying requirements for quality, safety and efficacy as assessed by drug regulatory agencies" (Eichler et al. 2010). In the future, ex ante FDA approval will cease (if it has not already) to be the most important regulatory event in a new therapeutic product's life cycle; far more important will be the decision by public and private payers to include that product on their reimbursable formulary.

In this new world, FDA's reluctance to embrace comparative effectiveness analysis places it at the margins of current policy thinking about comparative efficacy, which has become a central subject of inquiry in the health research and health care delivery systems in the United States and elsewhere. Exemplifying this disinclination to innovate, FDA in 2008, in issuing new regulations on the conduct of foreign clinical trials, flatly rejected a recommendation to require new drugs to be tested against existing therapies, explaining that "[w]e believe that [a requirement of drug–drug trials instead of placebo control] is inconsistent with U.S. law and policy because it would impose a standard for the design of clinical trials that is different from the standard of 'adequate and well-controlled investigations' which the act requires us to apply" (FDA 2008). As explained earlier, the strong-form endorsement of placebo control is more entrenched in FDA regulations than in the FDCA itself (O'Connor 2010), and FDA's reluctance to reconsider is perhaps explained by its statement in the same regulations that the issue of requiring comparator drugs as controls "invokes issues of health care policy that are not directly related to FDA's mission of ensuring that medical products are safe and effective" (ibid.). FDA's narrow interpretation of its mission contrasts with the embrace of comparative effectiveness analysis by other key public and private actors, resulting in "an asymmetric playing field: policy and reimbursement decisions will be made in one agency [the new federal CER body] based on data that a sister agency [FDA] judges too unreliable" (Commission Regulation (EC) 726/2004, 2004 O.J. (306) 22). Such bifurcation and inconsistency already characterizes the European pharmaceutical landscape, where market approval decisions are made for the entire European Union by a central European Medicines Agency (EMA) "on the basis of the objective scientific criteria of quality, safety and efficacy . . . to the exclusion of economic and other considerations" (Eichler et al. 2010) while simultaneously coverage decisions and allocative trade-offs are made at the national level by a different set of agencies, like the United Kingdom's National Institute for Health and Clinical Excellence (NICE), which quite explicitly applies cost-effectiveness metrics in making coverage recommendations for specific products.

This central dichotomy in pharmaceutical policy will only become more structurally entrenched in the next few decades of the twenty-first century as trends in cost containment, health system reorganization, and payer sophistication will greatly amplify the significance of the institutional coverage determinations that may or may not follow from ex ante market authorization. The ascendance of this bifurcated regime has several key implications for patients, providers, and the pharmaceutical industry going forward. First, the existence of sequential and independent gatekeepers controlling access to pharmaceuticals negatively affects both national uniformity and (relatedly) distributional equity in the allocation of expensive therapies. A measure of FDA's prestige historically derived from the fact that the agency was highly unusual within American medicine's diffuse authority structures—its drug and device approval decisions were national in scope even when most medical decisions remained at the bedside. This national standard-setting role will be blunted substantially as a multitude of public and private payers make (and are already making) disparate decisions about whether approved products are relatively effective enough to justify inclusion in a formulary mix sensitive to price concerns.

   This in turn may lead to new problems of distributional equity. For most of the late twentieth century, FDA's approval regime had a democratic character to it—lenient payer reimbursement policies meant that for most Americans with employment-based insurance, having private insurance was tantamount to having access to all but the most expensive or experimental new pharmaceutical therapies. This realm of patient and physician choice within existing insurance structures no longer exists, and going forward one's access to particular products will depend increasingly on the variable choices made by private insurers, state Medicaid administrators, and federal agencies. One predictable silver lining for the FDA in this new regime is that much of the public and congressional ire about suboptimal "access to medicines" that is often directed at FDA's approval process can be expected to be channeled in the future toward the public and private entities making these coverage decisions.

   The increasing bifurcation between market approval and coverage has other implications for industry product development, research, and marketing. The twentieth-century world of a highly centralized FDA approval regime coupled with a highly diffuse prescribing regime (driven by individual doctors and their patients) was instrumental in shaping the structures and practices of both R&D and marketing in the pharmaceutical industry. Industry research on new chemical entities (NCEs) was tailored primarily to fit the standards and demands of FDA, and the prodigious amounts spent on marketing were directed at influencing individual physician behavior and consumer choices. These modes will remain important in the future, but industry increasingly faces both research and marketing challenges in realigning their message to operate in a world of two gatekeepers rather than one. The research challenge is particularly pointed in that private payers are demanding precisely the kind of comparative studies that FDA has deemphasized. As the formulary director of one national insurer recently opined, "we've been really hungry for head-to-head data. . . . It will be difficult for drugs to get front-line consideration [for reimbursement] unless they have that kind of data available" (Wang et al. 2012).

How these dynamics play out in actual operation remains to be seen, and much uncertainty remains. What is certain is that the twenty-first-century world of pharmaceutical regulation looks to be a great deal more institutionally complex than the prior half century. FDA's traditional focal points of therapeutic safety and efficacy will continue to be necessary benchmarks for marketed products, but meeting such standards will no longer be sufficient to create a presumption of broad access. FDA is hardly relegated to the sidelines in this regime but to a greater extent than in past decades shares the field of play with several other increasingly important players. This regulatory multiplicity may ultimately be the right solution but only if other public and private entities achieve the success in these new functions that FDA has enjoyed in fulfilling its traditional ones.

III. POTENTIAL NEW DIRECTIONS IN FOOD AND DRUG REGULATION

In a world where patterns of consumption, pricing, and production of food and drugs have become imperatives, FDA will continue to maintain a role, but a diminished one. Two policy responses are already emerging that may prefigure the shape of a twenty-first century food and drug regulatory regime—one with a continuing but less predominant role for FDA.

The first of these, the phenomenon of multiplicitous and multimodal regulation, involves numerous public and private entities at the state and federal levels that work in loose concert toward a given regulatory end. The progress made in reducing youth smoking rates over the past fifteen years illustrates the power of such a coordinated, multiplicitous approach while also revealing FDA's relative marginality on the issue. In the 1990s, smoking was identified as the leading preventable cause of death and disease in the United States. Under Commissioner David Kessler, FDA famously took action to address this public health crisis, issuing new proposed regulations in 1995 that asserted agency jurisdiction over tobacco and ambitiously aimed to cut youth smoking in half in the seven years after promulgation of the new rules—only to be thwarted by the U.S. Supreme Court in the *Brown & Williamson* ruling (2000). Yet even though FDA was pushed to the side by the Court, the agency's stated goal of dramatically reducing youth smoking was almost fully realized in a decade (Centers for Disease Control 2008).

How did this happen? The answer lies in the range of institutional strategies and governmental interventions that emerged to fill the void created by FDA's ultimate lack of jurisdiction. Aggressive countermarketing campaigns by state governments, local bans on the sale or use of tobacco in public places and workplaces, employer disincentives for smoking, innovative state lawsuits, and continued Federal Trade Commission restriction of tobacco marketing all coalesced to achieve substantially the same goal that FDA had sought in its assertion of regulation. This story of multiple and decentralized institutions has clear implications for the pressing policy battle in the decades ahead against obesity and its related health complications.

The second new direction in regulation that will reduce FDA's primacy in pharmaceutical regulation is an already emerging dynamic of regulation by payment, in which rules for primary conduct are enforced not by traditional coercive regulation but rather by strings attached to government or private payment streams. In the United States and throughout the developed world, various public and private institutions are spending increasing amounts of money on food, pharmaceuticals, and medical care, and these institutional payers are becoming increasingly aggressive at attaching policy-related conditions to the funds they disburse.

Today in the health care field, although the federal government remains (for now) reluctant to use its ample purchasing power to force price concessions from manufacturers or to encourage shifting to cheaper generics, state governments have already asserted authority through their Medicaid reimbursement policies to shift physician and patient behavior in the area of prescription drugs. Such efforts have produced a tangible policy payoff: from 2005 to 2009 the percentage of total prescriptions nationally that were filled with brand-name drugs fell from 40 percent to 25.6 percent (National Conference of State Legislatures 2010). Similar reimbursement incentives and restrictions have been implemented by private health insurance companies and stemmed the tide of pharmaceutical cost growth.

In this world of restrictive and cost-conscious reimbursement policy, it is possible to imagine what would have been unthinkable only a few decades ago—namely, that within ten years the most important agency with jurisdiction over pharmaceuticals will not be FDA but rather the Centers for Medicare and Medicaid Services or some successor or state entities that decide which subset of approved drugs are cost-effective enough to include on the shrinking public and private formularies. This regulatory divergence and disuniformity is already on display in the European Union, which has centralized and standardized drug safety determinations but devolved coverage, payment, and cost-effectiveness determinations for pharmaceuticals to national or subnational units—leading to wide discrepancies in the public formularies of different nations.

Many of the same shifting priorities also apply in the realm of food. Although most individual food purchases are not directly subsidized by third-party payers, federal and state governments are actively involved in purchasing or subsidizing billions of dollars of food for segments of the population, and here too payers are beginning to exercise a more active or directive approach. As an example, the U.S. Department of Agriculture (USDA) spends 3.5 times more annually on subsidized school lunches ($14 billion) than the entire yearly budget of FDA ($4 billion), and with this huge funding stream comes the opportunity to compel healthier menus at tens of thousands of local schools. Under the Obama administration, the USDA has begun to require that recipient school districts offer a menu of healthy options for their students and remove foods and beverages that increase public health risks (Cohen et al. 2014). State and local governments are enacting similar policies for both school lunch and food stamp programs. In the years ahead, these funding restraints will likely perform functions relating to public health and food consumption that are as important as the work of FDA.

IV. CONCLUSION

Despite the increasing shift of policy relevance to institutions outside of FDA, on many issues touching crucial food and drug policy questions FDA retains significant discretionary authority, which it could use to remain an important player in addressing the major public health issues of the twenty-first century. However, the shifting imperatives of food and drug policy in the twenty-first century have created a policy world where FDA will cede at least some of its traditional predominance. Today's most pressing food and drug problems arise from the way we use, consume, and pay for such products, and these are variables outside of FDA's traditional jurisdiction and core competency. In the next century, FDA will be a key partner, but less of a leader, on the key regulatory choices the United States makes in food and drug policy.

NOTE

1. In rem proceedings are actions against physical objects (not persons or their conduct); the more common in personam actions are brought against individuals or entities.

REFERENCES

Baker, Danial D. 2010. "REMS—One Year Later." *Hosp. Pharm.* 45(5):348–51.
Carpenter, Daniel. 2010. *Reputation and Power: Organizational Image and Pharmaceutical Regulation at the FDA*. Princeton, NJ: Princeton Univ. Press.
Centers for Disease Control. 2008. "Cigarette Use Among High School Students—United States, 1999–2007." *Morbidity & Mortality Wkly. Rep.* 57:689–91.

Coglianese, Cary and Theodore Ruger. 2103. "Diagnosis and Cure for the FDA's Failure in Meningitis Cases." *Detroit Free Press* (January 5).
Cohen, Juliana F. W. et al. 2014. "Impact of the New U.S. Department of Agriculture School Meal Standards on Food Selection, Consumption, and Waste." *Am. J. Preventative Medicine* 46(4):388–94.
Cohen, Lizbeth. 2003. *A Consumers' Republic: The Politics of Mass Consumption in Postwar America*. New York: Random House.
Eichler, Hans-Georg et al. 2010. "Relative Efficacy of Drugs: An Emerging Issue Between Regulatory Agencies and Third-Party Payers." *Nature Rev. Drug Disc.* 9:277.
Finkelstein, Eric et al. 2009. Annual Medical Spending Attributable to Obesity: Payer- and Service-Specific Estimates, *Health Affairs* 28:w822–31. http://content.healthaffairs.org/content/28/5/w822.full.pdf+html.
Fischer, Elliott S. et al. 2003. "The Implications of Regional Variations in Medicare Spending, Part 2: Health Outcomes and Satisfaction with Care." *Annals Int. Medicine* 138:288.
Food and Drug Administration. 2008. "Human Subject Protection; Foreign Clinical Trials Not Conducted Under an Investigational New Drug Application." *73 Fed. Reg.* 20800–12.
Food and Drug Administration. 2010. Prescription Drug User Fee Act (PDUFA): Adding Resources and Improving Performance in FDA Review of New Drug Applications. http://www.fda.gov/ForIndustry/UserFees/PrescriptionDrugUserFee/ucm119253.htm.
Food and Drug Administration. 2014. Menu and Vending Machines Labeling Requirements, (May 22). http://www.fda.gov/Food/IngredientsPackagingLabeling/LabelingNutrition/ucm217762.htm.
Food and Drug Administration v. Brown & Williamson Tobacco Corporation, 529 U.S. 120 (2000).
Finkelstein, Eric et al. 2009. Annual Medical Spending Attributable to Obesity: Payer- and Service-Specific Estimates, *Health Affairs* 28:w822–31. http://content.healthaffairs.org/content/28/5/w822.full.pdf+html.
Fischer, Elliott S. et al. "The Implications of Regional Variations in Medicare Spending, Part 2: Health Outcomes and Satisfaction with Care." *Annals Int. Medicine* 138 (2003):288.
Garrison, Louis P. et al. 2010. "A Flexible Approach to Evidentiary Standards for Comparative Effectiveness Research." *Health Affairs* 29(10):1812–17.
Hammer v. Dagenhart, 247 U.S. 251 (1918).
Harris, Gardner. 2004. "F.D.A. Failing in Drug Safety, Official Asserts." *New York Times* (November 10).
Heron, Melonie. 2011. "Deaths: Leading Causes for 2007." *Nat'l Vital Stat. Rep.* 59:1–96.

Herper, Matthew. 2007. The Biggest FDA Reform in a Decade. *Forbes* (September 24). http://www.forbes.com/2007/09/21/drugs-safety-fda-biz-sci-cx_mh_0924fda.html.

Hilts, Philip J. 1989. "How the AIDS Crisis Made Drug Regulators Speed Up." *New York Times* (September 24):D5.

Kaiser Family Foundation. 2009. *Trends in Healthcare Costs and Spending* (March 19). http://www.kff.org/insurance/upload/7692_02.pdf.

Kessler, David. 2001. *A Question of Intent: A Great American Battle with a Deadly Industry*. Cambridge, Mass.: Perseus.

Merrill, Richard A. and Earl M. Collier, Jr. 1974. "Like Mother Used to Make: An Analysis of FDA Food Standards of Identity." *Colum. L. Rev.* 74:561–621.

National Conference of State Legislatures. 2010. *Health Cost Containment and Efficiencies: Use of Generic Prescription Drugs and Brand-Name Discounts*. 1–6 (June). http://www.ncsl.org/portals/1/documents/health/GENERICS-2010.pdf.

O'Connor, Alec B. 2010. "Building Comparative Efficacy and Tolerability Into the FDA Approval Process." *JAMA* 303(10):979–80.

Olson, Mary K. 2004. "Managing Delegation in the FDA: Reducing Delay in New-Drug Review." *J. Health Law & Policy* 29:397.

Regulations Restricting the Sale and Distribution of Cigarettes and Smokeless Tobacco Products, 60 *Fed. Reg.* 41,314 (proposed August 11, 1995) (codified at 21 CFR Parts. 801, 803, 804, 897).

Scallan, Elaine et al. 2011. "Foodborne Illness Acquired in the United States—Unspecified Agents." *Emerg. Infect. Dis.* 17:7–15.

United States v. Two Barrels Desiccated Eggs, 185 F. 302 (D. Minn. 1911).

Wang, Anthony et al. 2012. "U.S. Payer Perspectives on Evidence for Formulary Decision Making." *J. Oncol. Pract.* 8:22s–24s.

Washington Legal Foundation v. Henney, 56 F. Supp.2d 81 (D.D.C. 1999).

CHAPTER FIVE

The Future of Prospective Medicine Under the Food and Drug Administration Amendments Act of 2007

BARBARA J. EVANS

I. INTRODUCTION

The United States health care industry is sometimes characterized as a system of disease care that sits on its hands until people fall ill and then mounts a costly, crisis-oriented response, often after the patient is beyond help (Brigham and Johns 2012; Topol 2012). The Patient Protection and Affordable Care Act of 2010 modestly alters the way health care is financed but does very little to alter this essence of U.S. health care. After decades of investing more than other Organisation for Economic Co-operation and Development (OECD) nations spend per capita on health care, Americans qualify as some of the sickest, shortest-lived rich people in the world. A growing body of literature decries this "disease-oriented, reactive, and sporadic approach to care" and calls for a shift to "prospective medicine" that would help healthy people stay well (Snyderman and Yoediono 2008). This shift presents unique challenges for the U.S. Food and Drug Administration's (FDA's) regulation of medical products. This chapter explores these challenges. It describes how the Food and Drug Administration

Amendments Act of 2007 (FDAAA) helps address the challenges and inquires whether FDA is making full use of its new powers under FDAAA. Although it is too early to appraise the agency's overall success in implementing FDAAA, there are early signs that the agency may be underconstruing its statutory powers in ways that may impede the progress of twenty-first-century prospective medicine.

II. THE CHALLENGES OF PROSPECTIVE MEDICINE

Prospective medicine is sometimes described as "personalized, predictive, preventive, and participatory" (Snyderman and Yoediono 2008). Each of these attributes strains FDA's traditional regulatory paradigm. Predictive technologies involve testing gene variants or other biomarkers to identify future health risks or to forecast treatment responses, as in pharmacogenetic testing (Bookman et al. 2006; Institute of Medicine 2008). The ultimate goal is to be able to respond to predictive or presymptomatic testing with preventive interventions that can avert the development of disease or at least minimize burdens of illness and treatment.

Predictive and preventive technologies imply a shift from the curative model that now dominates health care. Rapid advances in diagnosis and treatment made it possible for twentieth-century doctors to diagnose and treat manifest disease, but prognosis—which had held an important place in medical practice from ancient times until late in the nineteenth century—fell into disrepute (Christakis 1997; Rich 2002). Predicting health outcomes at the level of individual patients was seen as ethically questionable and unscientific because twentieth-century science treated individual health outcomes as a product of random chance (Christakis and Iwashyna 1998; Evans 2010). Recent advances in the life sciences are giving prediction and prognosis an evidentiary basis that they previously did not have and are "catalyzing a revolution in healthcare focused around an informational view of medicine" (Institute for Systems Biology 2010).

As Richard Merrill has remarked, FDA's product-oriented regulations furnish an "odd-fitting framework for regulating what is basically an information service" (Merrill 2000), which is an apt characterization of predictive tests. To complicate matters further, participatory medicine envisions a "far greater role for patient involvement" (Snyderman and Yoediono 2008), which means patients will need greater access to health information than they traditionally have enjoyed. Even when a test is analytically valid, it may take many years to establish that the test results have clinical validity and utility, and when relationships between biomarkers and clinical conditions are "well established," this often implies a statistical rather than deterministic relationship (Bookman et al. 2006). Even when motivated by worthy health and safety objectives, governmental efforts to restrict flows of information can raise First Amendment questions as FDA has learned in a series of bruising encounters with the courts in recent years (see e.g., *Washington Legal Foundation v. Friedman* 1998; *Pearson v. Shalala* 1999; *Thompson v. Western States Medical Center* 2002; *Whitaker v. Thompson* 2002; *United States v. Caputo* 2003; *Alliance for Natural Health US v. Sebelius* 2010; *United States v. Caronia* 2012). It is one thing for FDA to deem a *product* insufficiently safe and effective to be allowed onto the market. It is more disturbing when a government agency declares that *information* is not safe and effective and thus must be suppressed (Evans 2014).

Personalized medicine negates the pivotal simplifying assumption of FDA's twentieth-century drug regulatory framework—that population-average statistics for safety and effectiveness are informative when treating individuals (Evans 2010; Topol 2012). Personalized medicine accepts that this assumption is a fiction and aspires to use medical products in ways that will be safe and effective at the individual level. "Over the past half century, biomedical science has developed randomized, controlled clinical-trial methods that can distinguish treatment effects from the noise of human variability" (Woodcock and Lesko 2009). In personalized medicine, unfortunately, the eliminated "noise" is precisely what is interesting. FDA's premarket clinical trials are too small, too brief in duration, and too unreflective of actual patients to detect and explain the differences in individual responses to medical products. Today's medical products routinely come to market trailing clouds of uncertainty about how patients will respond and with little or no practical information to help clinicians optimize individual health outcomes.

This problem only grows worse in a world of preventive medicine. Unlike today's medical products that regularly exhibit late-emerging safety problems, tomorrow's products will have late-emerging issues with efficacy as well as safety. Preventive medical products deliver their benefits over long time scales—from twenty to fifty years—that make it impractical to assess effectiveness using FDA's traditional premarket clinical trials (Evans 2010). The impossibility of proving "preventive effectiveness" in a compact premarket trial implies that many preventive interventions will be off-label uses of products that already are on the market for therapeutic indications, and there may be little data to support the preventive use of the product. Products that do receive FDA approval for a preventive indication often will have been approved using surrogate endpoints of effectiveness. Assessing whether preventive interventions actually work will require ongoing postmarketing studies—not only of safety but of effectiveness. Validating the effectiveness of preventive interventions is critical because prospective medicine envisions putting drugs into healthy people who, although susceptible to illness, are presently asymptomatic. Safety risks that would be acceptable when treating manifest disease may be entirely unacceptable unless there is clear evidence that the preventive intervention works (Evans 2010).

III. NEW REGULATORY POWERS UNDER FDAAA

FDAAA was a forward-looking statute that gave FDA crucial powers to regulate new kinds of medical products that are emerging in the age of prospective medicine. FDAAA represented the most profound modernization of FDA's drug regulatory framework since the 1962 Drug Amendments (Pub. L. No. 87-781, 76 Stat. 780), which first empowered FDA to require evidence of safety and effectiveness before approving new drugs. FDAAA did not change FDA's premarket drug approval process but supplemented it with an array of new powers after drugs are approved (Evans 2010). To summarize, FDAAA accepted that premarket clinical trials cannot fully answer questions about new drugs and expanded FDA's authority to require manufacturers to conduct postmarketing studies and clinical trials (21 USC § 355(o)(3)). It recognized that making drugs safe in clinical use requires ongoing, shared effort that draws on numerous methodologies, including large-scale observational studies to monitor how approved drugs perform in actual clinical use. It authorized FDA to develop a large-scale data infrastructure in the form of a 100-million-person postmarketing "risk identification and analysis system" (21 USC § 355(k)(3)). This data infrastructure, known as the Sentinel System, relies on administrative data (such as insurance claims information) and clinical records (Behrman et al. 2011; Robb et al. 2012). FDA's Sentinel pilot project, known as Mini-Sentinel, is already operating and responding to drug safety queries from FDA (Curtis et al. 2012; Platt et al. 2012).

FDAAA establishes a hierarchy of approaches for investigating safety issues with approved drugs (21 USC § 355(o)(3)(D)(i)-(ii)). First, FDA must study the problem itself using its traditional adverse event reporting system and the new Sentinel System. If these approaches are insufficient to answer the drug safety question, FDA may order the drug manufacturer to do a § 505(o)(3) "study," which is defined in a way that includes investigational methods *other than* clinical trials, for example, laboratory and animal studies and observational approaches (Food and Drug Administration 2011). The agency can order the manufacturer to conduct a randomized, controlled clinical trial during the postmarketing period only as a last resort if all of these other methods are deemed insufficient to answer the safety question.

Crucially, FDAAA lets FDA take concrete regulatory actions in response to new evidence developed after drugs are approved. The agency can require safety-related labeling changes after certain procedural steps (21 USC § 355(o)(4)), and it can distribute risk information to patients and physicians through a new Internet-based communication system (21 USC § 355(r)(2)(a)). FDAAA's Risk Evaluation and Mitigation Strategy (REMS) program offers a way for FDA to condition the sale of drugs on specific measures to manage their risks (21 USC §§ 355(p), 355-1). FDA can require a REMS at the time of a new drug's approval or later in the drug's life if emerging evidence shows that a REMS is necessary to ensure that the drug's benefits outweigh its risks (21 USC § 355-1(a)). Every REMS must provide for ongoing evaluation of the drug's risks and also may include additional discretionary elements (21 USC § 355-1(d)-(e)). These may be as simple as requiring a medication guide, patient package inserts, or warning letters to health care providers.

A REMS can include more stringent elements if a drug is effective but has known risks so serious that the drug otherwise would not be approvable (or, if already approved, would need to be removed from the market). Under these circumstances, FDA can condition sales of the drug on restrictions known as "elements to ensure safe use"

(21 USC §§ 355-1 (f)(1)(B), (3)). FDAAA identifies six such elements, including one that allows the agency to require that patients undergo laboratory tests before they can receive the drug (21 USC § 355-1(f)(3)(D)). To date, FDA has used this power to require pregnancy testing in conjunction with prescriptions for teratogenic drugs, but this provision also would lend itself to requiring pharmacogenetic tests in situations where a drug, without the testing, does not have an acceptable risk–benefit ratio. This provision, in effect, gives FDA a way to achieve the functional equivalent of cross-labeling the drug and a companion diagnostic. It thus could help resolve legal and practical problems that have made it hard to cross-label drugs and tests (Evans 2010). This is just one example of the ways FDAAA helps address barriers to the clinical translation of pharmacogenomics and personalized medicine; a more-detailed discussion has been provided elsewhere (Evans 2010).

IV. REGULATING EFFECTIVENESS DURING THE POSTMARKETING PERIOD

FDAAA's § 505(o)(3) lets FDA order postmarketing studies and clinical trials only if the agency has concerns about a drug's safety (21 USC § 355(o)(3)(B)). Safety concerns also are the trigger for requiring labeling changes and REMS elements to assure safe use. FDAAA equips FDA to address postmarketing safety questions, but what about efficacy questions that loom so large in prospective medicine? FDA can only order postmarketing investigations of drug safety issues, but this focus is not as narrow as it initially sounds.

FDAAA's § 505(o)(3) (21 USC § 355(o)(3)) adopts a set of definitions that include "failure of expected pharmacological action" as a type of serious drug safety risk, if the failure entails serious medical consequences for the patient (21 USC §§ 355-1(b)(1)(E), (b)(4), and (b)(5); 355(o)(2)(C)). For example, if an antibiotic fails to work because germs are now resistant to it, and if patients who take the drug are seeing their infections progress to gangrene, the drug's efficacy failure would qualify as a serious safety risk under FDAAA's definitions. FDAAA's approach of viewing failures of expected efficacy as potential safety problems is consistent with FDA's longstanding practice of framing "safety" as a favorable ratio of benefits and risks

(Food and Drug Administration 2011). If a drug fails to deliver its expected pharmacological action, then obviously its ratio of benefits and risks is diminished, and the drug is not as "safe" as it was originally thought to be.

FDA's power to address unexpected efficacy failures is an outgrowth of FDAAA's definition of serious drug safety risks. Specifically, the statute defines the term *adverse drug experience* as including events FDA traditionally has counted as drug-related adverse events plus "any failure of expected pharmacological action of the drug" (21 USC § 355-1(b)(1)). This last provision, for example, seems to count therapeutic nonresponse as an adverse drug experience. Nonresponse happens when a patient is neither poisoned nor helped by a drug. For such patients, the expected pharmacological action, which was inferred from population-average data from premarket clinical trials, fails to materialize.

Traditionally, FDA regarded nonresponse as an efficacy problem but not a safety problem. The risk–benefit methodology FDA employs when approving new drugs distinguishes harms that are directly caused by the drug itself (drug toxicity) from consequential harms (such as worsening of the patient's illness that occurs because the drug failed to produce its expected therapeutic response) (Food and Drug Administration 2004; Food and Drug Administration 2005). These consequential harms can be real and serious, for example, when patients have chronic or progressive diseases that worsen while the patient is taking an ineffective drug. The safety dimensions of treatment failure are all the more sobering because an estimated 30 to 60 percent of the prescriptions written in the United States result in nonresponse (Peakman and Arlington 2001; Henderson and Reavis 2008).

FDAAA defines a "serious" adverse drug experience as one that kills or puts the patient at immediate risk of death, that causes persistent or significant incapacity or substantial disruption of ability to conduct normal life functions, or that requires medical or surgical intervention to prevent such outcomes (21 USC § 355-1(b)(4)(A)). FDAAA equates "serious risk" to risk of a serious adverse drug experience (21 USC § 355-1(b)(5)). These definitions supply statutory authority for FDA to address nonresponse using the same powers the agency has for addressing problems with drug toxicity. Thus, FDAAA would allow FDA to order postmarketing studies and clinical trials, labeling changes, or REMS use conditions (such as ordering a pharmacogenetic test) to address unexpected efficacy failures. These powers potentially include the power of FDA to address heterogeneity of treatment effects, the phenomenon that causes individual responses to depart from the population-average effectiveness data that are the basis for FDA's new drug approvals. This new statutory authority to address unexpected efficacy failures in approved drugs is one of FDAAA's most crucial features as FDA adapts to an era of predictive, preventive, and personalized medicine.

V. IS FDA UNDERCONSTRUING ITS POWERS UNDER FDAAA?

In its 2006 *Future of Drug Safety* report, the Institute of Medicine remarked on FDA's record of poor follow-through on proposed initiatives (IOM 2006). Perhaps attentive to that criticism, the agency has displayed admirable speed in implementing key aspects of FDAAA such as the Sentinel System. There are isolated areas of concern, however, one of which involves the postmarket efficacy issues discussed in the previous section. While it is still too early to draw conclusions, FDA appears to be turning its back on the crucial statutory authority that FDAAA gave the agency to address efficacy failures during the postmarket period.

The agency published final guidance on Postmarketing Studies and Clinical Trials ("Guidance") in April 2011 (Food and Drug Administration 2011). This Guidance clarifies how the agency intends to exercise its powers to order manufacturers to conduct postmarketing studies and clinical trials under FDAAA's § 505(o)(3). As just discussed, the statute envisions that FDA can use this power to generate evidence about serious *adverse drug experiences* and defines this italicized phrase as meaning:

> any adverse event associated with the use of a drug in humans, whether or not considered drug related, including—
> (A) an adverse event occurring in the course of the use of the drug in professional practice;
> (B) an adverse event occurring from an overdose of the drug, whether accidental or intentional;
> (C) an adverse event occurring from abuse of the drug;

(D) an adverse event occurring from withdrawal of the drug; and
(E) any failure of expected pharmacological action of the drug.
—(21 USC § 355-1(B)(1))

FDA's April 2011 Guidance did considerable violence to this definition, shortening it to:

Adverse drug experience includes an adverse event in the course of the use of the drug in professional practice.
—(FOOD AND DRUG ADMINISTRATION 2011)

This rendering appears in an appendix to the Guidance that purports to provide definitions from 21 USC § 355-1(b), yet it completely jettisons prongs (B)–(E) of the statutory definition of *adverse drug experience*. Elsewhere, this same appendix faithfully renders the complete statutory definitions of other terms such as *serious adverse drug experience* and *serious risk*. Omitting subsections (B)–(E) eliminates the crucial clause at section 355-1(b)(1)(E) that mentions "any failure of expected pharmacological action of the drug." This is the clause that authorizes FDA to address efficacy failures during the postmarketing period.

This omission does not appear to have been a mere accident or an abbreviation of longer statutory text. In the main text of the Guidance, FDA affirmatively disclaims its authority to address unexpected efficacy failures during the postmarketing period. The Guidance asserts that FDA cannot require clinical trials

in which the primary endpoint is related to further defining efficacy, designed to:
- Evaluate long-term effectiveness or duration of response
- Evaluate efficacy using a withdrawal design
- Evaluate efficacy in a subgroup
—(FOOD AND DRUG ADMINISTRATION 2011)

Yet evaluating efficacy in a population subgroup is precisely the type of investigation that the statute at 21 USC § 355-1(b)(1)(E) *does* appear to authorize. In effect, the April 2011 Guidance rewrites the statute in a way that eliminates a power that the statute, in fact, appears to confer. Some of the pharmaceutical industry's comments on the proposed Guidance called for postmarketing efficacy studies to be classified as "commitments" that companies could agree to do rather than "requirements" that FDA could order them to do (see, e.g., Baranello 2009). The Guidance may have been trying to allay industry concerns that the agency will exploit FDAAA's definitions to order wide-ranging postmarketing investigations of efficacy-related issues. The agency disclaimed any intent to require randomized, controlled clinical trials (RCTs) in which the primary endpoint is an efficacy question. This seemingly includes trials of efficacy in a suspected nonresponding population subgroup (Food and Drug Administration 2011).

Subgroup efficacy studies are very important in the age of pharmacogenomics. The agency approves drugs based on average safety and efficacy statistics that may badly misrepresent a drug's actual clinical performance in people who happen to be cursed with nonaverage genes. FDAAA's definition of "serious risks" seemingly would allow the agency to require postmarketing investigations of subgroup efficacy if the agency chose to do so. The April 2011 Guidance suggests that FDA has no wish to do so. It is true that FDA's authority to order investigations under § 505(o)(3) is discretionary in nature, and federal agencies can choose not to regulate to the maximal extent that their statutes theoretically would allow. The April 2011 Guidance modestly suggests that clinical trials of subgroup efficacy might go forward as *voluntary* commitments between drug manufacturers and FDA. In contrast, the statute envisions that the agency can *require* such studies.

Federal agencies are allowed to exercise enforcement discretion, but they are not allowed to rewrite their statutes. A familiar example conveys this crucial distinction: A police officer is not required to ticket a motorist who is driving sixty miles per hour (mph) in a thirty-mph zone. The officer can choose to let the driver go with a warning, and this would amount to an exercise of the officer's enforcement discretion. On the other hand, the officer cannot declare that the speed limit is sixty mph when it is, in fact, thirty mph. That would be amending the statute, which only the legislature can do. When the April 2011 Guidance affirmatively disclaimed the agency's power to order clinical trials of subgroup efficacy effects, it arguably crossed the line into rewriting FDAAA.

FDA did not, however, completely wash its hands of addressing efficacy issues that pose potential risks to patient safety. The agency stood its ground when public comments asked it to delete certain types of efficacy-related investigations from the list of § 505(o)(3) *studies* (that is, investigations other than RCTs) that FDA can require. For example, public comments asked FDA to remove studies that explore mechanisms of drug resistance that might cause a drug to fail to deliver its expected level of efficacy (Baranello 2009). The final Guidance kept those studies in the list of postmarketing investigations that FDA can require under § 505(o)(3).

At the risk of overinterpreting what may have been inadvertent language in the April 2011 Guidance, it appears that FDA may be willing to require a § 505(o)(3) study (such as an observational study or a laboratory study) to investigate failures of a drug's expected pharmacological action, but the agency will stop short of ordering postmarketing RCTs to investigate such failures. Under the rubric of § 505(o)(3) studies, FDA could, for example, require a prospective registry study to identify subgroups that have failed to respond to a drug, or FDA could require a laboratory analysis of biospecimens obtained with the consent of such patients in the hope of identifying a gene variant that predicts their nonresponse to the drug. However, FDA apparently would stop short of requiring a clinical trial that randomizes patients from a discrete, genetically defined subpopulation to receive various therapies to determine which treatment works better for them.

Actual practice in coming years will clarify how actively FDA intends to use its powers under § 505(o)(3) to foster pharmacogenetic discovery to improve drug safety for *all* people, including those in atypical genetic subgroups. Although the April 2011 Guidance shrank from requiring *clinical trials* of subgroup efficacy, it did not slam the door shut on other valuable pharmacogenetic investigations, such as laboratory and observational studies that could be carried out as § 505(o)(3) studies.

The same is true of § 505(o)(3) observational studies to assess long-term efficacy and duration of treatment response—the April 2011 Guidance did not rule them out. It merely ruled out RCTs for assessing long-term efficacy. Observational studies may, in fact, be the superior alternative for assessing long-term efficacy and duration of treatment response. Long-term RCTs, apart from their cost, raise troubling ethical issues (for example, keeping patients unaware whether they are taking a test or control drug during a double-blind RCT that continues for many years) and logistical problems (e.g., high drop-out rates in prolonged RCTs; Evans 2010). In light of the many problems they pose, FDA may not have sacrificed a practical option by renouncing RCTs of long-term efficacy.

VI. CONCLUSION

FDAAA imbued FDA with important powers for adapting the agency's medical product regulations to the needs of twenty-first-century prospective medicine. To succeed, however, the agency will need to embrace its new powers and make skillful, aggressive use of them.

REFERENCES

Alliance for Natural Health US v. Sebelius, 714 F. Supp. 2d 48 (DDC 2010).

Baranello, Jr., Roy J. 2009. Comments submitted on September 29, 2009, on behalf of Wyeth Pharmaceuticals in U.S. Department of Health and Human Services, Food and Drug Administration Docket No. FDA-2009-D-0283. Available at http://www.regulations.gov/#!documentDetail;D=FDA-2009-D-0283-0005.

Behrman, Rachel E. et al. 2011. "Developing the Sentinel System—A National Resource for Evidence Development." *New England Journal of Medicine* 364:498–99.

Bookman, Ebony B. et al. 2006. "Reporting Genetic Results in Research Studies: Summary and Recommendations of an NHLBI Working Group." *Am. J. Med. Genet. Part A* (May 15):1033–40.

Brigham, Kenneth and Michael M. E. Johns. 2012. *Predictive Health: How We Can Reinvent Medicine to Extend Our Best Years*. New York: Basic Books.

Christakis, Nicholas A. 1997. "The Ellipsis of Prognosis in Modern Medical Thought." *Social Science and Medicine* 44(3):301–15.

Christakis, Nicholas A. and Theodore J. Iwashyna. 1998. "Attitude and Self-Reported Practice Regarding Prognostication in a National Sample of Internists." *Archives of Internal Medicine*. 158:2389–95.

Curtis, Lesley H. et al. 2012. "Design Considerations, Architecture, and Use of the Mini-Sentinel Distributed Data System." *Pharmacoepidemiology and Drug Safety* 21(Supplement S1):23–31.

Evans, Barbara J. 2010. "Seven Pillars of a New Evidentiary Paradigm: The Food, Drug, and Cosmetic Act Enters the Genomic Era." *Notre Dame Law Review* 85:419–524.

Evans, Barbara J. 2014. "The First Amendment Right to Speak About the Human Genome." *University of Pennsylvania Journal of Constitutional Law* 16:549–636.

Food and Drug Administration, Center for Drug Evaluation and Research. 2004. *Manual of Policies and Procedures 6010.3*, at § 1 and §§ 7.1,·7.1.5.5. http://www.fda.gov/downloads/AboutFDA/ReportsManualsForms/StaffPoliciesandProcedures/ucm080121.pdf.

Food and Drug Administration, Center for Drug Evaluation and Research. 2005. *Reviewer Guidance: Conducting a Clinical Safety Review of a New Product Application and Preparing a Report on the Review.* http://www.fda.gov/downloads/Drugs/GuidanceComplianceRegulatoryInformation/Guidances/ucm072974.pdf.

Food and Drug Administration, Center for Drug Evaluation and Research and Center for Biological Evaluation and Research. 2011. *Guidance for Industry: Postmarketing Studies and Clinical Trials—Implementation of Section 505(o)(3) of the Federal Food, Drug, and Cosmetic Act.* http://www.fda.gov/downloads/Drugs/GuidanceComplianceRegulatoryInformation/Guidances/UCM172001.pdf.

Henderson, Rebecca and Cate Reavis. 2008. *Eli Lilly: Recreating Drug Discovery for the 21st Century* 16 (Cambridge, Mass.: M.I.T. Sloan Sch. of Mgmt., Doc. No. 07-043, Mar. 13). https://mitsloan.mit.edu/LearningEdge/CaseDocs/07–043-Recreating-Drug-Discovery.pdf.

Institute for Systems Biology. 2010. *Annual Report: Pushing Boundaries.* Seattle: Institute for Systems Biology.

Institute of Medicine, Committee on the Assessment of the U.S. Drug Safety System. 2006. *The Future of Drug Safety.* Washington, D.C.: The National Academies Press.

Institute of Medicine, Roundtable on Translating Genomic-Based Research for Health. 2008. *Diffusion and Use of Genomic Innovations in Health and Medicine: Workshop Summary.* Washington, D.C.: The National Academies Press.

Merrill, Richard A. 2000. "Genetic Testing: A Role for FDA?" *Jurimetrics* 41(1):63–66.

Peakman, Tim and Steve Arlington. 2000. "Putting the Code to Work: The Promise of Pharmacogenetics and Pharmacogenomics." *Drug Discovery World* (Winter). http://www.ddw-online.com/winter-2000/p148609-putting-the-code-to-work:-the-promise-of-pharmacogenetics-and-pharmacogenomics.html.

Pearson v. Shalala, 164 F.3d 650, 658 (D.C. Cir. 1999)

Platt, Richard et al. 2012. "The U.S. Food and Drug Administration's Mini-Sentinel Program: Status and Direction." *Pharmacoepidemiology and Drug Safety* 21(Supplement S1):1–8.

Rich, Ben A. 2002. "Prognostication in Clinical Medicine: Prophecy or Professional Responsibility?" *Journal of Legal Medicine* 23:297–358.
Robb, Melissa A. et al. 2012. "The U.S. Food and Drug Administration's Sentinel Initiative: Expanding the Horizons of Medical Product Safety." *Pharmacoepidemiology and Drug Safety* 21(Supplement S1):9–11.
Snyderman, Ralph and Ziggy Yoediono. 2008. "Perspective: Prospective Health Care and the Role of Academic Medicine: Lead, Follow, or Get Out of the Way." *Academic Medicine* 83(August):707–14.
Thompson v. Western States Medical Center, 535 U.S. 357 (2002).
Topol, Eric. 2012. *The Creative Destruction of Medicine: How the Digital Revolution Will Create Better Health Care*. New York: Basic Books.
United States v. Caputo, 288 F. Supp.2d 912, 920–21 (S.D.N.Y. 2003).
United States v. Caronia, 703 F.3d 149 (2d Cir. 2012).
Washington Legal Foundation v. Friedman, 13 F. Supp. 2d 51 (DDC 1998).
Whitaker v. Thompson, 248 F. Supp. 2d 1, 2, 7 (DDC 2002)
Woodcock, Janet and Lawrence J. Lesko, 2009. "Pharmacogenetics: Tailoring Treatment for the Outliers." *New England Journal of Medicine* 360(8):811–13.

PART TWO

Preserving Public Trust and Demanding Accountability

Introduction

CHRISTOPHER ROBERTSON

THE PRESIDENT and CEO of the Pharmaceutical Research and Manufacturers of America (PhRMA) recently said that "there is one great problem that seriously challenges the ability of America's research-based pharmaceutical companies to . . . research and develop new cures and treatments. In a word, it is trust" (Pricewaterhouse Coopers' Health Research Institute 2006). This section includes four chapters that explore the role of the drug and device industries, in this time when public trust has eroded. In two chapters, Patrick O'Leary and Katrice Bridges Copeland address the strategies that the United States government uses to prosecute industry transgressions. Alla Digilova, Barbara Bierer, Mark Barnes, and Rebecca Li discuss trends toward transparency in clinical trial data, which end an industry monopoly on this information. Genevieve Pham-Kanter provides a novel empirical analysis, which shows how industry financial ties are related to the votes of U.S. Food and Drug Administration (FDA) advisory board members.

The O'Leary and Copeland chapters seem to be bookends, with O'Leary worrying that recent U.S. Department of Justice enforcement actions against the drug and device companies are simply not substantial enough to provide sufficient deterrence, while Copeland worries that when the U.S. government instead goes after the industry executives themselves, it may have an "over-deterrent" effect. Both of these points deserve close scrutiny.

When O'Leary describes the "massive settlements" entered by many of the major pharmaceutical companies, he gives as an example the $2.2 billion settlement against Johnson & Johnson in 2013. While $2.2 billion is a lot of money in absolute terms, it must be remembered that these are also massive companies. In that same year, Johnson & Johnson had $71.3 billion in revenues (Johnson & Johnson 2014). Thus the settlement was about 3 percent of income. Similarly, the $1.195 billion settlement against Pfizer in 2009 amounted to 2.4 percent of Pfizer's 2009 revenues of $50.0 billion (Pfizer Inc. 2009). In the context of a median American household, that would be like paying a $1,200 fine or paying as much as a deductible for health insurance. While sizeable, this sort of payment is hardly debilitating for Johnson & Johnson or Pfizer.

O'Leary worries that these large payments fail to deter future misconduct, which seems likely given their relative scale, though it may be too soon to tell. Nonetheless, O'Leary suggests that it may not be "tenable" to simply exact more pain in the next settlement: "penalties large enough to really 'hurt' would likely also be large enough to materially endanger a firm's financial well-being," he argues. Wherever that line between painful and devastating may be, the recent settlements seem far from it. As long as we have liquid capital markets and consumer demand for products, it seems unlikely that a substantial fine would actually cause a company to stop its operations. Even corporate bankruptcies frequently allow firms to reorganize in a way that enables them to continue operations. Yet, these major companies are nowhere near bankruptcy. Pfizer, for example, has more than $32 billion in cash on hand.[1]

O'Leary argues that "the most straightforward way to bridge the deterrence gap" is to go after corporate officers themselves. Regardless of whether this strategy is a complement or an alternative to corporate liability, O'Leary is prescient to put greater scholarly focus on the question of individual accountability.

Copeland adopts this same focus, but worries that the collateral consequences of such a conviction—administrative exclusion from federal programs for three years or more—can be "a grossly disproportionate remedy," at least in cases where there is no "showing of moral blameworthiness on the part of health care executives." Copeland shows that the federal rules for exclusion provide for a three-year period "unless there are mitigating circumstances that lengthen or shorten the period of exclusion."

Essentially, Copeland argues that the only aggravating circumstance that should be considered is whether the executives had personal knowledge of the misconduct and the intent that it be committed. Copeland argues that beyond three years, a longer period of exclusion is "the modern day equivalent of 'civil death' for health care executives." The term here is figurative, given that death is permanent while a five- or ten-year exclusion is temporary, and given that the executives are not excluded from jobs in the remaining 80 percent of the economy that is not health care. Copeland also leaves open the difficult question of why a three-year exclusion is a reasonable sanction but five or eight years is qualitatively so very different.

Copeland raises interesting philosophical questions about whether executives should be at all vicariously liable for the wrongdoing of their subordinates. In corporate culture, we have no problem recognizing that credit floats upward in the flowchart—a CEO can garner a large bonus simply because a unit of his company performs well, even if he had little or no role in that success. But should blame for wrongdoing likewise float upward when a CEO has no knowledge of its particulars? Copeland says no, and she has a lot of traditional doctrine and jurisprudence on her side.

In a country founded on notions of individualism, the law has never been comfortable with notions of collective responsibility, especially not in the domain of criminal law. Still, one danger of this position is that it creates an incentive for willful ignorance. If a corporate executive can escape liability by remaining ignorant of wrongdoing in his organization, then he has an incentive to do so. That ignorance will often be easier than identifying the wrongdoing and putting a stop to it. Thus, as long as the personal benefits of wrongdoing float upward, but the risks of wrongdoing can be externalized to the underlings, it remains rational for the executive to tolerate corporate lawlessness. Alternatively, in a regime with vicarious liability, an executive has an incentive to create a corporate culture and internal regulatory systems to ensure compliance. This insight has been recognized by the Supreme Court for forty years (*United States v. Park* 1975:667). Copeland's challenge to this doctrine—at least for exclusions beyond three years—is thus provocative.

The chapter by Digilova and colleagues deals not with industry wrongdoing but rather progress. The coauthors trace a remarkable change in the practices of biomedical science occurring in less than two decades. In 1997, it seemed like a landmark reform to simply get the pharmaceutical industry to disclose the very rudimentary facts about what clinical trials they were performing. Now, due to the leadership of the biomedical science journals, the regulatory agencies, and a few companies, we are seeing a trend toward companies disclosing not just the existence of clinical trials but actually sharing the patient-level data, so that other researchers, and maybe even competitors, can perform their own independent analyses of the data.

Why are companies doing this? As the coauthors' history makes clear, neither FDA nor the EMA are yet requiring it, and only one major journal (BMJ) is requiring it as a condition of publication. Interestingly, the company leading this charge, GlaxoSmithKline, paid a $3 billion settlement in 2012, which specifically turned on its failure "to report safety data" and failure to "make available data from two studies," which showed that a heavily promoted drug actually did not work as intended (Department of Justice Office of Public Affairs 2012). The Corporate Integrity Agreement does not actually require the full data transparency that GSK has adopted.[2] GSK may be altruistic or it may be rationally adopting these policies as something of a prophylactic. Rather than bearing the burden of deciding data point by data point whether to disclose, GSK has adopted a broader transparency, which may shield it from future prosecution. As Digilova and colleagues explain, other companies are following suit. If this analysis is correct, it suggests that the U.S. federal government prosecutions (and the qui tam claims on which they are based) may be succeeding in creating a real deterrent effect after all.

Pham-Kanter's chapter goes in another direction, examining conflicts of interest for the members of the advisory panels that inform FDA decisions about whether to approve new drugs or devices. It is important to understand that FDA itself depends on scientific expertise, drawn from the outside world, which is rife with financial relationships and ulterior interests. It is not uncommon for advisory board panelists to have financial relationships with the drug and device industries, and remarkably the panelists may even take money from the very same company that makes the product that the panel is tasked with evaluating. Does this money buy influence?

Pham-Kanter has painstakingly constructed a data set of more than 15,000 votes taken by almost 1,400 members of scientific advisory panels, a data set that is ten times as large as prior studies in this area. The findings are striking. Panelists who have a sole financial relationship with just the drugmaker whose product is being considered are one-and-a-half times as likely to favor the product. Interestingly, the correlation is weaker and not statistically significant for panelists who have relationships with multiple companies. Perhaps those with multiple financial relationships have divided loyalties (sometimes a new drug by one company undermines market share of another), or perhaps it is because they have greater scientific expertise, which both attracts the funding and makes them more discerning and skeptical consumers of biomedical science. In any observational study, it is difficult to tease apart causal effects (money buying influence) from selection effects (money following expertise), but this research is striking regardless of the interpretation.

These four chapters contribute to our understanding of the future of FDA, especially in its relation to the industry that it regulates. The goal is to create a regulatory system that incentivizes companies and their executives to compete on quality and compete on price, as they develop the next generation of innovative products, which will improve the public health by reducing the burdens of morbidity and mortality. A well-functioning and predictable regulatory regime serves the interests of patients and industry alike.

NOTES

1. This data was obtained through Thomson Reuters and is available at http://markets.ft.com/research//Markets/Tearsheets/Financials?s=PFE:NYQ&subview=BalanceSheet&period=a.

2. "Corporate Integrity Agreement Between the Office of Inspector General of the Department of Health and Human Services and GlaxoSmithKline LLC 2010."

REFERENCES

"Corporate Integrity Agreement Between the Office of Inspector General of the Department of Health and Human Services and GlaxoSmith Kline LLC." 2010. http://oig.hhs.gov/fraud/cia/agreements/GlaxoSmithKline_LLC_06282012.pdf.

Department of Justice Office of Public Affairs. 2012. "GlaxoSmithKline to Plead Guilty and Pay $3 Billion to Resolve Fraud Allegations and Failure to Report Safety Data." (July 2). http://www.justice.gov/opa/pr/2012/July/12-civ-842.html.

Johnson & Johnson. 2014. "Johnson & Johnson Reports 2013 Fourth-Quarter and Full-Year Results." *Acquire Media* (January 21). http://files.shareholder.com/downloads/JNJ/3361666437x0x719867/185e8986-d100-445a-88a1-6fd4691a8786/JNJ_News_2014_1_21_Earnings.pdf.

Pfizer Inc. 2009. "2009 Financial Report." http://www.pfizer.com/files/annualreport/2009/financial/financial2009.pdf.

PricewaterhouseCoopers' Health Research Institute. 2006. "Recapturing the Vision: Restoring Trust in the Pharmaceutical Industry by Translating Expectations into Actions." http://www.pwc.com/he_IL/il/publications/assets/11recapturing.pdf.

CHAPTER SIX

Global Trends Toward Transparency in Participant-Level Clinical Trials Data

ALLA DIGILOVA, REBECCA LI, MARK BARNES,
AND BARBARA BIERER

APPEALS FOR increased clinical trials data transparency date back at least a decade, but thanks to a series of catalytic events, efforts to broaden access to clinical trials data have accelerated in recent years. The complementary effects of regulatory trends in the United States and Europe, academic journals' new data disclosure requirements, and voluntary efforts by the industry have led to an unprecedented growth of data-sharing mechanisms within the clinical research enterprise. The resulting access to clinical trials information has important implications for academicians, pharmaceutical companies, other research funders, study participants, and the biomedical sciences. While transparency and data sharing are laudable goals, more open access may portend risks to privacy and commercial interests and may alter the behavior of regulatory agencies. This chapter will outline the features of a well-designed data-sharing model that would allow for the realization of associated benefits while protecting against the underlying risks and will review recent trends toward participant-level clinical trials data sharing.

As a preliminary matter, it is important to clarify that this chapter focuses on the most recent efforts in sharing clinical trials *participant-level* data sets as distinct from *summary level, aggregated,* or *pooled results.* Summary level results do not include a listing of individual level data and outcomes, thereby carrying a lower risk of reidentification but only offering limited utility for secondary research. While summary level data sharing is vital for full disclosure of both positive and negative trial results, data sharing intended to facilitate collaborative advancement of science is better served through access to participant-level data that allows secondary researchers to replicate previous analyses, generate new hypothesis, and conduct novel studies.

I. RISKS, BENEFITS, AND ISSUES OF INCREASED DATA TRANSPARENCY

Participant-level data sharing requires careful consideration of associated risks and benefits and implications for future system design. Participant-level data sharing needs to account for concerns regarding participants' privacy and the risk of reidentification, public trust, sponsors' commercial interests, research quality, the financial costs, and overall viability of the clinical research enterprise. Critical evaluation and analysis of modern data-sharing steps should, however, consider and value the underlying positive goals of broader data access—advancement of scientific knowledge and treatment innovation.

The difficulties inherent in broadening access to participant-level clinical trials data cannot be underestimated. Standard research practice, guided by the mandates of the Declaration of Helsinki (WMA 2013) and other applicable regulations, requires researchers to protect study participants' privacy and confidentiality. In the modern era of information, releasing participant-level data, as compared to pooled, summary level data, increases the probability that study participants may be reidentified; well-publicized studies illustrate the possibility of reidentification through the utilization of publicly available information (Narayanan and Shmatikov 2008). Fears of data misuse and actual instances of participant reidentification, even if rare, could cause significant adverse consequences for the clinical trial enterprise, reducing public trust and further decreasing future participation in clinical trials. However, while the risks of reidentification necessitate robust de-identification standards and methodologies, stripping any data set of too many identifiers, or grouping variables in wide ranges, will render the data less useful for secondary analyses, thus undermining the purpose of data sharing. A carefully balanced approach is, therefore, necessary. Supplementing data-protection safeguards with restrictions on data access, with potential penalties for misuse, is one efficient way to achieve the desired balance.

In addition to managing the risk of patient reidentification, controlling access to participant-level clinical trials data will also accommodate industry concerns regarding confidential information and competitive efforts. A robust data-sharing environment relies on the industry's willingness to participate. Yet, as exemplified by the June 2013 European Medicines Agency (EMA) proposal that would require industry to make participant-level clinical trials data public, expansive access can engender forceful opposition from industry, which counters that mandatory release of certain documents may interfere with commercial confidentiality and threaten disclosure of trade secrets (EMAa 2013).

Mandatory disclosure, as proposed by the EMA policy, raises the specter of competitive spying. Rival companies may use information contained in new drug applications submitted in one jurisdiction to seek regulatory approval of competing products in another jurisdiction without incurring associated costs (Mello et al. 2013). Competitors may also peruse disclosed data to learn of submitters' general scientific and commercial strategies, with the goals of informing their development of rival products. These consequences of unmediated, uncurated data sharing may damage profitability of new research and correspondingly decrease industry's willingness to invest in developing new treatments and medicines (Mello et al. 2013). Companies may also support agendas that are distasteful to industry and perhaps even harmful to public discourse, such as "fishing expeditions" seeking masses of data so that in any way possible evidence can be developed that would dispute treatment efficacy claims. These issues can be addressed by controlled access data-sharing models, providing sponsors the opportunity to redact data of commercially sensitive information. However, if a sponsor is also entrusted with final authority to grant or deny data access requests, the credibility of data-sharing systems will suffer. A neutral body that determines whether and how data can be shared, and the conditions of use, would resolve this tension.

Permitting wider access to participant-level clinical trials data also requires weighing the costs and benefits to research quality. Here, open and uncontrolled access to clinical trials data may result in studies conducted by unqualified researchers who reach erroneous conclusions with unfortunate risks to public health. Rather than achieving the intended goals of advancing knowledge and treatment of disease, poorly performed analyses will confound scientific and public understanding of disease and treatment. Further, already underfunded regulatory agencies will need to reexamine any new data with respect to their earlier conclusions and analyses. Limiting access to data to qualified investigators who have the requisite training, skills, and tools can be one step to assure quality analysis. Assuring scientific quality will foster progress of science and public health by generating new hypotheses, uncovering new relationships, and speeding the process of finding new treatments.

Lastly, broadening access to clinical trials data should take into account globalization of the clinical trials enterprise and competing regulatory requirements of various jurisdictions. Local differences in applicable legal regimes, including trade secrecy/commercial confidentiality laws, privacy regulations, and emerging data disclosure laws, can hinder sharing efforts. For example, if a clinical trial were conducted in a nation that restricts secondary use of data, the sponsor may be unable to file for marketing authorization in the European Union if the proposed EMA policy, mandating disclosure of clinical trials data for future research, takes effect. Development of common national sets of rules and practices is therefore necessary to foster a successful clinical trials data-sharing environment that encompasses the increasingly multinational clinical trials enterprise.

The interplay of risks and benefits associated with access to participant-level data reinforces the need for a well-designed data-sharing regime. The rigor of requirements for data access should balance the need to allow access to qualified researchers with the need to protect against data misuse and poor quality or even erroneous analyses. Data access and denial decisions should be free from competitive bias and real or perceived conflict of interest. Satisfactory standards of de-identification and data-access protection are essential and should be adequate to minimize the risk of identifying study participants while maintaining the utility of the data set for future research.

One example of a data-sharing model that can provide for appropriate management of risks and benefits is a learned intermediary model, consisting of an independent, third-party decision-making authority basing data access decisions on specific standards (Mello et al. 2013). Unlike models that rely on data sponsors to review the requests and/or retain custody of the data, by vesting control over decision making in a neutral party, the learned intermediary model can guard against biased determinations by the data sponsor and thus improve public trust in the system (Mello et al. 2013). At the same time, in contrast to open access models, the learned intermediary model can help prevent flawed analyses and can foster research quality through evaluating and approving proposals based on scientific merit and the requesting party's expertise.

II. HISTORY AND CATALYSTS OF INCREASED DATA TRANSPARENCY

In light of the risks and benefits described previously, the next section outlines historical trends in clinical trials data sharing and the various approaches undertaken by policy makers, academic journals, and voluntary efforts to broaden access to participant-level data. We pay particular attention to influential initiatives that have served as catalysts for greater transparency in clinical trials data.

A. Efforts by Policy Makers

Over the past decade, policy makers and governmental institutions in the United States and abroad have heard and responded to public demands for greater clinical trials data transparency stemming from mistrust of industry and perceived lack of information that has been made voluntarily available. Over time, regulators have exhibited increasingly greater receptiveness toward data sharing, shifting from more restrictive traditional approaches—demanding trial registration in public databases—to provisions of aggregate and participant-level data. This section will discuss some of the prominent catalysts in the regulatory field responsible for fostering broader access to clinical trials data.

1. Early Registration Requirements

Before the international debate to allow access to participant-level data gained force, the U.S. government undertook steps to provide the public with selected information from clinical trials. In 1997, as part of the Food and Drug Administration Modernization Act, Congress required trial sponsors to register certain clinical trials and directed the National Institutes of Health (NIH) to establish a public database resource of clinical trials conducted under investigational new drug applications (INDs) for drugs intending to treat serious or life-threatening diseases or conditions (NIH 2013a). In February 2000, NIH unveiled ClinicalTrials.gov, an online resource of open clinical trials for experimental treatments that includes trial objectives, eligibility requirements, location, and point of contact; the website was made available to the public and became one of the first trial registries of its kind. Starting in 2009, the ClinicalTrials.gov database was expanded in order to facilitate the required posting of summary results. While originally focused on trial registration and the public availability of information about possible clinical trials in which patients might enroll, ClinicalTrials.gov was the first step toward more comprehensive data sharing. Yet its initial goal was to improve patients' access to trials rather than to promote secondary research analyses.

In November 2004, the World Health Organization (WHO) identified the need to establish an international voluntary platform to register clinical trials; that platform became the International Clinical Trials Registry Platform (ICTRP). In establishing this resource, WHO underscored the importance of trial registrations for overcoming the negative effects of publication bias and selective reporting on evidence-based health care practices (WHO 2014). In September 2005, the International Committee of Medical Journal Editors (ICMJE) unveiled a new policy requiring trial registration prior to first patient enrollment in a publicly available database as a condition of later publication of study results in any of the ICMJE-affiliated journals. The ICMJE policy statement articulated a perceived obligation of the clinical trials community members to respect the altruism and trust of study participants in dedicating themselves and their time to advance scientific knowledge. According to the policy statement, the obligation demands ethical research conduct and honest reporting. While the ICMJE policy has not mandated more advanced data-sharing requirements, it recognized that "[r]egistration is only part of the means to an end; that end is full transparency with respect to performance and reporting of clinical trials" (De Angelis et al. 2004). ICJME laid the predicate for later, more comprehensive data-sharing initiatives and mechanisms.

Reinforcing these requirements, the 2008 revised Declaration of Helsinki mandated that "[e]very research study involving human subjects must be registered in a publicly accessible database before recruitment of the first subject" (WMA 2013). Trial registration thereby became effectively unavoidable, with registration databases spreading through all parts of the globe, reaching, among others, the European Union, Japan, Australia, and African nations.

2. NIH Sharing Research Data Policy

In March 2003, NIH released its Statement on Sharing Research Data. The policy required all NIH-grant applicants seeking a budget exceeding $500,000 in any single year to include a detailed *data-sharing plan*. Foreshadowing themes in the current data-sharing debate, the policy sought to balance competing interests of improved clinical trials data access with the need to protect patients' privacy, emphasizing that data intended for broader use must meet appropriate anonymization standards. The policy provided an important break from earlier transparency initiatives that had primarily focused on improving patients' access to clinical trials; instead, the new NIH initiative became one of the first concerted efforts to encourage data sharing in order to promote the flow of information among researchers for the benefit of science and public health.

3. Food and Drug Administration Amendments Act (FDAAA) of 2007

The U.S. Congress took significant additional steps to promote availability of clinical trials information with its passage of the Food and Drug Administration Amendments Act (FDAAA) of 2007. Pertinent to clinical trials transparency, FDAAA extended the applicability and scope of ClinicalTrials.gov registration requirements to mandate reporting of summary results and adverse events for certain trials (NIH 2013b). The effect of FDAAA on data sharing was further strengthened by a concurrent NIH policy requiring all future grantees in applicable clinical trials to certify compliance with FDAAA, including its clinical trial registration and results reporting requirements (NIH 2003).

4. Recent Trends Toward Participant-Level Data Sharing

As the possible benefits of access to raw clinical trials data gained broader recognition and voluntary data-sharing mechanisms emerged, in summer 2013, FDA and the European Medicines Agency (EMA) of the European Union introduced policy proposals aimed to allow public access to participant-level clinical trials data submitted by pharmaceutical (and in the case of FDA, medical device and biotechnology) companies to the agencies as part of a drug approvals process. Both the request for comment (FDA) and the proposed policies (EMA) stood in stark contrast to regulators' historical treatment of industry data as protected against disclosure in order to safeguard commercial interests and prompted a passionate debate among a broad range of stakeholders.

a. FDA's role in data transparency: FDA has traditionally played a circumscribed role in participant-level data-sharing efforts, based on the position that commercial confidentiality laws prevent voluntary public disclosure of data submitted to the agency (Mello 2013). Following a marketing authorization, FDA generally voluntarily releases only two kinds of clinical trials data received during a new drug approval process: (1) a summary basis of approval with limited description of results of the clinical trials in humans, and (2) a synopsis of reviews, including medical and pharmacological opinions, made by the agency staff during the approval process (Kesselheim and Mello 2007). Additional disclosures may be made through release of FDA advisory committee process reports. Releases are routinely reviewed internally before disclosure and are redacted for sensitive and commercially confidential information (Kesselheim and Mello 2007).

In view of the historically limited voluntary disclosure by FDA, researchers, consumer groups, and advocates have typically relied on the Freedom of Information Act (FOIA) requests to gain access to clinical trials data submitted to the agency (Kesselheim and Mello 2007). However, the utility of FOIA for data sharing is significantly impaired by FDA's legal responsibilities toward protecting drug manufacturers' commercially sensitive information, mandated by Exemption 4 to FOIA (5 USC § 552(b)(4)), and FDA's obligations to guard trial participants' privacy pursuant to international principles of good clinical practices (FDA 2010). When FDA refuses disclosure pursuant to a FOIA exemption, the only recourse of the requesting party is complicated, time-consuming, and expensive litigation with uncertain likelihood of success (Kesselheim and Mello 2007).

While respecting the mandates of industry confidentiality and participant privacy, FDA recently proposed for public comment some measures to promote greater accessibility to clinical trials data (FDA 2013). In June 2013, FDA published a proposal for public comment on *Availability of Masked and De-identified Non-Summary Safety and Efficacy Data*. The proposal would permit public release of pooled, de-identified, and "masked" clinical and preclinical data from marketing applications submitted to FDA. "Masked" data was defined by FDA as data removed from any link to a specific drug or device product or application. Though inherently less useful for some secondary research purposes than data identifying specific products, masked data can be used to build disease progression models, characterize natural history of illnesses, assess population responses to standard-of-care products, or characterize risk factors (FDA 2013). It currently remains unclear whether the agency will move forward with the proposal and how the process of data sharing would be structured.

b. European Medicines Agency's proposed policies: The European Union has signaled changes in its regulatory policies regarding data sharing and has gradually moved from a more restrictive to a more open approach. Since November 2010, in accordance with its new policy on access to documents related to medicinal products for human and veterinary use, the EMA has been releasing upon request certain contents of applications submitted for marketing authorization. Documents available for release include clinical study reports for medicines on which the agency has completed its decision-making process. Further extending its approach to transparency, in June 2013, the EMA published a new draft policy on *Publication and Access to Clinical-Trial Data* (EMA 2013a). Citing hopes that new policy will enable the "wider scientific community to make use of detailed and high-quality clinical trial data to develop new knowledge in the interest of public health," the proposal laid out a framework that would permit significantly expanded and unprecedented public access to participant-level clinical trials data.

The EMA's new approach faced stiff opposition from the drug manufacturing industry, among others. The European Federation of Pharmaceutical Industries and Associations (EFPIA), the trade association of the research-based pharmaceutical industry in Europe, submitted public comments on the draft policy, stressing the risks and likely "negative implications" associated with the EMA's proposal (EFPIA 2013). In a bid to block disclosure of clinical trials data, two pharmaceutical companies, AbbVie and InterMune, initiated legal actions against the EMA, arguing that data intended for sharing by the agency is commercially confidential information and that publication of such information would violate European Union (EU) privacy laws and the International Agreement on Trade-Related Aspects of Intellectual Property Rights (TRIPS) (*AbbVie v. EMA* 2013; *InterMune v. EMA* 2013). By July 2014, however, the EMA had settled the lawsuit filed by AbbVie by agreeing to release a more limited data set, and InterMune has dropped its lawsuit against the agency.

The draft policy sparked an animated public debate. More than 1,000 comments from a diverse group of stakeholders were submitted to the EMA, which postponed final action on its data-sharing proposal until June 2014 (EMA 2013b). In a parallel development, in April 2014, the European Parliament approved a new regulation regarding clinical trials on medical products for human use. Among other things, the new regulation extended trial registration requirements, mandated publication of study results within a year of trial's end, and required sponsors to make clinical study reports publicly available (EMA 2014).

III. *BRITISH MEDICAL JOURNAL* POLICY AND VOLUNTARY INDUSTRY AND OTHER EFFORTS

A. *British Medical Journal* Policy

Clinical trials data-sharing efforts gained significant traction in the years following the 2005 policy statement put forth by ICJME. Individual sponsor data-sharing models, precompetitive consortia, and public–private partnerships were developed for data sharing, each with different and diverse foci and varying degrees of access. For instance, GlaxoSmithKline (GSK), in a project known as "Share," began to permit controlled access to anonymized patient-level data under a data use agreement that allowed requesters unprecedented access to industry-sponsored clinical trials data (Harrison 2012). As the research community grew more accustomed to the concept of safeguarded clinical trials data access, including access to anonymized patient-level data, the *British Medical Journal* (BMJ) introduced new, robust data-sharing requirements in October 2012 (Godlee 2012).

Starting in January 2013, the BMJ required authors to commit to sharing, upon "reasonable request," "all anonymized data on individual patients on which the analysis, results, and conclusions reported in the paper are based" as a condition for publication (Godlee 2012). To ensure process transparency, the policy required all data requesters to make the requests public and submit reanalysis protocols to the primary authors. Authors denying data requests were required to provide BMJ with a brief explanation of their decision (Godlee and Groves 2012).

The movement from the ICMJE policy on trials registration to the more demanding policy of BMJ on patient-level data sharing reflects the ongoing evolution in the data-sharing environment. Journal policies fostering improved clinical trials data access have encouraged broader recognition of the need for open science and have moved the scientific community toward greater acceptance of data sharing. While such journal efforts undoubtedly have fostered the current climate of growing clinical research transparency, journal policies only reach studies whose results are intended for publication, thus indirectly favoring positive or noninferiority trial results at the expense of negative trial results. To take advantage of the full range of benefits offered by greater access to accumulated scientific knowledge, more inclusive data-sharing initiatives have been undertaken voluntarily by private industry, government, and nonprofit sponsors. Industry has played a prominent role in data sharing, in part as a response to media reports of industry's resistance to open data practices. Specific industry sponsors faced significant public censure after studies published in 2013 revealed instances of selective reporting and "cherry-picking" of results; indeed, according to one report, only half of clinical trial results were published in full, and positive-result trials were published twice as often as negative-result trials (Goldacre 2013). Media attention intensified following reports that some pharmaceutical companies had planned to enlist patient advocates to fight EMA transparency efforts (Sample 2013). At the same time, some industry sponsors have willingly taken steps toward clinical trials data sharing. With industry responding to public pressure and changing data-sharing expectations, voluntary mechanisms to permit controlled access to clinical trials data have proliferated, moving data-sharing initiatives from regulation induced to voluntary.

These data-sharing initiatives present a wide diversity in model structures as well as the underlying terms of access. There are single- and multisponsor initiatives, academic–private partnerships, and public–private models built either around specific sponsors' data or select conditions. The access requirements also vary, with some conducting a more rigorous evaluation of a requestor's proposal than others, and some requiring executed data-use agreements. A diverse sample of noteworthy initiatives, selected for their varying approaches to data sharing, is described further in the following sections.

B. Company-Specific Platforms

One of the earliest company-specific efforts to increase transparency in clinical trials was undertaken by GlaxoSmithKline. In 2004, GSK became the first pharmaceutical company to introduce a publicly accessible clinical trial database, providing data from GSK's clinical research studies (GSK 2013). In late 2012, GSK announced plans to make anonymized participant-level data available to qualified researchers through a controlled access model. To ensure that data is used for valid scientific purposes, the model requires interested researchers to submit a study protocol, certify the presence of a statistician on the research team, and commit to publish research results. Safeguarding against potential for bias in granting access to data, all decisions are made by an independent review panel consisting of experts appointed and compensated—but not controlled—by GSK. If a request is approved, the researchers gain access to raw data sets with specifications, updated study protocols, clinical study reports, annotated case report forms, reporting and analysis plans, and clinical study reports. Studies dating back to 2000 are made available if a drug is approved by regulators or has been terminated from development and the study has been accepted for publication (GSK 2014). In furthering its commitment to transparency, the GSK model informs requestors of the reasons for proposal acceptance/denial and makes public relevant statistics on the number of data requests submitted and actions taken.

GSK's approach proved popular among other industry entities, and by July 2014 the GSK platform had evolved to become a multisponsor initiative hosting data from Bayer, Boehringer Ingelheim, GSK, Lilly, Novartis, Roche, Sanofi, and ViiV Healthcare, accessed at https://www.clinicalstudydatarequest.com (ClinicalStudyDataRequest.com 2014). While the sponsors utilize different guidelines for the inclusion of a study in the database and vary in anonymization standards, the type of shared information, review criteria, process, and data use agreements for enquiries are generally similar. One limitation of the system is that data must be analyzed on the system platform and cannot be downloaded; therefore, only data sets within the platform, and not external data sets, can be combined.

C. Disease-Specific Platforms

While GSK's sharing model is an example of a platform built around a specific company, other initiatives have undertaken steps to organize and share data from clinical trials on specific diseases. For example, Parkinson's Progression Marker Initiative (PPMI), an international partnership between mostly university-based clinical sites working toward identifying progression biomarkers in Parkinson's disease, grants qualified researchers access to download clinical, biological, and imaging data from PPMI trials. Another disease-specific initiative gaining prominence is Project Data Sphere of the CEO Roundtable on Cancer, first announced in early 2013.

Unlike the GSK controlled-access initiative, Project Data Sphere can be characterized as a broad access model. Seeking to accelerate the pace of cancer research by facilitating broader sharing, Project Data Sphere attempts to create an independent online research repository of comparator arm data from both academic and industry-sponsored studies (Beetsch 2013). Data sets from various sponsors are integrated, including de-identified patient level data made accessible in a firewalled environment (Hugh-Jones 2013). Researchers granted access after meeting basic expertise requirements and agreeing to terms of use can search and filter studies based on such criteria as sponsor, tumor type, region, age group, study type, and cancer stage (Hugh-Jones 2013). Project Data Sphere does not utilize an applicant review panel and does not require a study proposal. It depends on voluntary data contributions from study sponsors, which are promised the benefits of precompetitive mutual data insights and the resulting improved productivity, cost savings, and increased citation rates (Hugh-Jones 2013). A significant limitation of Project Data Sphere is that it provides only nongenomic comparator arm data, at least in these initial stages. While Project Data Sphere generated significant support in its planning stages, it remains to be seen whether the initiative will enjoy sufficient popularity after its launch and gather the requisite critical mass of trials to make the database worthwhile.

D. Academic–Private Partnership Platforms

The Yale University Open Data Access (YODA) Project is an example of a successful data-sharing initiative centering on a partnership between academia and industry. YODA owes its start to a public controversy. In 2011, media reports disclosed that researchers with financial ties to the medical device company Medtronic had appeared to underreport adverse events and to exaggerate the efficacy results in clinical trials of Medtronic's rhBMP-2 spinal fusion surgery product marketed as "Infuse" (Stanton 2013). In the wake of these reports, Medtronic engaged two groups of researchers at Yale to conduct an independent review of complete data sets from the Infuse clinical trials. Following completion of the initial review, Yale announced that, in its continuous partnership with Medtronic, YODA would make available clinical trials data from seventeen rhBMP-2 trials to qualified researchers who wished to advance scientific knowledge. A controlled-access model, YODA requires interested researchers to submit a study protocol, a certification or waiver of IRB approval, a conflict of interest statement, and a signed data use agreement (CORE 2014). Approved requesters receive data via a secured file transfer system and have one year to complete their study or apply for an extension.

Utilizing the experience gained from its cooperation with Medtronic, in January 2014, YODA announced a new initiative with Johnson & Johnson (J&J). Under the new partnership, a J&J subsidiary, Janssen Research and Development LLC, will release its clinical trials data, including de-identified patient-level data and clinical study reports, to qualified researchers (J&J 2014). While J&J will participate in establishing guidelines for reviewing requesters, all decision making on data access will be independently carried out by the university team (Peart 2014). YODA's collaboration with J&J illustrates the potential of academic–industry partnerships.

E. Public Organization–Specific Platforms

A final category of platforms includes data-sharing projects initiated and run by public organizations. For example, the Immune Tolerance Networks's (ITN) TrialShare portal, operated by an NIH-sponsored international clinical research consortium, provides the research community access to clinical trials data and biological specimens from ITN-funded studies, as well as auxiliary tools for data visualization to assist in interactive exploratory analyses (ITN 2013). The National Heart Blood and Lung Institute (NHBLI) BioLINCC project similarly shares biological specimen and participant-level data from NHBLI-sponsored trials with researchers who meet basic scientific qualifications. Lastly, the Laboratory of Neuro Imaging (LONI) project, partially supported by NIH, is a large-scale collaborative effort aiming to provide researchers with an interactive visualization environment for safely integrating, archiving, and querying neuro imaging data and associated patient data (Dinov et al. 2006; LONI).

CONCLUSION

A critical examination of participant-level data transparency efforts reveals that data sharing is possible and desirable and that the inherent risks can be mitigated. The move toward broader clinical trials data access reinforces the need for a sharing regime that promotes the potential benefits of transparency while including adequate safeguards against risks. It is noteworthy that, to date, major pharmaceutical companies have made the most visible and robust efforts to provide access to clinical trial data. Over time, small biotechnology companies, academicians, not-for-profit sponsors, and government agencies should develop mechanisms and secure resources to implement data-sharing initiatives. Funders of clinical trials (e.g., NIH, foundations) that are performed by academic investigators should anticipate and resource the incremental financial costs of data sharing, including development of data standards, metadata, and hosting the data. As the real power of data sharing involves combining different data sets, an early commitment to data standards and a common taxonomy are important for such interoperability. Finally, the participants themselves, patient groups, and patient advocates are becoming increasingly sophisticated and capable of planning and executing their own research (often termed "citizen science" or "participant-led research"); data availability will be important for these efforts as well. Well-tailored data-sharing mechanisms, with careful protections for participant privacy and private commercial interests, have the potential to revolutionize the pace of clinical research, facilitating the finding of new treatments and the promotion of public health.

NOTE

Alla Digilova and Rebecca Li are co–first authors.

REFERENCES

AbbVie, Inc. v. European Medicines Agency, Case T-44/13. Action brought on January 29, 2013 (March 1, 2013).
Beetsch, J. 2013. "Project DataSphere." *Partnering for Cures.* http://partneringforcures.org/assets/Uploads/Innovators/ProjectDataSphere.pdf.

Center for Outcomes Research & Evaluation (CORE). 2014. "Request Medtronic rhBMP-2 Data." http://medicine.yale.edu/core/projects/yodap/datasharing/medtronic/requestrhbmp2data/requestrhbmp2.aspx.

ClinicalStudyDataRequest.com. 2014. "Study Sponsors." https://clinicalstudydatarequest.com/Study-Sponsors-Info.aspx.

De Angelis, Catherine, Jeffrey M. Drazen, Frank A. Frizelle, Charlotte Haug, John Hoey, Richard Horton, et al. 2004. "Clinical Trial Registration: A Statement from the International Committee of Medical Journal Editors." *New England Journal of Medicine* 351:1250–51.

Dinov, Ivo D., Daniel Valentino, Bae Cheol Shin, Fotios Konstantinidis, Guogang Hu, Allan MacKenzie-Graham, et al. 2006. "LONI Visualization Environment." *Journal of Digital Imaging* 19(2):148–58.

European Federation of Pharmaceutical Industries and Associations (EFPIA). 2013. "EFPIA Response to EMA Consultation on the Publication and Access to Clinical-Trial Data." http://www.efpia.eu/mediaroom/118/44/EFPIA-Response-to-EMA-Consultation-on-the-Publication-and-Access-to-Clinical-Trial-Data.

European Medicines Agency (EMA). 2013a. "Publication and Access to Clinical-Trial Data." http://www.ema.europa.eu/docs/en_GB/document_library/Other/2013/06/WC500144730.pdf.

European Medicines Agency (EMA). 2013b. "Release of Data from Clinical Trials." http://www.ema.europa.eu/ema/index.jsp?curl=pages/special_topics/general/general_content_000555.jsp&mid=WC0b01ac0580607bfa.

European Medicines Agency (EMA). 2014. "Regulation (EU) No 536/2014 of the European Parliament and of the Council on Clinical Trials on Medicinal Products for Human Use, and Repealing Directive 2001/20/EC."

Food and Drug Administration (FDA). 2013. "Availability of Masked and De-identified Non-Summary Safety and Efficacy Data; Request for Comments." https://www.federalregister.gov/articles/2013/06/04/2013-13083/availability-of-masked-and-de-identified-non-summary-safety-and-efficacy-data-request-for-comments#h-8.

GlaxoSmithKline (GSK). 2013. "Data Transparency." http://www.gsk.com/explore-gsk/how-we-do-r-and-d/data-transparency.html.

GlaxoSmithKline (GSK). 2014. "GSK Adds Detailed Clinical Trial Data to Multi-Sponsor Request System as Part of Continued Commitment to Data Transparency." http://www.gsk.com/media/press-releases/2014/gsk-adds-detailed-clinical-trial-data-to-multi-sponsor-request-s.html.

Godlee, Fiona. 2012. "Clinical Trial Data for All Drugs in Current Use." *British Medical Journal* 345:e7304.

Godlee, Fiona and Trish Groves. 2012. "The New BMJ Policy on Sharing Data from Drug and Device Trials." *British Medical Journal* 345:e7888.

Goldacre, Ben. 2013. "Trial sans Error: How Pharma-Funded Research Cherry-Picks Positive Results [Excerpt]." *Scientific American* (February 12). http://www.scientificamerican.com/article/trial-sans-error-how-pharma-funded-research-cherry-picks-positive-results/.

Harrison, Charlotte. 2012. "GlaxoSmithKline Opens the Door on Clinical Data Sharing." *Nature Reviews Drug Discovery* 11:891–92.

Hugh-Jones, Charles. 2013. "Project Data Sphere: Overview." *CEO Life Sciences Consortium* (January 29). http://ceo-lsc.org/projectdatasphere.

Immune Tolerance Network (ITN). 2013. "ITN Achieves Scientific Manuscript First—Provides Open, Interactive Access to Clinical Trial Data." (July 31). http://www.immunetolerance.org/sites/files/TrialShare%20PRESS%20RELEASE.pdf.

InterMune UK Ltd. v. European Medicines Agency, Case T-73/13. Action brought on February 11, 2013 (Apr. 5, 2013).

Johnson & Johnson (J&J). 2014. "Johnson & Johnson Announces Clinical Trial Data Sharing Agreement with Yale School of Medicine." http://www.jnj.com/news/all/johnson-and-johnson-announces-clinical-trial-data-sharing-agreement-with-yale-school-of-medicine.

Kesselheim, A. S. and Mello, M. M. 2007. "Confidentiality Laws and Secrecy in Medical Research: Improving Public Access to Data on Drug Safety." *Health Affairs* 26(2):483–91.

Laboratory of Neuro Imagine (LONI). "Data Archive." http://www.loni.usc.edu/about_loni/resources/data_archive/.

Mello, Michelle M., Jeffrey K. Francer, Marc Wilenzick, Patricia Teden, Barbara E. Bierer, and Mark Barnes. 2013. "Preparing for Responsible Sharing of Clinical Trial Data." *New England Journal of Medicine* 369(17):1651–58.

Narayanan, Arvind and Vitaly Shmatikov. 2008. "Robust De-Anonymization of Large Sparse Datasets." *IEEE Symposium on Security and Privacy*. http://www.profsandhu.com/cs6393_s13/ns_2008.pdf.

National Institutes of Health (NIH). 2003. "Final NIH Statement on Sharing Research Data." http://grants.nih.gov/grants/guide/notice-files/NOT-OD-03-032.html.

National Institutes of Health (NIH). 2013a. "History, Policies and Laws." http://clinicaltrials.gov/ct2/about-site/history.

National Institutes of Health (NIH). 2013b. "FDAAA 801 Requirements." http://clinicaltrials.gov/ct2/manage-recs/fdaaa.

Parkinson's Progression Markers Initiative (PPMI). "Access Data & Specimens." http://www.ppmi-info.org/about-ppmi/who-we-are/.

Peart, Karen N. 2014. "Yale Program's Agreement with Johnson & Johnson Allows Broad Access to Clinical Trial Data." *Yale News* (January 30). http://news.yale.edu/2014/01/30/yale-program-s-agreement-johnson-johnson-allows-broad-access-clinical-trial-data.

Sample, Ian. 2013. "Big Pharma Mobilising Patients in Battle Over Drugs Trials Data." *The Guardian* (July 21). http://www.theguardian.com/business/2013/jul/21/big-pharma-secret-drugs-trials.

Stanton, Terry. 2013. "Will YODA End Debate Over rhBMP-2?" *AAOS Now* (August). http://www.aaos.org/news/aaosnow/aug13/cover1.asp.

U.S. Food and Drug Administration (FDA). 2010. "Science & Research: Protection of Human Subjects; Informed Consent." http://www.fda.gov/ScienceResearch/SpecialTopics/RunningClinicalTrials/ucm113818.htm.

U.S. Food and Drug Administration (FDA). 2013. "Availability of Masked and De-Identified Non-Summary Safety and Efficacy Data; Request for Comments." *78 Fed. Reg. 33, 421–23.*

World Health Organization (WHO). 2014. "International Clinical Trials Registry Platform (ICTRP): About Trial Registration." http://www.who.int/ictrp/trial_reg/en/index.html.

World Medical Association (WMA). 2013. "World Medical Association Declaration of Helsinki: Ethical Principles for Medical Research Involving Human Subjects." *Journal of the American Medical Association* 310(20):2191–94.

CHAPTER SEVEN

Conflicts of Interest in FDA Advisory Committees

The Paradox of Multiple Financial Ties

GENEVIEVE PHAM-KANTER

I. INTRODUCTION

In medicine, there is currently robust debate about the ways in which physician-industry relationships can best be managed. On the one hand, clinicians and physician researchers have experience that can contribute to the greater public good if they lend their expertise to help develop clinical therapies. On the other hand, financial ties between physicians and industry can create incentives for physicians to, either intentionally or unintentionally, behave in ways that are not in the best interest of their patients. In addition, the perception of conflicts of interest created by these financial ties, even if these ties do not affect behavior, can lead to diminished public trust in physicians and in medicine (Grande et al. 2012; Perry, Cox, and Cox 2014).

Medical centers and policy makers thus face an empirical challenge in distinguishing the conditions under which physician-industry relationships can be harmful rather than helpful. In addition, to the extent that the same financial relationship can have both good and bad consequences, institutions and regulators must determine the criteria to be used in evaluating and weighting these opposing effects.

A. FDA Advisory Committees

At the U.S. Food and Drug Administration (FDA), industry financial interests held by external experts who serve on FDA advisory committees have been, since at least the 1990s, a source of policy concern (Glodé 2002; Institute of Medicine 2002). Although the agency itself issues final decisions on the approval of pharmaceuticals that can be sold and marketed in the United States, it relies heavily on its advisory committees for guidance. These advisory committees are composed of biomedical researchers, clinicians, pharmacologists, and biostatisticians who can provide independent expert evaluation of the scientific evidence related to the safety and efficacy of products under review. At product approval meetings, committee members listen to hours of testimony and review thousands of pages of technical reports on the product being considered for approval. At the end of these meetings, members vote on questions that elicit members' broad assessments of the safety and efficacy of the product under review. For example, members might be asked to vote yes or no on the question: "Should the sponsor [manufacturer] conduct further controlled clinical trials to assess whether and how drug X should be used in patients with condition Y?" Members are also asked to vote for or against a recommendation for product approval. The approval recommendations of advisory committees are typically (although not always) adopted by the FDA.

Advisory committee members typically serve two-year terms, and during their service, receive special government employee appointments. Because committee members are not regular employees of the FDA—many members' main appointments are at academic medical centers—the financial conflict-of-interest rules governing the conduct of federal employees are not as stringently applied to them. More precisely, although the level and type of financial interest some members hold would, according to the published rules, disqualify them from participating and voting in meetings, they are often able to obtain waivers to participate despite their financial interests. These waivers are justified on the basis of the "essential expertise" provided by these members or that the benefits of a member's participation are deemed by the FDA to outweigh the risks (Food and Drug Administration 2008).

B. FDA Advisory Committees as Research Focus

Because of the gravity of the decisions made by FDA advisory committees, the possibility of committee members holding financial interests that might affect the impartial scientific review of prescription products has far-reaching health and public trust implications. For this reason, conflicts of interest among advisory committee members are not merely a scholarly curiosity but a matter with potentially large health and social welfare consequences.

In addition, because of how the waiver system operates, FDA advisory committees are also a particularly good site for the study of how financial ties might work to influence physician and researcher behavior. In other contexts, individuals with known financial interests in firms that are potentially affected by a decision in which these individuals are involved would be disqualified or would recuse themselves (for example, in presiding over legal proceedings). In the case of FDA advisory committees, however, individuals with relevant financial interests can often participate and vote in approval decisions. Moreover, because the process for obtaining a waiver involves a well-defined process of disclosure, it is possible to obtain detailed information about these financial interests. We can thus, for instance, differentiate among various kinds of financial ties. Being able to draw these kinds of distinctions is important because we might think that, for example, the effect of an ownership stake in a firm whose product is under review is very different from the effect of employment at a university that receives research grants of the same dollar value. Because of the waiver disclosure process, it is also possible to distinguish individuals who have a financial tie to a single firm—and whether that tie is to the product sponsor or a competitor—from those who have financial ties to many different firms.

Finally, in studying FDA advisory committees, we have behavioral measures that are good first-order signals of bias associated with financial interests. For example, it is possible to examine members' voting patterns in committee and see if they are associated with members' financial interests to determine if committee members vote in a way that benefits the firms to which these members have financial ties.

Note that, based on this association alone, we cannot make definitive causal claims; that is, we cannot claim that members vote in favor of a firm *because* they have a financial interest in that firm. We can say, however, that if we observe an association between financial ties and voting, then individuals who have financial ties to a firm systematically vote in a way that is favorable to that firm.

This chapter reports findings from an analysis of the association between the financial ties of FDA advisory committee members and their voting. The findings reported here relate to a facet of conflicts of interest that has not been empirically well studied—differences between exclusive (single) financial ties and nonexclusive (multiple) financial ties. In the medical literature, there appears to be some consensus that having many financial ties is unambiguously undesirable and worse than having a single financial tie (Henry et al. 2005; Lurie et al. 2006; Campbell et al. 2007). Having many financial relationships suggests that a physician is primarily driven by financial motives, or worse, is a hired gun, willing to change his testimony to suit his financial underwriter. In the context of FDA advisory committees, a member who has many financial ties may be more likely to favor industry interests—because he or she is financially dependent on them—rather than the public interest.

II. PREVIOUS LITERATURE

Previously, there have been two quantitative studies that have looked at the relationship between the financial interests of FDA advisory committee members and member voting behavior. In 2009, the FDA commissioned the Eastern Research Group (ERG) to examine the relationship between the financial ties of committee members and member voting in drug or device approval recommendations. That report found no statistically significant relationship between individuals' financial ties to the firm sponsoring the drug or device under review and these individuals' likelihood of voting in favor of approval (Eastern Research Group 2009). These results were similar to those from an earlier published study looking solely at drug approvals; in that study, Lurie et al. (2006) found no relationship between individuals' financial ties to the firm sponsoring the drug under review and these individuals' voting in favor of the drug's approval.

One difficulty with these two studies, however, is their small sample sizes. In the Lurie et al. study, as the authors themselves note, although approximately 200 meetings were in the original sample, only eleven meetings could be analyzed for the relationship between voting and financial interest in the drug sponsor. In the ERG report, at most thirty-nine meetings could be analyzed for sponsor relationships. Small sample sizes mean that underlying associations may go undetected. Statistical theory tells us that it can be very difficult to statistically detect an effect in small samples even if there is a relationship in the underlying population.

A second difficulty with the previous studies is that the reported estimates between financial ties and drug approval voting were the result of simple two-variable relationships. In other words, the estimates were obtained first by looking at the link between voting and financial tie to a sponsor, and then separately, examining the relationship between voting and financial tie to a competitor. This type of analysis, which does not control for both key variables at the same time, is problematic because the lack of association between voting and financial tie to the sponsor does not also account for the possibility of someone having ties to *both* the sponsor and its competitors. In particular, it is entirely plausible, as discussed in more detail below, that having financial interests in many different firms might cancel each other out in terms of preferential voting. Economic theory predicts that a single firm may benefit from a financial tie, but if there are multiple competing firms, no single firm has an advantage over the others. It is therefore possible that none of the firms will preferentially benefit. By failing to control for the possibility of multiple relationships (which account for approximately 14 percent of conflicts in our sample), the estimates from these previous studies could lead to a misleading conclusion of no association between voting and financial interests in the sponsoring firm, especially if there are small sample sizes.

III. THEORETICAL ACCOUNTS OF MULTIPLE PHYSICIAN-INDUSTRY TIES

Comprehensively accounting for multiple relationships and exclusive relationships is important for clarifying the kinds of mechanisms through which conflicts of interest might operate. A common presumption, as noted previously, is that having more financial ties is worse than having fewer financial ties. This could be because, with multiple ties the physician is more beholden to commercial interests, which could create more bias, or it could be because the many ties are a signal of more worrisome but unobservable motives, which themselves are the source of bias in decision making. Although this account is the dominant narrative in the conflicts-of-interest literature, it is not the only theory available to us for interpreting the behavior of physician researchers and their financial ties to industry. An alternative account based in economic theory is that if multiple firms are establishing financial ties to the same researcher, no single firm will have a particular advantage. Each individual firm may seek initially to generate preferential treatment through its tie to a physician researcher. If, however, other firms see one firm developing a financial relationship, they too will want to develop financial ties to the same researcher. In competing to create individual advantage through financial ties, firms can enter into an arms race in which all firms are expending money to generate advantages over other firms. In the end, no individual firm "wins" in terms of holding an advantage because all firms are using the same strategy. Informally speaking, their expenditures cancel out.

A third theory related to multiple financial ties is that physicians who are very good at what they do are sought out by many different firms. These types of researchers are well-respected in their fields and may be highly influential. It may come as no surprise that high-status physician researchers, because of their ability to influence others, are pursued by many firms; but the crucial factor in this third narrative is not the influence of researchers with multiple ties but rather the researchers' expertise. Because these researchers are extremely proficient in their specialty, they are of course attractive to many firms that would like to use their expertise in drug development. At the same time, however, these researchers are also expert at detecting problems with scientific evidence and product claims. These researchers will consequently take a more skeptical stance toward evidence presented to them, regardless of the source. Their exceptional expertise and skepticism mean that they will be *less* likely to be influenced by any particular firm.

These three theoretical accounts are developed in more detail elsewhere (Pham-Kanter 2014), but for now, the most important point to note is that these accounts generate three different empirical implications. In the first account, where physician researchers are hired guns, committee members who have multiple, nonexclusive financial ties should tend to vote in ways that are favorable to industry because they financially benefit from any and all firms. In this context, "favorable to industry" refers to whether individuals vote in favor of firms whose products are under review. For example, an individual who votes to recommend product approval would be considered to vote in a way favorable to the sponsoring firm and to industry. In the second account, where firms are in an arms race to gain a voting advantage, individuals with multiple financial ties should not vote differently from individuals who have no financial ties. The financial ties of competing firms in effect cancel each other out so no single firm—and in particular, the sponsor—is able to generate a voting advantage. In the third account, where physician researchers with multiple ties are highly skilled experts, individuals with many financial relationships should behave more skeptically than other members toward the scientific evidence presented in committee. Those individuals with multiple ties would therefore be less likely to vote in favor of the sponsor—for example, less likely to vote in favor of drug approval—than individuals with no financial ties.

Because these three accounts have different predictions for the effects of multiple financial interests on voting, they can be empirically distinguished from each other. In particular, if the experts are simply pro-industry hired guns, we should observe a positive association between having multiple financial ties and voting in favor of the sponsor. If the arms race mechanism is dominant, we should observe no association between having multiple ties and voting in favor of the sponsor. If the expert selection mechanism is important (selecting for top-performing but more skeptical experts), we should observe a negative relationship between having multiple ties and voting in favor of the sponsor.

IV. DATA AND METHODS

To analyze the first-order validity of these alternative accounts, the frequency with which members voted in favor of the drug sponsor was modeled. Predictor variables in this model included: (1) an indicator of whether a member had an exclusive tie to the sponsor; (2) an indicator of whether a member had an exclusive tie to a competitor; and (3) an indicator of whether a member had ties to both the sponsor and at least one competitor. The base comparison group was therefore the set of committee members who had no financial ties to either the sponsor or to competitors. This model was estimated using data collected on the financial interests and voting behavior of all individuals who attended and voted in meetings of advisory committees convened by the FDA Center for Drug Evaluation and Research between 1997 and 2011. The final sample consisted of fifteen years of meetings held for sixteen committees; 1,168 questions and 15,739 question-votes from 379 meetings were analyzed. The full data panel included 1,379 unique persons who cast at least one vote during the fifteen years. Because this data set is ten times larger than previous data sets, this study significantly improves the degree to which a link between voting and financial interests can be detected. Data collection protocols and statistical models and methods are reported in detail elsewhere (Pham-Kanter 2014).

V. RESULTS

A. Voting Behavior with Multiple Ties Versus Exclusive Ties

In brief, the study found that contrary to conventional wisdom, individuals with multiple, nonexclusive financial relationships do not vote differently from committee members who have no financial ties. Whereas individuals with an exclusive relationship to the sponsor have an increased odds of voting in favor of the sponsor when compared to individuals with no financial ties to any firms (odds ratio = 1.49, p = 0.032; reference group = 1.00), individuals with nonexclusive ties to multiple firms are not more likely to vote in favor of the sponsor (odds ratio = 1.16, p = 0.481). In other words, if one compares the voting behavior of individuals with exclusive ties to the sponsoring firm with the voting behavior of individuals with no financial ties, one observes that members with sponsor-only ties are more likely to vote in favor of the sponsor. When one looks at the voting behavior of individuals with ties to multiple firms (i.e., those with ties to the sponsor and to at least one competitor) and compares this behavior to the voting behavior of individuals with no financial ties, one observes that those with multiple nonexclusive ties do not vote differently from those who have no financial ties at all. This finding of no difference in voting between those with multiple ties and those with no financial ties supports the second of the three narratives presented—multiple financial ties reflect multiple firms competing in an arms race in ultimately futile efforts to secure a voting advantage.

Similar results are observed when the sample is restricted to only nonunanimous votes (about 60% of the original sample). Unanimous votes occur when there is overwhelming evidence in favor of the efficacy of a product or unambiguous and alarming safety evidence against a product. In these cases, committee members are in general agreement regarding the efficacy or safety of a product based on the evidence presented. In contrast, nonunanimous votes occur when safety or efficacy evidence is more ambiguous. In these cases, there is room for more subjectivity and for nonscientific factors to affect voting. By analyzing a sample that includes only nonunanimous votes, one can focus on situations in which the scientific evidence is more mixed and there is more room for discretion and bias. Interestingly, when the original model is estimated using this restricted sample, one finds that—in terms of point estimates—individuals who have multiple ties have reduced odds of voting in favor of the sponsor compared to members who have no financial ties (odds ratio = 0.79, p = 0.369; reference group = 1.00). That is, members with multiple ties were *less* likely to vote in favor of the sponsor than members who had no financial ties at all, although the difference was not statistically significant. These results are thus statistically supportive of the arms race account where multiple ties cancel out. At the same time, the point estimates are suggestive of the third account, in which individuals with many different ties are expert researchers who bring a more skeptical eye to the evidence than other committee members.

In general, these empirical findings suggest that, contrary to the conventional narrative described in the conflicts-of-interest literature, individuals who have many financial ties do not appear to vote differently from members with no financial ties. They do not consistently vote in pro-industry ways, and under some conditions, such as when the scientific evidence is more ambiguous, tend to react more skeptically, voting more frequently against the sponsor than individuals with no ties do.

B. Voting Behavior and Types of Financial Ties

The findings reported so far are estimates of overall associations between financial ties and voting, aggregating over all types of payments. Because the type of payment may matter (e.g., a research grant may have a different effect than an ownership or royalty financial interest), a more elaborate model was estimated using indicators of financial ties specific to the payment type. In particular, the sponsor-only ties, competitor-only ties, and multiple ties were also characterized by the type of financial relationship: (1) research (investigator or grant/contract recipient); (2) employer grant or contract; (3) ownership interest such as equity or bond holdings as well as income from royalties or licenses; (4) consulting; (5) member of scientific or other advisory board or steering committee; (6) blinded endpoint reviewer or member of data safety monitoring board; or (7) paid speaker.

Separating payments in this way, one finds interesting patterns with sponsor-exclusive versus multiple ties for some categories. For example, individuals with ownership interest (e.g., stocks, income from royalties or licenses) only in the sponsoring firm had greater odds of voting in favor of the sponsor than those with no financial ties, but those with ownership interests in multiple firms voted no differently from those with no financial ties. Here, again, there is empirical support for the second account—multiple payments reflect payments arms races and result in no net voting effect.

When one looks at speaking payments rather than ownership interests, one finds that individuals who have sponsor-only financial interests have greater odds of voting in favor of the sponsor, whereas those with multiple financial interests have *lower* odds of voting in favor of the sponsor. Thus, when one focuses on financial interests related to paid speaking, one finds that those with nonexclusive ties are less likely to vote in favor of the sponsor than those with no financial ties. This is evidence in favor of the third account, where individuals with multiple ties are more skeptical than other members of the advisory committee. It would appear that in this case multiple financial ties are paradoxically more innocuous than single, exclusive financial ties.

VI. CONCLUSION

It is important, however, to emphasize that the estimates reported here are of overall effects of multiple financial relationships across many individuals and different committees. There are likely to be specific individuals with strong pro-industry preferences among those researchers with multiple industry ties. In addition, we know that committees vary a great deal in their average levels of conflictedness (Pham-Kanter 2014) and in the degree to which financial ties are associated with bias (results available upon request). Thus, the negative consequences of multiple ties may be more manifest in some committees and product markets than others. More detailed theorizing and further investigation of the factors underlying heterogeneities and of the conditions under which multiple financial relationships can be detrimental will be important.

The findings reported here also suggest that policy makers might do well to focus on the exclusivity of financial ties. Contrary to popular belief, a single financial tie need not be a signal of weak industry bias. The results described in this chapter suggest quite the opposite— a single exclusive tie could be associated with substantial bias.

This analysis shows that the number of financial ties held by a physician researcher can be a misleading measure of conflictedness and that exclusivity appears to be an important neglected dimension of conflictedness. Overall, this chapter argues for a more granular and nuanced approach to managing financial conflicts of interest in FDA advisory committees and, more broadly, in medical activities.

NOTE

Author Genevieve Pham-Kanter has no conflicts of interest to disclose. This research was supported by the Edmond J. Safra Philanthropic Foundation through a grant to the Edmond J. Safra Center for Ethics at Harvard University. Pham-Kanter had full access to all of the data in the study and takes responsibility for the data and the accuracy of the data analysis. Laquesha Sanders, Igor Gorlach, Magdalina Gugucheva, Kenneth Oshita, and John Barnes provided valuable research assistance.

REFERENCES

Campbell, Eric G., Russell L. Gruen, James Mountford, Lawrence G. Miller, Paul D. Cleary, and David Blumenthal. 2007. "A National Survey of Physician-Industry Relationships." *New England Journal of Medicine* 356:1742–50.

Eastern Research Group. 2009. "Financial Conflict-of-Interest Disclosure and Voting Patterns at FDA Advisory Committee Meetings." Unpublished report.

Food and Drug Administration. 2008. Guidance for the Public, FDA Advisory Committee Members, and FDA Staff on Procedures Determining Conflict of Interest and Eligibility for Participation in FDA Advisory Committees. http://www.fda.gov/downloads/RegulatoryInformation/Guidances/UCM125646.pdf. Accessed March 8, 2014.

Glodé, Elizabeth R. 2002. "Advising Under the Influence?: Conflicts of Interest Among FDA Advisory Committee Members." *Food and Drug Law Journal* 57:293–322.

Grande, David, Judy A. Shea, and Katrina Armstrong. 2012. "Pharmaceutical Industry Gifts to Physicians: Patient Beliefs and Trust in Physicians and the Health Care System." *Journal of General Internal Medicine* 27(3):274–79.

Henry, David, Evan Doran, Ian Kerridge, Suzanne Hill, Paul M. McNeill, and Richard Day. 2005. "Ties That Bind: Multiple Relationships Between Clinical Researchers and the Pharmaceutical Industry." *Archives of Internal Medicine* 165(21):2493–96.

Institute of Medicine. 1992. *Food and Drug Administration Advisory Committees.* Washington, D.C.: National Academy Press.

Lurie, Peter, Cristina M. Almeida, Nicholas Stine, Alexander R. Stine, and Sidney M. Wolfe. 2006. "Financial Conflict of Interest Disclosure and Voting Patterns at Food and Drug Administration Drug Advisory Committee Meetings." *Journal of the American Medical Association* 295(16):1921–28.

Perry, Joshua, Dena Cox, and Anthony D. Cox. 2014. "Trust and Transparency: Patient Perceptions of Physicians' Financial Relationships with Pharmaceutical Companies." *Journal of Law, Medicine & Ethics* 42(2):475–91.

Pham-Kanter, Genevieve. 2014. "Revisiting Financial Conflicts of Interest in FDA Advisory Committee Meetings." *Milbank Quarterly* 92(3):446–70.

CHAPTER EIGHT

The Crime of Being in Charge

Executive Culpability and Collateral Consequences

KATRICE BRIDGES COPELAND

I. INTRODUCTION

The pharmaceutical industry has long been an enforcement headache for the Food and Drug Administration (FDA). In particular, FDA has struggled with enforcing the laws restricting the promotion of drugs. For many years, FDA pursued a strategy of cooperative enforcement with pharmaceutical companies to avoid the collateral consequences of criminal conviction. If a pharmaceutical company is convicted of a felony, it would be automatically excluded from participation in federal health care programs, such as Medicare and Medicaid, for a period of at least five years. Exclusion has been described as a "death sentence" for pharmaceutical manufacturers because it prevents the federal government from reimbursing patients for any drug produced by the excluded pharmaceutical manufacturer (Girard 2009:137). To avoid the potentially devastating consequences of exclusion for patients, FDA opted not to pursue felony criminal charges against pharmaceutical manufacturers that engaged in off-label promotion. Instead, FDA entered into corporate integrity agreements that required the pharmaceutical companies to pay large fines and enact compliance measures designed to prevent the illegal promotional activity from recurring (Copeland 2012). In return, the government agreed not to pursue felony criminal charges or debar the pharmaceutical manufacturers from participation in federal health care programs.

FDA's approach seemed like a reasonable response to a complicated enforcement problem. The harsh reality, however, was that pharmaceutical companies viewed the fines and compliance measures as nothing more than the cost of doing business. Thus, the corporate integrity agreements were not enough to deter pharmaceutical companies from engaging in illegal promotional activity. As a result, FDA began to see repeat offenders of the pharmaceutical marketing rules.

In March 2010, FDA announced that it intended to use individual misdemeanor prosecutions "to hold responsible corporate officials accountable" (Hamburg 2010). FDA also made it clear that it intended to enhance and make better use of its debarment and disqualification procedures (Hamburg 2010). Thus, the focus shifted from simply pursuing pharmaceutical companies to pursuing misdemeanor criminal charges against individual executives for the misconduct of their subordinates. To support its theory of indirect criminal liability, the government dusted off the responsible corporate officer doctrine. The responsible corporate officer doctrine permits the government to prosecute an executive for a misdemeanor violation of the Federal Food, Drug, and Cosmetic Act (FDCA, 21 USC §§ 301–92), regardless of the officer's lack of awareness of misconduct if by reason of the officer's position in the company she had the responsibility and authority either (1) to prevent the misconduct in the first place, or (2) promptly to correct the violation, but failed to do so (*United States v. Park* 1975:673–74). The government has already successfully obtained guilty pleas based on the responsible corporate officer doctrine (see infra part III.B).

The misdemeanor convictions, however, are not the end of the story. Just as there are collateral consequences that flow from convicting pharmaceutical manufacturers, there are collateral consequences that flow from convicting executives as well. Following a misdemeanor conviction, the government has the discretion to exclude executives from participation in federal health care programs. The baseline period of exclusion is three years but can be increased based on aggravating factors. For an individual, exclusion means that the executive is virtually unemployable in the health care industry because there are only limited circumstances where a company that directly or indirectly receives money from federal health care programs can employ an excluded individual without jeopardizing its own participation in those programs.

Despite its use by FDA as a means to punish pharmaceutical executives, exclusion is not meant to be punitive. The goal is supposed to be to protect the health care system from unscrupulous individuals (67 Fed. Reg. 11,928 (Mar. 18, 2002)). This chapter argues that absent a showing of moral blameworthiness on the part of health care executives who were in charge at the time the misconduct occurred, a period of exclusion longer than the three-year base level for permissive exclusions is a grossly disproportionate remedy.

II. BACKGROUND

FDA regulates the pharmaceutical industry through the FDCA. Violations of the FDCA are considered to be public welfare offenses. Public welfare offenses are a special category of regulatory offenses that involve dangerous activities or materials. Because a reasonable person should be aware of the risks involved with activities proscribed by public welfare offenses, these crimes are an exception to the fundamental requirement that individuals should not be punished unless the government can demonstrate wrongful conduct along with a guilty mind (*Liparota v. United States* 1985:433).

The responsible corporate officer doctrine, which has only been applied to public welfare offenses, grew from two Supreme Court cases dealing with violations of the FDCA. In the first case, *United States v. Dotterweich*, the government charged Buffalo Pharmacal Company, Inc. and Joseph H. Dotterweich, the president and general manager of Buffalo Pharmacal Company, with violating the FDCA by shipping misbranded and adulterated drugs in interstate commerce (1943:278). Although Dotterweich was not personally involved in packaging or shipping the drugs, the jury convicted him of violating the FDCA (*United States v. Buffalo Pharmacal Co., Inc.* 1942:501). The Supreme Court found that Dotterweich was liable under the act because he had a "responsible share" in the shipment of the misbranded and adulterated drugs despite the fact he did not have "consciousness of wrongdoing" (*United States v. Dotterweich* 1943:284). In the Court's view, Dotterweich, as an executive in the company, was in a better position than the public to ensure that the drugs met the requirements of the FDCA. Therefore, he had to bear the risk of those laws being violated, not the public (ibid., 284–85).

In the second case, the government charged Acme Markets, Inc. (Acme) and John R. Park, Acme's chief executive officer, with violating the FDCA by causing adulteration of food (*United States v. Park* 1974:660). The evidence showed that an FDA inspector found evidence of rodent infestation of food on multiple occasions and sent a letter to Park detailing the violations (ibid., 662–63). In a follow-up inspection, an FDA inspector found that there was still rodent activity in the warehouse (ibid., 662). Park testified that all Acme employees were under his general control but the job of sanitation was handled by somebody else at Acme (ibid., 663). The jury convicted Park.

The Supreme Court upheld Park's conviction (*United States v. Park* 1975:667). The Court explained that the government makes a prima facie case by demonstrating that "the defendant had, by reason of his position in the corporation, responsibility and authority either to prevent in the first instance, or promptly to correct, the violation complained of, and that he failed to do so" (ibid., 673–74). Thus, the burden of complying with the law is placed on "a person otherwise innocent but standing in responsible relation to a public danger" (ibid., 668–69). The Court recognized that there is a defense to liability when it would be "objectively impossible" for the executive to prevent or correct the misconduct (ibid., 673). In the cases following *Park*, however, the impossibility defense has never been successfully raised.

III. EXCLUSION OF RESPONSIBLE CORPORATE OFFICERS

Pharmaceutical executives face misbranding charges under the FDCA as responsible corporate officers.[1] A drug or device is misbranded if its label is false, misleading, or does not contain "adequate directions for use" (21 USC § 352(a), (f)). Misbranding is a misdemeanor offense punishable by less than one year in jail, unless it is undertaken "with the intent to defraud or mislead," in which case it is a felony punishable by up to three years in jail (21 USC § 333(a)(1), (2)).

Following a guilty plea or conviction and sentencing, the executives are faced with the potential collateral consequences, or civil restrictions, of their criminal convictions (Demleitner 1999:154). For health care executives, the most severe collateral consequence of conviction is the Department of Health and Human Services' (HHS) discretionary authority to exclude the officers from participation in federal health care programs.

A. The Exclusion Authority

The HHS Office of Inspector General (OIG) exercises the exclusion authority. If the OIG excludes an individual or entity from federal health care programs, those programs may not pay for any item or services furnished directly or indirectly by the excluded individual or entity. The OIG has extensive authority to debar individuals and executives for a wide range of offenses. The OIG's exclusion authority is mandatory in some cases and within the OIG's discretion in other cases. Exclusion is mandatory if the individual or entity is convicted of a felony offense "relating to" health care fraud (42 USC § 1320a-7(a)(3), (4)). The mandatory exclusion period is a minimum of five years (42 USC § 1320a-7(c)(3)(B)). Exclusion is permissive if the individual or entity is convicted of a misdemeanor offense "relating to" fraud in connection with the delivery of a health care item or service (42 USC § 1320a-7(b)(1), (3)). Permissive exclusions are for a period of three years unless there are mitigating or aggravating circumstances that lengthen or shorten the period of exclusion (42 USC § 1320a-7(c)(3)(D)). While there are base periods of exclusion, there are no outer limits on the OIG's ability to increase the exclusion period beyond the base period through aggravating factors. In situations where the individual is convicted as a responsible corporate officer, the authority to extend the base period of exclusion is potentially subject to misuse or abuse.

B. The Purdue Pharma Case

A prominent example of the OIG's use of its broad exclusion authority is the Purdue Pharma case. FDA investigated Purdue Pharma for marketing violations concerning its pain drug OxyContin (Information, *United States v. Purdue Frederick Co., Inc.* 2007:10). Purdue Pharma supervisors and employees marketed and promoted OxyContin as less addictive than other pain medications (ibid., 5–6, 15). Their own studies, however, demonstrated that those claims were not true (*Goldenheim v. Inspector General* 2009:6).

After a five-year investigation, the government entered into a global settlement with Purdue Pharma that included a corporate integrity agreement, a fine, and a plea of guilty by Purdue Frederick Company, Inc. (a subsidiary of Purdue Pharma) to felony misbranding with intent to defraud or mislead. In addition, the government charged three Purdue Pharma executives—Michael Friedman (chief executive officer), Howard Udell (chief legal officer), and Paul D. Goldenheim (chief scientific officer)—as responsible corporate officers, with misdemeanor counts of misbranding.[2] All three of the executives entered guilty pleas. The district court accepted the plea agreements even though they did not impose prison sentences because there was no "proof of knowledge by the individual defendants of the wrongdoing" (*United States v. Purdue Frederick Company, Inc.* 2007:576).

Following these convictions, the OIG sent notices to the executives informing them that they would each be excluded from participation in federal health care programs for twenty years. According to the OIG, the exclusion was proper because "the misdemeanor offense[s] relat[ed] to fraud . . . in connection with the delivery of a health care item or service" (*Goldenheim v. Inspector General* 2009:7). Because these were permissive exclusions, the base period of exclusion would have been three years.

The executives sought review of the inspector general's decision. After briefing was complete, the administrative law judge (ALJ) reduced the period of exclusion to fifteen years based on the executives' cooperation with law enforcement officials (a mitigating factor) (*Goldenheim v. Inspector General* 2009:7). The executives argued that they had no intent to defraud and that their convictions rested solely on their status as responsible corporate officers rather than their own misconduct (ibid., 2, 13). The ALJ found that the executives' misbranding offenses were offenses "relating to fraud" because of the relationship between the executives' conduct and Purdue's fraudulent misbranding (ibid., 8).

On appeal, the executives argued that their convictions were not related to fraud because they were not convicted of fraudulent conduct. The Departmental Appeals Board (DAB) found that based on the plain language of the exclusion provision, the statute "does not restrict exclusions to only offenses constituting or consisting of fraud, but requires merely that the offense at issue be one 'relating to' fraud" (*Goldenheim v. Inspector General* 2009:10). The DAB found that an offense "relating to" fraud is one that has some "nexus" or "common sense connection" to fraud (ibid., 8). In the DAB's view, the misdemeanor misbranding offense clearly related to fraud because without Purdue Frederick's fraudulent misbranding there would not have been any charges against the executives (ibid., 10–12). Thus, the executives' lack of knowledge concerning or role in the fraudulent conduct was irrelevant (ibid., 13). Ultimately, the DAB upheld the exclusion but reduced it to a period of twelve years because it found that the ALJ relied on an aggravating factor that was not supported by substantial evidence (ibid., 25–26). The executives sought review of the final decision, but the federal district court held that the secretary's decision was supported by substantial evidence (*Friedman v. Sebelius* 2010:105). Similarly, the D.C. Circuit found that the convictions related to fraud because the executives' were in charge when the felony misbranding took place (*Friedman v. Sebelius* 2012:816). The D.C. Circuit was convinced, however, that the length of exclusion was arbitrary and capricious because it was a significant departure from prior agency decisions and was not sufficiently justified (ibid., 828).

IV. THE MORAL BLAMEWORTHINESS OF RESPONSIBLE CORPORATE OFFICERS

The Purdue Pharma case perfectly illustrates the fundamental problem with imposing significant collateral consequences on individuals convicted as responsible corporate officers—neither their prosecution nor their collateral consequences are based on their intent or conduct. It makes sense to permit the conviction and perhaps even the three-year period of exclusion because with violations of the FDCA, the concern is the protection of society (*Morrisette v. United States* 1952:257). The Supreme Court has explained that in these cases, the "penalties commonly are relatively small, and conviction does no grave damage to an offender's reputation" (ibid., 256). Given that violations of the FDCA could potentially lead to serious injury or death, the interests of the public are paramount and outweigh an executive's lack of intent or moral blameworthiness in a given criminal case. The question becomes whether that justification holds when there are significant and long-lasting collateral consequences that attach to that conviction. This section contends that it does not.

A. Exclusion Is a Harsh Remedy

Exclusion from participation in federal health care programs for a prolonged period of time, while technically a civil penalty, is far harsher than the potential criminal penalty for a misdemeanor misbranding violation. Using the example of the Purdue executives, their criminal penalties included fines, disgorgement, and three years of probation. Even if they had received prison time, it would have been for less than a year. In contrast, the OIG initially wanted to debar the executives for a period of twenty years. Thus, the collateral consequences of their convictions would have lasted seventeen years after their probation ended. The exclusions effectively ended the careers of the Purdue Pharma executives in the health care industry and forever labeled them as untrustworthy criminals.

Professor Gabriel Chin argues that these types of overly restrictive collateral consequences bear a striking resemblance to a form of punishment called "civil death" (Chin 2012:1790). In the nineteenth century, civil death statutes took away the civil and political rights of felons "on the theory that they ceased to exist as legal persons after their conviction" (Labelle 2008:85). Although civil death statutes fell out of favor in the 1960s, they returned in the 1980s and 1990s (Love 2011:770). Federal laws barred people with convictions from public benefits, government employment, and government contracts (ibid., 771). Professor Chin argues that these collateral consequences are the new form of "civil death" (2012:1790). As he explains it, "[f]or many people convicted of crimes, the most severe and long-lasting effect of conviction is not imprisonment or fine. Rather, it is being subjected to collateral consequences . . ." (ibid., 1791). Further, he argues that "for a person who must work for a living, [the] loss of the right to do business with the government—or work in any regulated industry—could result in exclusion as complete as civil death under the nineteenth-century statutes" (ibid., 1802). It is these collateral consequences of conviction that become the "most important part" of the conviction (ibid., 1806).

There is no question that for health care executives "the most important part" of their conviction is exclusion from participation in federal health care programs. Debarred health care executives cannot contract with the government and have very limited ability to work for any health care company that contracts with the government. The health care industry accounts for approximately 18 percent of the United States' gross domestic product (Radnofsky 2012). Thus, excluded health care executives are unable to access employment in a large segment of the economy. As Professor Demleitner has argued, exclusion from "vast segments of the labor market . . . parallels the effect of restrictions on the ex-offender's right to contract in the nineteenth and early twentieth centuries" (1999:156). Therefore, a long period of exclusion is not just a harsh remedy, it is also the modern day equivalent of "civil death" for health care executives.

B. The Government Should Prove Moral Blameworthiness Before Imposing a Prolonged Period of Exclusion

1. Exclusion Is Being Misused as Criminal Punishment

The collateral consequence of a long period of exclusion is far more devastating than the direct criminal consequences of conviction. Indeed, as Margaret Colgate Love has argued, "collateral consequences are increasingly understood and experienced as criminal punishment, and never-ending punishment at that" (2011:114). To avoid problems of over deterrence, it is important that the sanction fits the crime. Exclusion is not meant to be punitive. As such, it should not be used to transform a misdemeanor offense with minor punishment into a harsh penalty. This section argues that if the OIG wants to impose a period of exclusion greater than the base level of three years, it should be required to demonstrate that the officer is morally blameworthy for the misconduct. Otherwise, the debarment may be seen as unjust criminal punishment instead of a civil sanction.

The purpose of exclusion is essentially risk prevention (67 Fed. Reg. 11,928 (Mar. 18, 2002)). The government needs to be able to safeguard the federal health care programs and their participants from individuals who have defrauded the government or caused some other program-related harm. In that sense, the exclusion is for utilitarian reasons because the excluded individual's loss from the inability to participate in federal health care programs is outweighed by the need to prevent an untrustworthy individual from harming the program in the future. The problem is that when a responsible corporate officer is excluded, the decision is based on the harm caused by a subordinate. Even if one believes that harsh sanctions are appropriate to encourage health care executives to be more responsible when they supervise their subordinates, there still needs to be proportionality between the harm caused by the executive and the period of exclusion. The critical question becomes how much of the harm should be attributed to the executive for the purpose of setting a period of exclusion. If there is a strong belief that health care executives should face harsher criminal punishment for misbranding, that should be addressed by the legislature. In the absence of that, however, the OIG should not be permitted to use a long period of civil exclusion as a criminal sanction without some showing that the long period of exclusion is proportional to the harm caused by the executive. Thus, if an *individual* is to be excluded, the exclusion period should be proportional to the harm inflicted by the *individual* and the potential for future harm by the *individual*. As Professor Ewald has argued, "[w]hen the punitive elements of a sentence are premised on proportionality, the collateral consequences of conviction should be held to the same standard" (2002:1099).

In the Purdue Pharma case, the prosecutor and judge recognized that the individual executives had no knowledge of or involvement in the misconduct. Accordingly, the executives received three years of probation and paid large fines but did not receive prison time. But, the OIG initially imposed a twenty-year exclusion based on aggravating factors—financial losses to government programs, duration of the offenses, and the impact of those offenses on individuals—that did not relate to the failure of the executives to discover and remedy the misconduct. While exclusion for three years is likely warranted under these circumstances, it is unclear how an extended period of exclusion is warranted without any consideration of factors that relate directly to the failures of the individual executives. It seems that the OIG should exclude an *individual* based on the risk that the *individual* poses to federal health care programs. At a minimum, the length of exclusion for the *individual* should be tied to the *wrongdoing and risk that the individual* will violate the law in the future.

Thus, while exclusion may not technically be a criminal sanction, its imposition is much more serious than the criminal penalties associated with a misdemeanor misbranding conviction. If the government is going to impose "civil death" on an executive, then the exclusion should either be based on a conviction for the executive's personal misconduct or the executive's awareness of or participation in his or her subordinate's misconduct. Therefore, a responsible corporate officer should not be subjected to a prolonged period of exclusion unless the OIG can demonstrate the executive's moral blameworthiness for the misconduct.

2. Responsible Corporate Officers Are Not Always Morally Blameworthy for the Actions of Their Subordinates

Responsible corporate officers may be held criminally accountable for the conduct of their subordinates, but criminal accountability does not demonstrate moral blame for the misconduct. Moral blameworthiness is necessary because strict liability public welfare offenses are justified in part by the fact that they have minor sanctions associated with their violation. If the OIG intends to use long periods of exclusion as additional criminal punishment to scare executives into being responsible actors, as they did with the Purdue Pharma executives, then the justification no longer applies. A mere showing that the executive failed to detect and prevent or correct the wrongdoing is not the same as saying that the executive had a hand in the misconduct and is morally blameworthy for it. Moral blameworthiness is not demonstrated simply by virtue of the officer's position in the company. There needs to be some measure of culpability.

While there are many theories of moral responsibility, this chapter accepts the formulation by Professor Peter Arenella. Professor Arenella explains that in order to assign moral blame to an individual's conduct, four conditions must exist: "(1) moral agent must be implicated in (2) the breach of a moral norm that (3) fairly obligates the agent's compliance under circumstances where that (4) breach can be fairly attributed to the agent's conduct" (Arenella 1992:1518). Responsible corporate officers are certainly moral agents who have the capacity to make moral judgments and take actions in conformity with those judgments. While regulatory offenses such as misbranding under the FDCA may not be based on morality or have moral content, they are morally wrongful in the sense that they are done "in violation of a legal norm" (Green 1997:1575). Therefore, the sole question with respect to responsible corporate officers is whether the breach of a moral norm by a subordinate can be fairly attributed to the officer's conduct.

As Professor Arenella explains, to satisfy the fair attribution principle, first "there must be a voluntary act as well as causation between a defendant's act and any resulting harm" (1992:1523). But, the responsible corporate officer doctrine lacks the "traditional requirement . . . to show a connection between the individual and the particular wrong" (Petrin 2012:299). It does not conform to ordinary notions of causation and blame because the executive is being held accountable for the misconduct of others rather than herself. Professor Stanford Kadish explains that

> we are responsible for ourselves and for what our actions cause in the physical world, and we may cause things to happen unintentionally as well as intentionally. However, what other people choose to do as a consequence of what we have done is their action and not ours. Our actions do not cause what they do in the sense that our actions cause events.
>
> —(1985:355)

In many responsible corporate officer cases, the only way to show causation is by omission. As Professor Brickey explains, "proof that the officer had the responsibility and the power to prevent the violation and that he failed to fulfill the duty to do so establishes the required causal link between the officer and the violation" (Brickey 1982:1363). But, the notion that one person causes the actions of another runs counter to our idea that each person is an autonomous actor. "We regard a person's acts as the products of his choice, not as an inevitable, natural result of a chain of events" (Kadish 1985:333). While causation based on a factual connection between the officer and the violation may be sufficient for a misdemeanor conviction, it is insufficient to show that an action is fairly attributable to an executive.

Second, for an action to be fairly attributable there must be proof of mens rea with respect to the individual's act or risked harm (Arenella 1992:1523). A finding that a responsible corporate officer is guilty of a strict liability misdemeanor offense, such as misbranding, does not establish culpability because there is no requirement of mens rea. As Professor Brickey makes clear, "while one might well conclude that the [Park] Court's language . . . would support the proposition that liability must be predicated upon a finding of minimal culpability or negligence, within the context of the Park opinion *the premise is unsound*" (1982:1364; emphasis added). There is no explicit mens rea requirement in either the offense of conviction or in the responsible corporate officer doctrine. Therefore, a conviction as a responsible corporate officer does not demonstrate culpability. Some additional showing will be necessary to demonstrate that the wrongdoing can be fairly attributed to the responsible corporate officer's conduct.

To prove that the actions of subordinates are fairly attributable to the responsible corporate officer, there needs to be some showing that the executive in charge intended for the wrongdoing to occur. As Professor Kadish explains, "[w]e become accountable for the liability created by the actions of others . . . only when we join in and identify with those actions by intentionally helping or inducing them to do those actions; in other words, by extending our wills to their action" (1985:355). Without some showing of intentionality, through negligence, recklessness, knowledge, or purpose, there can be no finding that the actions of the subordinate are attributable to the executive. Thus, in the ordinary case there would be no showing of moral blameworthiness.

This is not to discount, of course, the fact that it would be difficult to prove the moral blameworthiness of a responsible corporate officer. This is particularly true if the executive is a high level official such as a chief executive officer. It is unlikely that the high level official's participation in the wrongdoing would be documented. Further, the executive might become aware of misconduct but choose not to intervene. The executive's discovery of the misconduct and subsequent decision not to get involved may similarly go unmemorialized. In many cases, the prosecutor would have to rely upon circumstantial evidence to prove culpability. While both of these scenarios are realistic in a large corporation, the fact that these situations could occur is not enough to relieve the prosecutor of the burden to prove that the officer is morally blameworthy before imposing the harsh remedy of a long period of exclusion. While the public certainly has an interest in these proceedings just like it does in criminal misdemeanor misbranding cases, the public's interest no longer outweighs the executive's interest because the punishment is no longer minor and there is great harm to the executive's reputation. In short, the justification for holding responsible corporate officers accountable without a showing of moral blameworthiness does not hold up when the executive is facing a career-ending exclusion.

V. CONCLUSION

FDA should be commended for its efforts to raise the stakes in cases of pharmaceutical fraud. Targeting executives who were in charge at the time that misconduct occurred at the pharmaceutical company sends a strong message that FDA does not take violations of the drug marketing rules lightly. It holds the executives accountable and gives them strong incentives to monitor their subordinates. After obtaining a conviction, however, FDA should not impose the collateral consequence of exclusion for more than three years without first demonstrating that the executive is morally blameworthy for the misconduct. While exclusion may be a civil remedy, it is the most damaging remedy that an executive can face. A long period of exclusion could amount to "civil death" that takes away an executive's livelihood and should not be imposed without sufficient justification.

NOTES

This chapter is adapted from a 2014 law review article of the same name that originally appeared in *American Criminal Law Review* 51: 799–836.

1. 21 USC § 331(a) (prohibiting "the introduction or delivery for introduction into interstate commerce of any . . . drug . . . that is adulterated or misbranded").

2. Plea Agreement, *United States v. Michael Friedman*, No. 1:07CR29 (W.D. Va. May 10, 2007); Plea Agreement, *United States v. Howard Udell*, No. 1:07CR29 (W.D. Va. May 10, 2007); Plea Agreement, *United States v. Paul Goldenheim*, No. 1:07CR29 (W.D. Va. May 10, 2007); 21 USC § 333(a)(1).

REFERENCES

Arenella, Peter. 1992. "Convicting the Morally Blameless: Reassessing the Relationship Between Legal and Moral Accountability." *UCLA Law Review* 39:1511–1622.

Brickey, Katheleen F. 1982. "Criminal Liability of Corporate Officers for Strict Liability Offenses—Another View." *Vanderbilt Law Review* 35:1337–81.

Chin, Gabriel. 2012. "The New Civil Death: Rethinking Punishment in the Era of Mass Conviction." *University of Pennsylvania Law Review* 160:1789–1833.

Copeland, Katrice Bridges. 2012. "Enforcing Integrity." *Indiana Law Journal* 87:1033–86.

Demleitner, Nora V. 1999. "Preventing Internal Exile: The Need for Restrictions on Collateral Sentencing Consequences." *Stanford Law and Policy Review* 11:153–63.

Ewald, Alec C. 2002. "'Civil Death': The Ideological Paradox of Criminal Disenfranchisement Law in the United States." *Wisconsin Law Review* 2002:1045–1137.

Friedman v. Sebelius, 755 F. Supp. 2d 98, 105 (D.D.C. 2010).

Friedman v. Sebelius, 686 F.3d 813, 820 (D.C. Cir. 2012).

Girard, Vicki W. 2009. "Punishing Pharmaceutical Companies for Unlawful Promotion of Approved Drugs: Why the False Claims Act Is the Wrong Rx." *Journal of Health Care Law & Policy* 12:119–58.

Goldenheim v. Inspector General, DAB No. 2268, at 6 (Dep't of Health & Human Servs. Aug. 28, 2009) (final admin. review).

Green, Stuart P. 1997. "Why It's a Crime to Tear the Tag Off of a Mattress: Overcriminalization and the Moral Content of Regulatory Offenses." *Emory Law Journal* 46:1533–1615.

Hamburg, Margaret A. 2010. Letter from Hamburg, FDA Commissioner, to Charles E. Grassley, Ranking Member, Senate Committee on Finance.

Information, *United States v. Purdue Frederick Company, Inc.*, No. 1:07CR00029 at 1 (W. D Va. May 10, 2007).

Kadish, Stanford A. 1985. "Complicity, Cause and Blame: A Study in the Interpretation of Doctrine." *California Law Review* 73:323–410.

Labelle, Deborah. 2008. "Bringing Human Rights Home to the World of Detention." *Columbia Human Rights Law Review* 40:79–133.

Liparota v. United States, 471 U.S. 419, 433 (1985).

Love, Margaret Colgate. 2011. "Paying Their Debt to Society: Forgiveness, Redemption, and the Uniform Collateral Consequences of Conviction Act." *Howard Law Journal* 54(3):753–93.

Morissette v. United States, 342 U.S. 246, 274 (1952).
Petrin, Martin. 2012. "Circumscribing the 'Prosecutor's Ticket to Tag the Elite'—A Critique of the Responsible Corporate Officer Doctrine." *Temple Law Review* 84(2):283–324.
Radnofsky, Louise. 2012. "Steep Rise in Health Care Costs Projected." *Wall Street Journal* (June 12). http://online.wsj.com/article/SB10001424052702303768104577462731719000346.html.
United States v. Buffalo Pharmacal Co., Inc., 131 F.2d 500, 501–2 (2d Cir. 1942), *rev'd sub nom. United States v. Dotterweich*, 320 U.S. 277 (1943).
United States v. Park, 421 U.S. 658 (1975).
United States v. Purdue Frederick Company, Inc., 495 F. Supp. 2d 569 (W.D. Va. 2007).

CHAPTER NINE

Recalibrating Enforcement in the Biomedical Industry

Deterrence and the Primacy of Protecting the Public Health

PATRICK O'LEARY

THE FOOD and Drug Administration's (FDA's) stated mission is to promote and protect the public health (Federal Food, Drug, and Cosmetic Act (FDCA), 21 USC § 393(b)). In the agency's own words, this mission has multiple aspects: "protecting the public health by assuring the safety, efficacy and security of [regulated products]," on the one hand, but also "advancing the public health by helping to speed innovations that make medicines more effective, safer, and more affordable. . . ." (FDA 2013). One of the key challenges that the agency faces is how to achieve the former goal—guaranteeing product safety and efficacy—with minimal negative impact on the important interests reflected by the latter—viz, industry's continued ability and incentive to develop and distribute innovative medical products. Further complicating this challenge is the fact that in health care enforcement in particular, regulators must place a premium on deterrence—when the cost of violations can be measured in human lives, making victims whole after the fact is often an impossibility. Although this chapter addresses only one small piece of this complex balancing act, it is a vital and increasingly urgent one, particularly in light of an escalation in enforcement that in recent years has imposed substantial costs and uncertainty on life-sciences firms without achieving meaningful deterrence.

This chapter makes two assertions. The first is that the current federal approach to deterring misconduct in the biomedical industry is broken. Prosecutors' emphasis on ever-larger monetary settlements and corporate integrity agreements (CIAs) has imposed significant costs on industry without achieving real deterrence, and threats of increased reliance on administrative and criminal sanctions against corporate officers at offending firms have raised practical and philosophical concerns. The second assertion is that our broken system can be fixed using existing enforcement tools—including monetary penalties and compliance programs, criminal prosecution, and administrative exclusion from federal health care programs—but only if they are used intelligently and responsibly. By coordinating and aligning enforcement policies at the various agencies involved and tying enforcement decisions at all of these agencies more directly to the objective of promoting and protecting the public health, I posit that existing tools can be deployed to effectively deter misconduct, ameliorate the flaws of the current regime, and meaningfully address apprehensions relating to the prosecution and exclusion of individuals.

I. THE SYSTEM IS BROKEN

A. The Futility of Escalating Monetary Penalties

According to a report compiled by Public Citizen,[1] from 1991 to July 2012, the federal government entered into 104 settlements with pharmaceutical manufacturers for a total of $26.9 billion (Almashat and Wolfe 2012:10). Well over half of that sum was collected in just a three-and-a-half-year period beginning in 2009, a period that saw multiple record-breaking settlements, including a $2.3 billion settlement against Pfizer and a $3 billion settlement against GlaxoSmithKline. Since the report was published, there have been additional large settlements, including a $762 million settlement against Amgen in 2012 and a $2.2 billion settlement against Johnson & Johnson in 2013 (Groeger 2014). Massive settlements like these are generally accompanied by CIAs, agreements imposing broad compliance obligations on firms as a condition of settlement (Copeland 2012:1050–52; Paulhus and Richter 2013). This reliance on ever-larger penalties and more demanding compliance programs reflects a "strategy aimed at financial recovery and organizational reform" (Copeland 2012:1075).

Notwithstanding the headline-making size of settlements and the increasingly comprehensive character of CIAs, the current enforcement approach has struggled to achieve meaningful deterrence (Copeland 2012:1075; Rodwin 2013). The commonly accepted explanation envisions industry as Justice Holmes's "bad man," who treats monetary fines as merely the prescribed cost of engaging in proscribed conduct (Holmes 1897:461). The idea is that these firms treat penalties and fines as simply the "cost of doing business" and thus treat the decision to engage in misconduct (or not to take greater steps to prevent it) as an economic one. When the expected revenues from misconduct exceed the risk-adjusted anticipated penalty, it becomes rational to break the rules (Boozang 2012:86–87; Copeland 2012:1064).

Logically, the response for regulators is to try to make misconduct economically unattractive—wrongdoers will be deterred if "the costs of their conduct, multiplied by the probability of their punishment, outweigh the net expected benefits of such conduct" (Baer 2012:629). That is, to deter, we need only increase either the cost of misconduct (by raising penalties) or the likelihood of punishment. Although increased focus from law enforcement and strong compliance programs have improved the likelihood of punishment to an extent, significant increases are ultimately constrained by limited resources and the inherent difficulty of investigating sophisticated organizations. Perhaps for this reason, the government has so far mostly sought additional deterrence by increasing penalties. But this approach suffers from a basic flaw: the evidence suggests that effective monetary penalties would have to be truly massive. Consider the government's 2009 settlement with Pfizer. The criminal fine in that case—$1.195 billion, which at that time was the largest criminal fine ever imposed—exceeded the profits resulting from the charged conduct by a factor of 1.8. The government explained this massive penalty by arguing that it would "deter such conduct in the future and make sure that the fine is not treated by such companies as merely a cost of doing business" (*United States v. Pharmacia & Upjohn Co.* 2009). But violations have persisted, and the fact that such settlements have been insufficient to achieve meaningful deterrence suggests that penalties may have to be even larger and exceed firms' ill-gotten profits by still more significant margins before they can be effective deterrents. At some point, this becomes untenable; penalties large enough to really "hurt" would likely also be large enough to materially endanger a firm's financial well-being. If that firm makes drugs or devices that lack other readily available sources, then the additional deterrence may come at the expense of public health.[2] The same problem prevents regulators from exercising their authority to exclude entire firms from federal health care programs, a punishment so severe it is sometimes referred to as the "death penalty" (Bucy 1995:720). Because firm-wide exclusion could deny access to essential drugs or devices, Boozang (2012:87–88) has characterized life-sciences firms as "too big to nail."

Faced with the prospect that penalties strong enough to meaningfully deter misconduct would have unacceptable social costs, the agencies principally responsible for enforcement in this area—FDA, the Department of Health and Human Services (HHS) Office of the Inspector General (OIG), and the Department of Justice (DOJ)—have faced a choice: either accept that some firms can afford and will choose to pay the maximum "price" regulators can impose to engage in misconduct, or determine that this misconduct is not just *legally* wrong but ethically so,[3] and find another way to deter it. Their controversial response in recent years has been to "up the ante" (Boozang 2012:89) by employing two powerful enforcement tools: prosecution of individual corporate officers under the *Park* doctrine and exclusion of these officers from federal health care programs (i.e., from employment in the industry) under OIG's permissive exclusion authority.

B. Targeting Corporate Officers

In 2010, federal officials began making public statements indicating an increased willingness to pursue enforcement actions targeting individual officers of biomedical firms. These statements came from high-ranking officials at all of the relevant agencies, including FDA Commissioner Margaret Hamburg and Deputy Chief Counsel for Litigation Eric Blumberg, OIG Chief Counsel Lewis Morris, and multiple lawyers from the DOJ's Consumer Protection Branch (CPB), which has a formal role in coordinating enforcement efforts under the FDCA. What many in industry found most alarming in these comments was the suggestion that the government was increasingly open not only to going after individuals in general but in pursuing at least some of these cases under the strict liability theory known as the *Park* doctrine.

A few words about the *Park* doctrine may explain the concern. The most severe individual sanction provided for by the FDCA is criminal prosecution. In fact, the statute creates two different criminal causes of action: a felony and a misdemeanor. The misdemeanor provision is more controversial because under the "responsible corporate officer doctrine" articulated in the seminal Supreme Court cases *United States v. Dotterweich* (1943) and *United States v. Park* (1975), individual managers can in principle be held criminally liable for any violation committed at their firm, even if they were unaware of the violation or took reasonable steps to prevent it, so long as they stood in a "responsible relationship" to the violation (i.e., if by reason of their position in the corporation, they had the responsibility and authority either to prevent the violation or to promptly correct it).

The prospect of more aggressive use of *Park* concerns critics for a number of reasons. On a philosophical level, the main criticism is that the doctrine authorizes criminal prosecution and even incarceration of individuals with no knowledge of, let alone intentions regarding, the underlying misconduct. This objection is as old as the doctrine itself, as Justice Murphy's oft-quoted dissent in *Dotterweich* (1943:286) illustrates: "It is a fundamental principle of Anglo-Saxon jurisprudence that guilt is personal and . . . ought not lightly to be imputed to a citizen who . . . has no evil intention or consciousness of wrongdoing." More directly, potential defendants are justifiably concerned about the serious consequences of conviction. Indeed, some commentators have argued that the consequences of a *Park* conviction are so fundamentally disproportionate to the conduct required for conviction that they raise due process concerns (Breen and Retzinger 2013:109–10).[4] The OIG's willingness to exclude individuals on the basis of *Park* convictions—a strategy upheld as legal by a divided panel of the D.C. Circuit in *Friedman v. Sebelius* (2013)—has only exacerbated these concerns. And, though the number of prosecutions and exclusions has in fact remained fairly modest (Savage and Klawiter 2013:32), regulators have to some extent made good on their threats, pursuing a nontrivial number of high-profile *Park* misdemeanor prosecutions, felony prosecutions, and exclusion actions over the last five years

(O'Leary 2013:165–74, discussing cases against Purdue Frederick, Synthes, KV Pharmaceutical, InterMune, and Stryker Biotech).

II. A PROPOSAL FOR REFORM

If the shortcoming of the current enforcement model is that it has proved impossible to impose a strong enough financial penalty to deter misconduct without also imposing unacceptable harm on society in the form of medicines delayed and denied, the objective going forward must be to find ways (practically feasible and philosophically defensible) to raise the stakes for misbehaving firms without imposing costs on the public. Used in a coordinated enforcement framework prioritizing public health, criminal prosecution and exclusion of individuals, including prosecution under the *Park* doctrine, can be valuable tools in achieving this goal.

A. The Case for Individual Enforcement

There is good reason to believe that enforcement actions targeting individuals are a particularly effective deterrent. Indeed, this conclusion follows from two simple premises. First, we assume that to at least some extent "[s]ystematic management failures and corporate cultures that emphasize profits over safety reflect the values and conduct of individual actors within the organization. . . . [P]eople—not a fictional entity—make the choices and decisions that translate into conduct" (Barrett 2011:313). Second, we assume that individuals will be more motivated by direct harms than indirect ones, and by harms they cannot externalize rather than ones they can. That is, "[p]ersonal accountability, which creates a risk to an individual that he might go to jail [or be excluded] as a result of decisions he makes, can change behavior and drive deterrence" (Barrett 2011:313) in a way that harm to a firm's bottom line (which might indirectly harm the firm's employees in the form of reduced compensation or termination) or individual monetary penalties (often subject to indemnification agreements and insurance) cannot. Because "[b]eing named in a pleading, put on trial, forced to make a public apology, required to pay a fine, or serve time in jail, are often expensive, professionally damning and personally humiliating consequences [that] most individuals, including corporate officers, prefer to avoid" (Wise 2002:285–86), we can infer (as regulators clearly have) that corporate officers will be more effectively deterred by the prospect of such consequences.[5]

Another reason individual liability adds deterrent value over corporate liability alone is that to the extent monetary penalties are tied to the profits earned through misconduct, certain proceeds of misconduct are unlikely to be reflected in these penalties. This could be the case, for example, when the benefits of misconduct accrue over a longer time horizon than the misconduct itself, such as where a firm is able to capture a significant share of a new product market through misconduct. In this scenario, the firm might benefit from the misconduct over the long term even if all of the profits directly attributable to the period of misconduct are nullified through a monetary penalty. Though difficult to reach through monetary penalties, the use of strategic misconduct to gain competitive advantage may be more susceptible to deterrence through individual liability because misconduct calculated to produce long-term benefits is likely to be driven by individual decision makers.

Beyond the deterrent effect of individual liability in general, the strict liability offense under *Park* offers an additional, important advantage through what might be described as the doctrine's "burden lowering" effect (Clark 2012:5; Ellis 2013:1007–8; O'Leary 2013:167 n.207). Unlike what Ellis describes as the "vicarious liability perspective" on *Park* (what I have elsewhere described as the "true strict liability" application of *Park* (O'Leary 2013:148)), which permits a prosecutor to impute misconduct onto an officer even where there is no reason to believe the officer was "in any way complicit in committing the crime" (Ellis 2013), the burden-lowering aspect of *Park* makes it easier to obtain a conviction (or plea) where there *is* reason to believe an officer is culpable. The U.S. Attorney's Manual expressly embraces this burden lowering conception of *Park*, noting that while "CPB attempts wherever possible to bring felony charges to deal with fraudulent behavior[,] . . . [*Park*] misdemeanor liability can attach to behavior that, due to lack of proof . . . , may not merit felony prosecution" (U.S. Attorney's Manual § 4–8.210).

As noted before, the *Park* doctrine has been the subject of considerable criticism both because of philosophical concerns over the relationship between culpability and criminal punishment and because of concerns involving the interaction of prosecution and the collateral consequence of exclusion. But fears that the *Park* doctrine will be abused to any great extent, in the sense of prosecutors using it to pursue convictions of individuals who they have no reason to believe possessed any knowledge of wrongdoing, are vastly exaggerated. For one thing, prosecution of such individuals would quickly run up against formidable political and legal obstacles. With respect to political barriers, it is worth observing that, over the years, political opposition to the *Park* doctrine based on fairness concerns has even come from otherwise staunch advocates of regulatory enforcement, like Senator Ted Kennedy and Secretary Joseph Califano, who led HHS in the Carter administration (O'Leary 2013:151–52). It is hard to believe the pharmaceutical industry has so few friends in Washington today that genuinely abusive use of *Park* would go unchallenged in Congress and at the highest levels of the executive branch. As for the legal safeguards, courts have from the very beginning expressed doubts about the propriety of imposing criminal sanctions on nonculpable defendants—even the Supreme Court majorities that upheld the doctrine did so only on the belief that "the good sense of prosecutors, the wise guidance of trial judges, and the ultimate judgment of juries must be trusted" (*United States v. Dotterweich* 1943:285). Indeed, some commentators have argued that in light of changes over the four decades since *Park* was decided, "it is no stretch to imagine that today's [responsible corporate officer] doctrine would not pass muster before the Supreme Court, whether before the bench of 1943, 1975, or 2010" (Bragg et al. 2010:534). Similar doubts arguably constrain the OIG's use of its permissive exclusion authority, which notwithstanding the D.C. Circuit's opinion in *Friedman*, stands on tenuous legal footing. In addition to political and legal constraints on the abuse of *Park*, such use is unlikely for the simple reason that federal prosecutors—who have the ultimate authority over the decision to employ *Park* in any particular case—have a variety of incentives to "pursue the most egregious types of violations" rather than "conduct that is illegal for reasons that are . . . cloudy, overly technical, or arcane" (Schiffer and Simon 1995:2532–33).

Concerns about the interaction between *Park* and OIG exclusion (and between criminal prosecution and exclusion in general) are another matter. There are legitimate reasons to wonder whether the recent emphasis on excluding individual officers as a mechanism of corporate regulation is sound enforcement policy. Not only does OIG's aggressive use of its permissive exclusion authority in connection with *Park* convictions raise questions about the appropriate scope of collateral consequences for strict liability crimes (recall that the eponymous Mr. Park, whose case divided the court, received only a $250 fine), but OIG's use of exclusion has the potential to undermine the effective use of criminal prosecution as a deterrence tool, e.g., by creating a strong disincentive to guilty pleas and cooperation (O'Leary 2013:171–74; Glasner 2011:4). These concerns are serious, and the collateral consequences problem undoubtedly casts a shadow over the otherwise promising role that (conservative) use of corporate officer liability could have in achieving effective deterrence without relying on ballooning corporate monetary penalties. Fortunately, these concerns are not irremediable, as the final portion of this chapter explains.

B. Interagency Coordination and Prioritizing Public Health

Putting aside anxieties about an aggressive expansion of "true strict liability" prosecutions under *Park*—which for the reasons expressed earlier, is unlikely—a federal enforcement regime that more heavily relies on holding corporate officers personally accountable for their companies' misconduct would still raise three substantial concerns absent reform. First, there is a plausible concern that case selection may be driven, at least in some instances, by principles not directly tied to protecting the public health. If the ultimate objective of the enforcement regime is to protect the public health, case selection should be ambivalent about the *size* of potential penalties. There is nothing inherently wrong with prosecuting large firms with deep pockets, but if we justify using harsh penalties by invoking the public welfare, decisions about imposing these penalties should be tied to that idea, not to the potential for treasury and reputation-enhancing settlements. Second, a troubling hallmark of the current framework for case selection, charging, and settlement decisions is its unpredictability and opacity. Each U.S. Attorney's office has the independent authority to pursue food- and drug-related cases without or in spite of FDA input, and many offices choose to do so using more general criminal statutes instead of health care–specific causes of action.[6]

This has the effect of (1) creating the possibility for disparate charging and settlement standards for comparable offenses and (2) making it difficult or impossible to track criminal enforcement in this area (O'Leary 2013:153 n.117). Finally, as noted previously, there is a real concern that the OIG's aggressive approach to permissive exclusion undermines the efficient resolution of criminal proceedings by discouraging defendants from entering plea agreements.

Although these problems with individual enforcement are significant, they can be substantially mitigated through greater coordination between the various agencies responsible for enforcement and through a shared commitment to prioritizing the public health goals underpinning the food and drug laws. By coordination, I mean to propose formal interagency policies guiding decision making in individual cases. The present reality is that criminal prosecutions for FDA-related misconduct—whether charged as a *Park* misdemeanor, an FDCA felony, or a more general Title 18 offense—are controlled by the DOJ. Although FDA investigates many of these cases, and many arrive in U.S. Attorneys' offices as a result of FDA referral, others are investigated by different law-enforcement agencies, or arise out of civil False Claims Act actions. In every case, the key decisions—whether to prosecute and on what charges, whether to settle and on what terms—belong to lawyers at the DOJ. Whether those lawyers are specialists in food and drug crimes working at the CPB in Washington or generalists at any of the ninety-three U.S. Attorneys' offices depends on how a case originates and on what charges it ultimately proceeds. Either way, the consequence is that strategic FDA-related enforcement decisions are routinely made by lawyers at the DOJ, which, unlike FDA or HHS, has no statutory obligation to prioritize public health. Particularly where FDA-related conduct is prosecuted under more general criminal theories—and much of it is, partially because charges like those under the fraud and false-statement statutes are more familiar to generalist prosecutors, more easily explained to jurors, and often comparatively easy to prove—officials from FDA may have at most a minimal role in the prosecution. Moreover, because many cases tried on more general theories (but based on conduct that *could* be prosecuted under a more specific statute) are not classified as food and drug prosecutions for tracking purposes, there is presently no straightforward way to develop a complete picture of how and how often the DOJ prosecutes this conduct.

Glasner (2011:7) has proposed remedying the FDA-DOJ coordination problem, at least with respect to the decision to proceed under *Park*, by "establish[ing] DOJ guidelines that reflect FDA guidelines in order to keep the application of the *Park* doctrine consistent even when . . . prosecutions are referred to DOJ from whistleblowers outside FDA." Consistency is obviously of paramount importance with respect to the *Park* doctrine, but it is also important for enforcement in this area more generally. Consistent but independent written policies, moreover, do not necessarily translate into decision making that reflects the shared values underlying those policies, particularly when the agency whose authorizing statute contributes the values— the public health–oriented FDA—and the agency with final decision-making authority differ. Rather than simply aligning the principles articulated in the agencies' policy manuals, the agencies could more fully address the concerns expressed above by adopting a binding interagency agreement that (1) formally recognizes that no matter *which* government agency acts to enforce the food and drug laws, it must do so in a manner that prioritizes public health impacts; (2) sets a uniform policy guiding prosecutors on how to charge (and track) cases where the underlying offense conduct is a violation of the food and drug laws; and (3) gives FDA a formal role in deciding what cases are brought and dropped. As Girard (2009:153) puts it: "If the goals of promoting and protecting the public health under the [FDCA] are truly paramount, FDA should be afforded an early and substantive role in the decision of whether prosecution is warranted."

The issues raised by exclusion can be addressed in much the same way. The decision whether to seek exclusion in any given case should not be divorced from decisions about whether to prosecute, whether to settle, and on what terms to settle. Whatever the differences between exclusion (technically a tool of incapacitation rather than deterrence, though not used that way) and criminal liability, the ultimate goal of both sanctions is to protect the public, and it makes no sense that two agencies of the same government, pursuing this goal in relation to the same actors for the same conduct, would not coordinate. The OIG must be on the same page with the DOJ and FDA, not only so that the risk of exclusion does not *undermine* negotiations between these agencies and actual or potential defendants but also so that these agencies can effectively employ the powerful stick that is exclusion in these negotiations in the first instance.

III. CONCLUSION

The federal enforcement regime for the biomedical industry is broken. Prosecutors have become addicted to a lightning-strike enforcement approach that emphasizes headline-making billion-dollar settlements despite growing evidence that these massive fines (and the CIAs that accompany them) do not effectively deter the misconduct they punish. With larger fines and firm-wide exclusion substantially off the table due to the social costs of shutting down the developers and suppliers of essential medical products, the most straightforward way to bridge the deterrence gap is by imposing accountability on the individual corporate officers who shape the culture and drive the decision making at these firms. While targeting these individuals using the full range of enforcement tools available—including *Park* prosecutions and exclusion from federal health care programs—can be an important element of a credible deterrence regime, these forms of enforcement also raise valid concerns. For this reason, it is vital that the agencies applying them coordinate and use their authority responsibly, according to shared principles emphasizing, above all, the public health mission that justifies such authority in the first place.

NOTES

1. Public Citizen is a consumer-rights advocacy group based in Washington, D.C.

2. This concern could theoretically be addressed in part through reforms, like some already being discussed in response to drug shortages, geared toward creating alternate sources for vital products. For example, Copeland (2012:1077–78) suggests that compulsory licensing could be employed in some cases as an intermediate sanction for off-label promotion, which would require the subject firm "to permit another manufacturer to produce and sell the patented drug."

3. Boozang (2012:91) explains that while "efficient breach of public law" can be seen as ethically acceptable "in the case of matters that are *malum prohibitum*" this is not the case for acts that are "*malum in se*." With the understanding that not all food and drug offenses can be fairly characterized as such, it seems fair to say that from the regulators' perspective, at least those prohibitions in the FDCA pertaining to safety and efficacy of regulated products are of the sort that cannot be ethically breached for a price.

4. As Baer (2012:632–33) observes, however, the notion "that offenders should be punished proportionally and in relation to their culpability" is really the "core claim" of retributive justice, not deterrence. While by "happy coincidence[]" this notion "bears some resemblance to the concept of 'marginal deterrence' in economics," in fact "[t]he amount of culpability someone bears for a given act may differ in translation from the specific sanction necessary to internalize and optimally deter that same act."

5. Petrin (2012:312–14) notes, however, that this inference is still just that, given the paucity of empirical evidence about the effectiveness of individual liability for corporate criminality.

6. A wide variety of Title 18 (and other) criminal charges can be used to prosecute conduct that arises in FDA-regulated industry, including mail and wire fraud, obstruction of justice, conspiracy, and violations of the Anti-Kickback Statute (Drake et al. 2013:1177–87; O'Leary 2013:161–62).

REFERENCES

Almashat, S. and S. Wolfe. 2012. "Pharmaceutical Industry Criminal and Civil Penalties: An Update." *Public Citizen* (September 27).

Baer, M. H. 2012. "Choosing Punishment." *Boston University Law Review* 92:577–642.

Barrett, J. F. 2011. "When Business Conduct Turns Violent: Bringing BP, Massey, and Other Scofflaws to Justice." *American Criminal Law Review* 48:287–334.

Boozang, K. M. 2012. "Responsible Corporate Officer Doctrine: When Is Falling Down on the Job a Crime?" *Saint Louis University Journal of Health Law and Policy* 6:77–112.

Bragg, J., J. Bentivoglio, and A. Collins. 2010. "Onus of Responsibility: The Changing Responsible Corporate Officer Doctrine." *Food and Drug Law Journal* 65:525–38.

Breen, G. B. and J. D. Retzinger. 2013. "The Resurgence of the Park Doctrine and the Collateral Consequences of Exclusion." *Journal of Health and Life Sciences Law* 6:90–120.

Bucy, P. H. 1995. "Civil Prosecution of Health Care Fraud." *Wake Forest Law Review* 30:693–758.

Clark, M. E. 2012. "The Responsible Corporate Officer Doctrine: A Re-Emergent Threat to General Counsel and Corporate Officers." *Journal of Health Care Compliance* 14(1):5–15.

Copeland, K. B. 2012. "Enforcing Integrity." *Indiana Law Journal* 87:1033–86.

Drake, T., A. Kanu, and N. Silverman. 2013. "Health Care Fraud." *American Criminal Law Review* 50:1131–98.

Ellis, A. R. 2013. "The Responsible Corporate Officer Doctrine: Sharpening a Blunt Health Care Fraud Enforcement Tool." *NYU Journal of Law and Business* 9:977–1036.

Food and Drug Administration. 2013. "About FDA: What We Do." *FDA*. http://www.fda.gov/aboutfda/whatwedo.

Friedman v. Sebelius, 686 F.3d 813 (D.C. Cir. 2013).

Girard, V. W. 2009. "Punishing Pharmaceutical Companies for Unlawful Promotion of Approved Drugs: Why the False Claims Act Is the Wrong RX." *Journal of Health Care Law and Policy* 12:119–58.

Glasner, A. 2011. "Are Misdemeanor Prosecutions Under the *Park* Doctrine an Effective Mechanism for Deterring Violations of the Federal Food, Drug and Cosmetic Act?" *FDLI Forum* 1(14):1–12.

Groeger, L. 2014. "Big Pharma's Big Fines." *ProPublica* (February 24). http://projects.propublica.org/graphics/bigpharma.

Holmes, O. W., Jr. 1897. "The Path of the Law." *Harvard Law Review* 10:457–78.

O'Leary, P. 2013. "Credible Deterrence: FDA and the Park Doctrine in the 21st Century." *Food and Drug Law Journal* 68:137–76.

Paulhus, M. E. and J. C. Richter. 2013. "New Government Strategies for Rooting Out Health Care Fraud." In *Health Care Law Enforcement and Compliance*, 63–86. Thomson Reuters/Aspatore.

Petrin, M. 2012. "Circumscribing the 'Prosecutor's Ticket to Tag the Elite'— A Critique of the Responsible Corporate Officer Doctrine." *Temple Law Review* 84:283–324.

Rodwin, M. A. 2013. "Rooting Out Institutional Corruption to Manage Inappropriate Off-Label Drug Use." *Journal of Law, Medicine & Ethics* 41:654–64.

Savage, J. F. and M. Klawiter. 2013. "The Revival of the Responsible Corporate Officer Doctrine?" *Health Lawyer* 26(1):32–39.

Schiffer, L. J. and J. F. Simon. 1995. "The Reality of Prosecuting Environmental Criminals: A Response to Professor Lazarus." *Georgetown Law Journal* 83:2531–38.

United States v. Dotterweich, 320 U.S. 277 (1943).

United States v. Park, 421 U.S. 658 (1975).

United States v. Pharmacia & Upjohn Co., No. 1:09-cr-10258-DPW-1 (D. Mass. Sept. 2, 2009).

Wise, N. 2002. "Personal Liability Promotes Responsible Conduct: Extending the Responsible Corporate Officer Doctrine to Federal Civil Environmental Enforcement Cases." *Stanford Environmental Law Journal* 21:283–344.

PART THREE

Protecting the Public Within Constitutional Limits

Introduction

I. GLENN COHEN

IN THE last set of Republican primary debates before the 2012 presidential election, primary contender Rick Perry famously stumbled in trying to name the three agencies he would eliminate as president, naming Commerce and Education (technically departments not agencies, but let's not be picky) but unable to name a third (Terkel 2011). Judging from the three chapters discussed in this part, for some libertarian-leaning conservatives, the Food and Drug Administration (FDA) might easily be high up on the list for that third spot, at least as to some parts of FDA's mission.

All three of the chapters center on *United States v. Caronia* (2012), a decision of the Court of Appeals for the Second Circuit stemming from a Department of Justice criminal misdemeanor prosecution of a pharmaceutical sales representative caught on tape promoting a drug for conditions not approved by FDA. Caronia successfully defended himself by arguing that FDA's prohibition of manufacturers and their representatives engaging in so-called "off-label" promotion as applied to him violated his First Amendment rights. Not coincidentally, the case drew an amicus brief from the Washington Legal Foundation, a prominent libertarian litigating organization that defines its goal as seeking to "preserve and defend America's free enterprise system by litigating, educating, and advocating for free market principles, a limited and accountable government, and individual and business liberties" (Institute for Justice 2014).

It may be surprising that this was the first major challenge along these lines, but this was the rare case where a prosecution was brought against the sales representative himself and not his drug company; the latter, it would seem, had a very strong incentive not to challenge FDA's authority in this area and instead "take its lumps."

The chapters in part 3 evaluate the merits of the Second Circuit's First Amendment analysis and FDA's options for regulating off-label promotion in the wake of this decision. But first, let us reflect upon the "long game" perspective and the origins of this decision, considering the ways in which *Caronia* seems to represent a way point for a larger libertarian constitutional project focused on governmental restrictions on health care that has grown in prominence in the 2000s.

FDA is no stranger to constitutional litigation. Much of this litigation, discussed in these chapters, centered on the First Amendment and drew on the Supreme Court's increasing solicitude for commercial speech. But there is also a distinct set of non-First Amendment challenges against FDA authority that have been raised recently, indeed sometimes by none other than the Washington Legal Foundation. Stepping back we can see this as a smart, slick, libertarian movement to rein in FDA on at least two fronts—a First Amendment front and a Fifth Amendment Due Process and Equal Protection front.

The most notable case on this second front is *Abigail Alliance v. Eschenbach*.[1]

The case was brought by the family of Abigail Burroughs, a young woman who died when she ran out of FDA-approved treatments for her cancer. Her oncologist had recommended she try several experimental cancer drugs, but she was unable to get into a clinical trial. Her father formed the Abigail Alliance for Better Access to Developmental Drugs, a nonprofit association representing several terminally ill patients who would like to have access to Phase I drugs. After pursuing a citizen's petition and not receiving relief from FDA, represented by the Washington Legal Foundation, the Alliance sued the agency to enjoin FDA from enforcing its ban on unapproved drugs insofar as it prohibits terminally ill patients with no other treatment options from purchasing investigational drugs. The Alliance claimed a fundamental right under the Fifth Amendment Due Process Clause to obtain access to potentially beneficial investigational new drugs that have completed Phase I trials (*Abigail Alliance for Better Access to Developmental Drugs v. von Eschenbach* 2007; Menikoff 2008). After losing before a district court judge, it initially scored a victory in the D.C. Circuit, which found (2-1) that the Due Process clause creates a fundamental right for "mentally competent, terminally ill adult patients who have no alternative government-approved treatment options" to access "potentially life-saving investigational new drugs that the FDA has yet to approve for commercial marketing but that the FDA has determined, after Phase I clinical human trials, are safe enough for further testing on a substantial number of human beings." The majority found this right to be "deeply rooted in this Nation's history and tradition," as reflected in various common law doctrines (the defense of necessity, the freedom from battery, the tort of intentional interference with a rescue) (*Abigail Alliance for Better Access to Developmental Drugs v. von Eschenbach* 2006). That holding, however, was later reversed by the en banc D.C. Circuit finding no such right (*Abigail Alliance for Better Access to Developmental Drugs v. von Eschenbach* 2007).

This chink in FDA's armor, though temporary, has prompted further Fifth Amendment libertarian challenges to parts of the federal government's health infrastructure abutting FDA.

One such case is *Flynn v. Holder*, brought by a different libertarian litigation organization, the Institute for Justice, which pursues "four pillars of litigation"—cases relating to "private property, economic liberty, free speech and school choice." In *Flynn* it helped bring a challenge on behalf of parents of children with leukemia and aplastic anemia (which can be fatal without bone marrow transplantation), parents of mixed-race children (for whom sufficiently matched donors are especially scarce), and MoreMarrowDonors.org (a California non-profit corporation that wanted to offer awards in the form of scholarships, housing allowances, or gifts to charities selected by donors of bone marrow) (Cohen 2012). They challenged the ban on selling "bone marrow" that is part of the National Organ Transplant Act (NOTA) of 1984 (42 USC § 274e; *Flynn v. Holder* 2012). They won a narrow victory in the Ninth Circuit Court of Appeals, which held that as a matter of statutory interpretation the prohibition does not encompass "peripheral blood stem cells" obtained through apheresis, such that the court did not reach a more radical claim they had presented earlier in the litigation: that the prohibition on selling bone marrow violates the Equal Protection Clause of the U.S. Constitution, which prohibits the federal and state governments from denying any person the equal protection of the law (*Flynn v. Holder* 2012; Cohen 2012).

Neither *Caronia*, nor *Abigail Alliance*, nor *Flynn* received Supreme Court review. One case that did, though, that sounds similar themes is the initial challenge to the Affordable Care Act, *NFIB v. Sebelius*, which challenged among other things the constitutionality of the individual mandate requiring many individuals to purchase health insurance (*National Federation of Independent Business v. Sebelius* 2012). Although decided on Commerce Clause and taxing power grounds, the posturing and briefing of the case had an undeniably libertarian flavor, with questions debated about whether the state could force you to eat (or perhaps more accurately "buy") broccoli (Elhauge 2012; Blackman 2013). To be clear this case was not a direct or even indirect challenge to FDA, but it nonetheless may be viewed as part of this larger libertarian constitutional attack on the regulation of health, perhaps even opening up a third front.

What does all this mean for the future? The takeaway is that the First Amendment scuffle that is the focus of these chapters is an important part of a larger movement of libertarian constitutional attack on the agencies responsible for health care. My own view is that this First Amendment front is likely even more appealing to a libertarian litigation army than the Fifth Amendment front for a few important reasons. First, we have a Supreme Court filled with Justices who (even the liberal ones) have shown themselves willing to enforce a very robust conception of the First Amendment, so this route is more likely to succeed. Second, First Amendment challenges can often be depoliticized from the left-right axis in that there are prominent left-leaning libertarian organizations, like the ACLU, who bring similar claims. Third, success as to substantive due process and equal protection theories is more likely to create tension with nonlibertarian conservative elements who have sough to fight this expansion in the abortion, gay rights, and affirmative action arenas. If FDA can be restrained without poking the bear of the rest of the conservative movement, this may be a promising way of doing it.

NOTE

1. Full disclosure: I represented FDA at the en banc stage of this case while an attorney at the Department of Justice, Civil Division, Appellate Staff.

REFERENCES

Abigail Alliance for Better Access to Developmental Drugs v. von Eschenbach, 445 F.3d 470, 472, 486 (D.C. Cir. 2006).
Abigail Alliance for Better Access to Developmental Drugs v. von Eschenbach, 495 F.3d 695 (D.C. Cir. 2007) (en banc), cert. denied, 128 S. Ct. 1069 (2008).
Blackman, Josh. 2013. "The Libertarian Challenge to Obamacare." *Reason.com* (September 24).
Cohen, I. Glenn. 2012. "Selling Bone Marrow—*Flynn v. Holder*." *New England Journal of Medicine* 366:296.
Elhauge, Einer. 2012. "The Irrelevance of the Broccoli Argument Against the Insurance Mandate." *New England Journal of Medicine* 366:e1.
Flynn v. Holder, 684 F.3d 852 (9th Cir. 2012).
Institute for Justice. 2014. "About IJ: Our Mission." http://www.ij.org/about.
Menikoff, Jerry. 2008. "Beyond Abigail Alliance: The Reality Behind the Right to Get Experimental Drugs." *University of Kansas Law Review* 56:1045, 1045–55.
National Federation of Independent Business v. Sebelius, 132 S. Ct 2566 (2012).
Terkel, Amanda. 2011. "Rick Perry Forgets Which Three Agencies He Would Eliminate as President (VIDEO)." *Huffington Post* (November 9).
United States v. Caronia, 703 F.3d 149 (2d Cir. 2012).

CHAPTER TEN

Prospects for Regulation of Off-Label Drug Promotion in an Era of Expanding Commercial Speech Protection

AARON S. KESSELHEIM AND MICHELLE M. MELLO

IN DECEMBER 2012, a three-judge panel of the U.S. Court of Appeals for the Second Circuit set down a decision that shook the very foundation of how the U.S. Food and Drug Administration (FDA) regulates pharmaceutical marketing and promotion. The case, *United States v. Caronia,* involved the Department of Justice's criminal misdemeanor prosecution of a pharmaceutical sales representative named Alfred Caronia who was caught on tape promoting a drug for conditions not approved by FDA (*United States v. Caronia* 2012). FDA generally prohibits manufacturers and their representatives from engaging in so-called "off-label" promotion, considering intent to sell a product for a non-FDA-approved use to violate the Federal Food, Drug, and Cosmetic Act (FDCA), which requires all prescription drugs sold in the United States to be supported by substantial evidence from adequate and well-controlled investigations (21 USC § 355(d)). At trial, Caronia argued that his First Amendment right to free speech protected him from being prosecuted for his statements. A majority of the Second Circuit panel agreed. Although only formally controlling in the states covered by the Second Circuit, the decision calls into question FDA's ability to regulate off-label promotional speech going forward.

This chapter first reviews the main principles guiding FDA's regulation of pharmaceutical promotion. It then discusses the *Caronia* case and the legal basis for the Second Circuit court's decision, as well as its public health implications. Finally, the chapter analyzes the range of options available for FDA to maintain oversight of off-label prescription drug promotion if the *Caronia* reasoning were to become the dominant judicial perspective. We discuss three strategies available to investigators and policy makers interested in preventing unsafe off-label marketing and focus on two that we believe hold the greatest promise for protecting the public health from non-evidence-based promotional speech about health care products: making a stronger case that FDA's regulations on off-label promotion satisfy the Supreme Court's interpretation of the First Amendment and modifying FDA regulations concerning off-label promotion to rely on reasonable "safe harbors."

I. OFF-LABEL USE AND FDA REGULATION OF PHARMACEUTICAL PROMOTION

FDA certifies the efficacy and safety of new drugs before they reach the market by examining the evidence that has been collected by manufacturers, but FDA does not directly regulate the practice of medicine, including how physicians actually prescribe drugs. If "adequate and well-controlled investigations" (21 USC § 355(d)) show that a drug's benefits outweighs its risks, FDA will approve the drug and authorize a label that describes the clinical trials it reviewed and the indication it has authorized. However, physicians can then prescribe the approved drug for any medical condition or patient group, or at any dose, even outside of the parameters of the label. Many off-label uses have a reasonable biological basis. For example, if a drug was approved to treat colon cancer, there is a strong basis for thinking it might also be successful in treating cancer of the appendix because the appendix is an outgrowth of the colon and made up of similar cells. Off-label uses may also be supported by considerable evidence (Walton et al. 2008). The drug gabapentin (marketed under the trade name Neurontin), for instance, was approved to treat epilepsy, but subsequent studies showed its efficacy in specific types of chronic pain. The drug was widely used for those purposes by the medical community even though the manufacturer had never submitted these studies to the FDA as part of a formal supplemental new drug application (NDA) package.

However, off-label uses may also be dangerous and non-evidence based. One widely cited study of prescribing patterns for 160 commonly used drugs found a 21 percent off-label prescription rate and concluded that 73 percent of such off-label uses lacked evidence (Radley et al. 2006). Antipsychotic drugs are commonly used off label in elderly patients with dementia and affective disorders (Alexander et al. 2010), but such use has been associated with a higher mortality risk (Wang et al. 2005).

Thus, a major concern about off-label drug use is that it may involve considerable risk for patients without proven benefit (Levi et al. 2010). A second concern is that high spending on drugs used for non-evidence-based, off-label purposes strains the budgets of payers like Medicaid, leading them to tighten eligibility requirements or reduce the services they provide (Pear 2011).

Although FDA does not restrict physicians from prescribing drugs off label, it does forbid pharmaceutical manufacturers from promoting the use of drugs for off-label purposes (Kesselheim 2011). This stance is not a capricious attempt to restrict manufacturers from communicating to physicians about their products. Rather, FDA's approach was a response to major problems caused by the lack of such regulation. Widespread manufacturer promotion of the safety of medical products in the absence of formal regulatory approval of those products led to cases in which patients died after taking products with poisonous constituents (sulfanilamide elixir, 1938), gave birth to babies with devastating congenital anomalies (thalidomide, 1962) (Avorn 2012), or used contraceptive devices that caused bacterial sepsis (Dalkon Shield, 1974) (Wood 2005). Even more common was the promotion of drugs to treat conditions for which they lacked efficacy (Waxman 2003). Thus, consensus grew that it was in the public's interest for manufacturers to prove that a medication actually worked and was safe for that use before it could be sold and promoted.

Notwithstanding FDA's prohibition on off-label promotion, it persists. Recent examples of pharmaceutical promotion of non-evidence-based off-label uses leading to widespread patient morbidity and mortality include rofecoxib (Vioxx) (Krumholz et al. 2007), rosiglitazone (Avandia) (Moynihan 2010), paroxetine (Paxil) (Wadman 2004), fenfluramine/phentermine (Fen-Phen) (Kolata 2007), and telithromycin (Ketek) (Ross 2007).

FDA considers off-label promotion to violate the provision of the FDCA that bars pharmaceutical manufacturers from introducing a new drug into interstate commerce unless it and its label have secured FDA approval (21 USC § 355(d)). Manufacturer-distributed materials that explain the uses of the drug are considered part of the labeling even if not packaged with the drug (21 USC § 321(m)). A second provision prohibits manufacturers from introducing into interstate commerce "misbranded" drugs (21 USC § 352). Drugs can be misbranded for false or misleading labeling information or labeling that does not bear "adequate directions for use" (21 USC § 352(f)(1)). The combination of the requirements for approval and the misbranding provision provides two avenues for FDA to argue that it is illegal for a drug's labeling to discuss uses of the drug that FDA has not validated as being supported by substantial evidence.

Notably, FDA's prohibition on manufacturers' ability to discuss off-label uses is not absolute. Companies can respond to unsolicited questions from health care professionals and others about unapproved uses. Companies can also distribute reprints of medical journal articles and support continuing medical education programs in which off-label uses are discussed. These "safe harbors" have been set forth in clear, explicit FDA guidance documents (Food and Drug Administration 2014) and the 1997 FDA Modernization Act (FDAMA).

II. COMMERCIAL SPEECH AND THE COURTS

Drug manufacturers and some affiliated activists have objected to FDA control over the scope of their marketing, arguing that it unfairly limits manufacturers' ability to disseminate truthful information about their products. They have found legal grounds for their objections in the Supreme Court's expanding protection of commercial speech or speech that primarily proposes a commercial transaction.

The First Amendment to the U.S. Constitution states that "Congress shall make no law . . . abridging the freedom of speech." Although the First Amendment has always been understood to protect speech regarding political or religious views and other forms of social speech, corporate advertising was not initially considered worthy of protection because it was seen as having less benefit for the public and a more distant relationship to the reasons the First Amendment was adopted. However, in a 1976 case, the Supreme Court struck down a law that prevented pharmacies from advertising the prices of prescription drugs, finding value in the free flow of commercial information that can allow consumers to make "intelligent and well informed" decisions (*Virginia State Board of Pharmacy v. Virginia Citizens Consumer Council, Inc.* 1976:765).

The Supreme Court did not grant commercial speech the same level of protection as noncommercial speech. The Court established a framework to evaluate regulation of commercial speech in *Central Hudson Gas & Electric Corp. v. Public Service Commission of New York* (1980), directing courts to answer four questions. First, is the speech false or misleading, or does it concern unlawful activity? If so, it will receive no protection. Second, is the government's interest in regulating the speech substantial? Third, does the regulation directly and materially advance the government's interest? Finally, is the regulation narrowly tailored, meaning no more extensive than necessary to serve that interest?

The Supreme Court most recently applied the *Central Hudson* framework to drug product promotion in *Sorrell v. IMS Health* (2011). The underlying dispute in that case concerned a Vermont law that restricted the commercial sale, disclosure, and use of prescribing records revealing prescribers' identities without prescribers' consent and prohibited pharmaceutical manufacturers from using that information for marketing or promotion. The law aimed to curtail the practice of pharmaceutical "detailing," in which pharmaceutical sales representatives used data about each physician's prescribing practices to tailor their sales messages to each physician. The Supreme Court struck down the law, holding that it was not narrowly tailored to achieving any of the asserted interests. Justice Kennedy, writing for the majority, characterized the law as a speaker- and viewpoint-based restriction because it was limited to pharmaceutical manufacturers and was adopted because of the Vermont legislature's disagreement with the message that the sales staff of manufacturers were trying to convey: that the physician should prescribe expensive, branded drugs more often.

In *Sorrell*, as in previous commercial speech cases, the Court voiced suspicion about speech restrictions that smacked of "paternalism" or a governmental attempt to keep consumers in the dark for what it has determined to be their own good. The Court found such overtures especially objectionable when directed at physicians, "sophisticated and experienced" consumers who are highly capable of critically weighing the information provided by pharmaceutical sales representatives (*Sorrell v. IMS Health* 2011:2658). As in prior cases, the Court stressed the role of commercial speech rights in supporting manufacturers' direct contribution to the marketplace of ideas, which leads to optimal economic decision making. Indeed, Kennedy cited the principle that "[t]he commercial marketplace, like other spheres of our social and cultural life, provides a forum where ideas and information flourish. . . . But the general rule is that the speaker and the audience, not the government, assess the value of the information presented." (*Sorrell v. IMS Health* 2011:2671–72). The Court did not examine whether the marketplace of ideas actually functions well in the case of prescription drugs, except insofar as it judged that the market's functioning would be worse in a world where pharmaceutical-related speech was restricted. Embedded within this affection for the marketplace-of-ideas concept is an important presumption: that the information at issue is true and, therefore, useful. However, the Supreme Court in *Sorrell* and other courts have generally declined to evaluate the quality of the information at issue.

III. THE *CARONIA* CASE

Alfred Caronia was a sales representative for the brand-name drug company Orphan Medical. In 2005, his job was to promote the drug sodium oxybate (Xyrem) to physicians. Sodium oxybate—also known as gamma hydroxybutyrate, a chemical that when used for recreational purposes has been associated with drug-assisted sexual assault—had been approved by FDA in 2002 for use in the rare clinical condition of narcolepsy with severe cataplexy, a condition marked by sudden onset of lethargy, sleepiness, and full loss of muscle tone. To grow the revenues and position for a lucrative acquisition, Orphan Medical encouraged sales representatives to suggest non-FDA-approved uses for the product (*United States ex rel. Lauterbach v. Orphan Medical* 2006).

In one physician's office, Caronia suggested that sodium oxybate was effective for conditions including insomnia, fibromyalgia, restless leg syndrome, chronic pain, Parkinson's disease, and multiple sclerosis. He also claimed that the drug could safely be used in elderly and pediatric patients despite clear statements on FDA-approved labeling that the drug had not been tested in those populations. Caronia's statements came to light because the physician to whom he was speaking was wearing a wire—he had earlier been recruited by the Department of Justice, which was investigating Orphan Medical for improper promotion of its drugs.

Based on this evidence, Caronia was prosecuted and convicted by a jury for conspiracy to introduce a misbranded drug into interstate commerce. Caronia sought to overturn his conviction in an appeal led by the Washington Legal Foundation (WLF), a libertarian advocacy group.

Caronia's case proved to be an excellent opportunity for the WLF to test the constitutionality of FDA's restrictions on off-label marketing. Most previous prosecutions of alleged off-label marketing had targeted drug companies and led to settlements rather than courtroom adjudication, because the companies could not take the business risk of losing a case and facing treble damages or the "corporate death sentence" penalty of exclusion from participation in federal health care programs such as Medicare and Medicaid (Schiff 1995). Caronia may have been more comfortable pursuing the appeal because his original conviction carried the modest penalty of probation, community service, and a $25 fine.

The Second Circuit overturned Caronia's conviction. Applying the *Central Hudson* test, Judge Denny Chin—writing for himself and Judge Reena Raggi—averred that the speech at issue was not false or misleading because the government, in presenting its case, had not tried to paint Caronia's statements as such. He also agreed that the government had substantial interests at stake, which he identified as "preserving the efficacy and integrity of the FDA's drug approval process and reducing patient exposure to unsafe and ineffective drugs" (*United States v. Caronia* 2012:166).

However, Judge Chin ruled that restriction on off-label promotion did not directly advance the government's interest in reducing unsafe drug use because it prohibited the "free flow of information" that would inform the legal practice of off-label use (*United States v. Caronia* 2012:167). Rather than advancing patient safety, "paternalistically" interfering with information dissemination about legal uses of a drug inhibited "informed and intelligent treatment decisions" (*United States v. Caronia* 2012:166). Judge Chin emphasized that physicians were a skilled and sophisticated audience, and it was their—not the government's—role to determine which information was useful.

Finally, the court held that restricting manufacturers' off-label promotion was not a narrowly tailored intervention because other options could achieve the same goal without inhibiting speech. The court proposed that the government could educate physicians and patients about how to distinguish between "misleading and false promotion, exaggerations and embellishments, and truthful or non-misleading information"; require disclaimers for off-label uses; limit the number of off-label prescriptions a physician may write; or warn physicians and manufacturers about their potential exposure to malpractice claims if adverse outcomes resulted from off-label treatment decisions (*United States v. Caronia* 2012:167–68).

IV. THE FUTURE OF REGULATIONS RESTRICTING OFF-LABEL PROMOTION

The *Caronia* decision has been hailed by advocates of expanded commercial speech rights for the pharmaceutical and medical device industries (Hall 2013). These commentators have argued that FDA's restrictions on off-label promotion are too broad because they cover potentially truthful speech about off-label uses and because those off-label uses are lawful and can be clinically indicated. They claim the combination of these factors "prevents promotion of valuable off-label uses of which doctors otherwise may well be unaware" (Klasmeier and Redish 2011).

However, the *Caronia* decision is also a setback in the government's effort to avoid the dangerous public health outcomes that may arise from non-evidence-based industry marketing. FDA insisted that *Caronia* would not "significantly affect the agency's enforcement of the drug-misbranding provisions of the Food, Drug and Cosmetic Act," as it "does not strike down any provision of the . . . act or its implementing regulations, nor does it find a conflict between the act's misbranding provisions and the First Amendment or call into question the validity of the act's drug approval framework" (Sell 2013). FDA's characterization of the scope of the opinion is correct: the *Caronia* court barred a particular avenue of criminal prosecution rather than striking down any statutory or regulatory provision per se. Notably, in the first year after the *Caronia* decision, pharmaceutical manufacturers Amgen and Ista settled prosecutions in the Second Circuit alleging improper off-label marketing for civil and criminal fines of $762 million and $33.5 million, respectively.

Despite those enforcement victories, it is hard to imagine the *Caronia* case not hampering the government's ongoing enforcement efforts. If the government seeks to prosecute other individuals like Caronia in the future in other jurisdictions, the effort will inspire similar appeals, and the prospect of other courts following the Second Circuit's lead will cast a shadow over future settlement negotiations with manufacturers, potentially reducing the government's ability to seek punitive fines or changes in corporate behavior. The Second Circuit court's prohibition on prosecuting manufacturer representatives for making off-label promotional statements therefore muddies the waters in terms of regulatory possibilities going forward.

At least three potential pathways remain, however. First, FDA could focus on the false or misleading nature of promotional materials or statements. Second, in defending its policies against future claims of commercial speech infringement, FDA could make a stronger case that its regulations meet the criteria of the *Central Hudson* test. Finally, FDA could reconfigure its regulatory regime to adapt its restrictions on commercial speech in ways that still permit oversight of public safety.

A. Off-Label Promotion as False or Misleading Speech

For many statements that sales representatives make to promote off-label uses, the government could make a reasonable argument that the speech qualifies as misleading and therefore lies outside the ambit of First Amendment protection. In previous Supreme Court cases, commercial speech has been found to be misleading if it omits essential information that can lead to "deception" of consumers (*Zauderer v. Office of Disciplinary Counsel of Supreme Court of Ohio* 1985:650) or where "most commercial uses" of a statement "are likely to be confusing" (*San Francisco Arts & Athletics Inc. v. United States Olympic Committee* 1987:539).

Among well-trained sales representatives, the tactic of communicating "non-demonstrably false information" (*United States v. Caronia* 2012:181) is common. Most off-label uses of drugs have little or no scientific support (Radley et al. 2006), but sales representatives may omit that material fact or misrepresent the strength of evidence. For example, a psychiatrist, Daniel Carlat, acting as a representative on behalf of a Wyeth antidepressant, addressed a physician's concern about a dangerous side effect of the drug by downplaying its significance. He later reflected, "I knew I had not lied—I had reported the data exactly as they were reported in the paper. But still, I had spun the results of the study in the most positive way possible, and I had not talked about the limitations of the data" (Carlat 2007). Such presentations may not be false, but they are misleading.

This approach has clear disadvantages. First, not all off-label promotion cases involve the pristine evidence available to the government in the *Caronia* case because of the willingness of a cooperating physician to record conversations. Second, even if the nature of the promotional statements can be clearly established, the government must prove that the statements are false or misleading, which requires a case-by-case evaluation. The boundary line between protected speech and false or misleading speech in the area of off-label promotion is unclear and likely to be heavily contested if the government pursues this strategy. Establishing that promotional speech is false or misleading will require assessing the strength, validity, and appropriateness of evidence for each claim. Consequently, it will also require a considerable amount of complex expert testimony and pose heavy cognitive demands on lay jurors (Silverman 2013). The misleading-speech strategy for prosecuting off-label promotion will be useful only in a narrow range of cases.

B. Readdressing the Remaining Requirements of *Central Hudson*

A second potential strategy for the FDA is simply to continue to prosecute off-label promotion on the theory that—debates over the nature of the speech as true, false, or misleading aside—its regulatory approach satisfies the *Central Hudson* test. This strategy relies on a belief that the Second Circuit's reasoning in *Caronia* is unsound and may not be followed by other courts, particularly if the government strengthened its argument concerning each of the three main prongs of the test: (1) substantiality of the government interest; (2) direct advancement; and (3) narrow tailoring.

1. Substantial Government Interest

The broader the interests that courts accept as legitimate justifications for restricting commercial speech, the broader the government's latitude will be in designing the regulatory scheme. Thus, it is worth honing the argument that the government has an important interest in ensuring that physicians receive accurate, unbiased information to support informed treatment choices. Even when off-label communications do not rise to the level that courts would define as false or misleading, they may fall well short of accurate, unbiased information. In particular, oral conversations in the confines of physician offices involve few mechanisms to ensure accountability for the accuracy of the information. The government has a substantial interest in preventing the communication of unsubstantiated information in such settings and should seek to disrupt courts' apparent assumption that off-label communications are truthful.

In making such an argument, the government could also emphasize that courts have had greater confidence than is warranted in physicians' ability to evaluate claims about off-label uses. Some promotional claims may be inherently impossible for physicians to verify, such as a claim that other physicians are already widely prescribing the drug for an off-label use and have encountered no serious safety problems. Even when documents supporting claims about the safety or effectiveness of off-label uses are offered, they may convey only a slice of the full empirical picture. There may be countervailing study findings or important study limitations that were omitted. The FDA approval process serves as a bulwark against such problems by deploying highly skilled scientists to analyze all of the available information about a drug's use, verify analyses conducted by the drug's sponsor independently, and scrutinize the design of the studies offered to support the new use. Individual physicians cannot approach this level of evaluation—even if they have the time and inclination to explore the veracity of claims made in off-label promotional communications, they lack access to the information necessary to do so thoroughly.

For these reasons, it is worthwhile for the government to continue to assert an interest in imposing reasonable restrictions on off-label communications for the purpose of ensuring that the information communicated to physicians is accurate and unbiased. It should also, of course, assert the other interests that courts have had less difficulty accepting as substantial, such as protecting the integrity of the drug approval process and protecting the public from unsafe prescribing (Spurling et al. 2010).

2. Direct Advancement

The next component of a reinvigorated defense of restrictions on off-label promotion under *Central Hudson* is establishing that the regulation directly advances the asserted government interests. First, the government should make clear that if off-label prescribing is to be allowed, it should be as informed as possible. The *Caronia* court assumed that obstructing the free flow of information about off-label uses works against this goal. But that conclusion is premised on the unchallenged assumption that off-label promotional communications are sufficiently accurate and unbiased to support informed decision making. Because there is reason to doubt this assumption, regulating off-label promotional communications is by no means incompatible with the decision to permit off-label prescribing.

In the context of this argument, it is worth recalling that FDA's current regulatory approach does not impose a blanket prohibition on off-label promotion but instead focuses on forms of communication most amenable to corruption. Unprompted oral communications from sales representatives in the personal confines of a physician's office are not permitted. By contrast, peer-reviewed journal articles and independent continuing medical education programs may discuss off-label uses, and manufacturers can respond to physician-initiated questions about off-label uses. These safe harbors bolster the case that FDA's regulation of off-label promotion is not a paternalistic attempt to keep physicians in the dark about off-label uses. Rather, FDA's intent is to permit the flow of truthful information while preventing companies from disseminating unsubstantiated claims through nontransparent mechanisms. The risk against which FDA is trying to guard is not that consumers will respond irrationally to the truth but that they will not be told the whole truth.

In arguing the direct-advancement prong, the government should also underscore the ways in which unfettered off-label promotion destabilizes its long-accepted drug-approval system. If companies are free to promote their products for any use or any population, once the product has been approved for one use or population, the incentive to invest in clinical studies to secure approval for new uses is dramatically undercut. A world in which off-label promotion is unregulated is likely to be one in which poorly substantiated marketing claims about unapproved uses proliferate, with attendant effects on prescribing decisions and patients' health and safety. The literature is clear that there is a strong, consistent, specific, and independent association between physician prescribing and exposure to pharmaceutical marketing messages (Manchanda and Honka 2005).

3. Narrow Tailoring

Two lines of argument could help establish that FDA's regulatory scheme for off-label promotion is narrowly tailored. First, the creation of several safe harbors represents thoughtful, adequate tailoring. Second, the policy approaches proposed by the *Caronia* majority as alternatives to speech restrictions fall well short of the mark.

On the latter point, the suggestions offered by the *Caronia* majority run the gamut from the merely ineffectual to the baldly ridiculous. For example, the *Caronia* majority proposed to train physicians and patients in differentiating misleading from truthful promotional statements. Even if didactic strategies for distinguishing among the types of claims made in off-label promotion and understanding the evidence base underlying them could be identified, along with strategies for effectively reaching every physician with this information, it is inconceivable that the government would appropriate funding at a level sufficient to create an effective counterweight to the $50 billion that pharmaceutical companies spend each year on promotion to physicians (Kesselheim et al. 2013).

Alternatively, the *Caronia* majority proposed to let FDA "develop its warning or disclaimer systems or develop safety tiers within the off-label market" (*United States v. Caronia* 2012:168). But drug risk communications emerging from FDA have been shown to have little impact on U.S. health care utilization or health behaviors (Dusetzina et al. 2013), and FDA does not commonly engage in de novo review of the evidence supporting off-label uses of drugs unless a manufacturer formally files a supplemental application requesting such a review. FDA thus would not be able to assign "safety tiers" to off-label uses of all prescription drugs without congressional action specifically authorizing it to do so and providing sufficient resources to support the effort. Legislators have been loath to expand FDA's regulatory authority in recent years even in the face of major public health crises (Outterson 2013). Even if FDA's ability to issue disclaimers were not at issue, the utility of such a strategy is doubtful. Consumers spend billions of dollars on vitamins and minerals that have no proof of efficacy, even in the presence of warning labels that alert consumers that FDA has not validated the health claims made about the products (Guallar et al. 2013).

The court also suggested that FDA impose "ceilings or caps on off-label prescriptions" or other mechanisms to restrict the amount of off-label prescribing directly—or prohibit off-label prescribing entirely (*United States v. Caronia* 2012:168). As previously stated, direct regulation of the practice of medicine is outside FDA's jurisdiction. FDA also recognizes that off-label prescribing can be helpful and, indeed, lifesaving in some clinical situations and has no apparent wish to deprive the public of those benefits (Schultz 1996).

Because no non–speech-restricting alternative policy could achieve the government's interest even marginally and because FDA has taken reasonable steps to allow companies to communicate truthful information about off-label uses in a responsible way, FDA's regulatory scheme for off-label promotion should be considered narrowly tailored.

C. Expanded Use of "Safe Harbors" for Off-Label Promotion

A third strategy that FDA could pursue to preserve its ability to regulate off-label promotion is to provide expanded pathways for such promotion while providing mechanisms to help ensure the reliability of the communications. This might bolster the argument that the regulatory scheme is narrowly tailored to advancing the goal of protecting the public from unsafe prescribing.

One mechanism would be to permit off-label communications in circumstances in which the manufacturer can document that the statements reflect substantial clinical experience. FDA rules stipulate that advertisements about a drug's efficacy for uses described in the labeling may rest on documentation of "substantial clinical experience" as an alternative to "substantial evidence" (evidence from "well-controlled investigations") (21 CFR § 202.1(e)(4)(ii)). For instance, a 2011 guidance document permitted manufacturers to promote antihypertensive drugs as efficacious in preventing deaths from cardiovascular events based on a "substantial clinical experience" rationale (Food and Drug Administration 2011). The endpoints in the clinical trials testing the drugs had only included intermediate outcomes such as blood pressure control, not cardiovascular deaths, but FDA explained that "blood pressure control is well established as beneficial in preventing serious cardiovascular events."

FDA's 2011 decision was based on decades of clinical experience with hypertension, antihypertensive drugs, and patient outcomes. The announcement also came in an official, prospective FDA guidance document, which involves months of internal deliberation and may involve input from outside experts (Colman 2012). These two features increase the likelihood that information disseminated about off-label uses based on "substantial clinical experience" will be evidence based and reliable. FDA could modify its regulation to apply the same standard to promotion of off-label uses and take steps to publicize the availability of this avenue of off-label promotion. The main advantage of the process is that it would be less rigid—and less expensive—than a formal supplemental NDA.

Second, FDA could permit more evidence-based off-label marketing by expanding its view of what constitutes "substantial evidence." Specifically, it could allow manufacturers to receive prospective approval for off-label promotion if the manufacturer submitted well-controlled observational research (in lieu of clinical trials) establishing the safety of the use (Silverman 2009). Indeed, large-scale observational studies are particularly well suited for detecting and quantifying drug safety problems, when they are conducted properly (Avorn 2007).

There is already flexibility within the FDCA to permit FDA to consider observational studies or other evidence short of prospective, randomized trials as offering "substantial evidence" for a given finding because the FDCA does not specify that substantial evidence needs to be supported by randomized controlled trials but only "adequate and well-controlled investigations." However, because of the risk that such studies can yield unreliable results if not carefully designed and conducted, FDA's current wariness about authorizing off-label marketing based on them is understandable. To clarify its expectations, FDA could issue a formal guidance document describing the features of high-quality observational research and the circumstances under which such research might be sufficient to support off-label marketing claims. For example, use of high-quality observational research could be initially limited to support claims about the safety of off-label uses and not be allowed to support claims about effectiveness. This approach would permit manufacturers to engage in truthful, evidence-based communications about certain types of off-label uses that observational research could validly support, while helping to assure that unsubstantiated marketing claims do not provoke inappropriate, unsafe prescribing.

V. CONCLUSION

The Second Circuit's decision in *Caronia* evinces judicial distaste for attempts to regulate pharmaceutical promotion, based in part on the claimed public health importance of the free flow of truthful information about medical products and in part on the role that physicians, as sophisticated intermediaries, play as receptors of pharmaceutical promotional communications. In the case of off-label promotion, neither of these assumptions withstands scrutiny. Years of experience with industry marketing practices leading to dangerous, non-evidence-based, off-label uses of medical products establishes the need for regulation in this arena.

Unfortunately, FDA did not pursue an appeal in *Caronia*. Its decision may reflect FDA's curious conclusion that *Caronia* will not significantly affect its enforcement activities, but the decision may have also been a strategic calculation. With the Supreme Court and Courts of Appeals showing ever stronger support for commercial speech rights, it is a treacherous time to pursue such an appeal.

As it stands, the *Caronia* decision would appear to preclude some of FDA's traditional enforcement practices, at least in the Second Circuit states. Significant retrenchment in the agency's enforcement activity would likely lead to adverse health consequences for patients. Companies would likely feel emboldened to engage in more promotional communications with fewer incentives to ensure that promotional statements have an evidentiary basis, and they would find less reason to seek FDA approval for new uses of their products. Such moves would heighten the prevalence of prescribing decisions that put patients at undue risk without clear benefits.

To help avoid such consequences, we have suggested several avenues along which FDA might continue to prosecute off-label promotion and defend its regulatory framework against future challenges. Fighting to ensure that a regulatory regime is in place to promote accurate and unbiased promotional communications is a public health imperative.

NOTE

This work is derived from an article previously published in the *North Carolina Law Review*: Kesselheim, Aaron S. and Michelle M. Mello. 2014. "Prospects for Regulation of Off-Label Drug Promotion in an Era of Expanding Commercial Speech Protection," *North Carolina Law Review* 92: 101–62.

REFERENCES

Alexander, G. Caleb et al. 2010. "Increasing Off-Label Use of Antipsychotic Medications in the United States, 1995–2008." *Pharmacoepidemiology & Drug Safety* 20:177–82.

Avorn, Jerry. 2007. "In Defense of Pharmacoepidemiology—Embracing the Yin and Yang of Drug Research." *New England Journal of Medicine* 357:2219–21.

Avorn, Jerry. 2012. "Two Centuries of Assessing Drug Risks." *New England Journal of Medicine* 367:193–96.

Carlat, Daniel. 2007. "Dr. Drug Rep." *New York Times*. (November 25).

Central Hudson Gas & Electric Corporation v. Public Service Commission of New York, 447 U.S. 557 (1980).

Colman, Eric. 2012. "Food and Drug Administration's Obesity Drug Guidance Document: A Short History." *Circulation* 125:2156–64.

Dusetzina, Stacie B. et al. 2012. "Impact of FDA Drug Risk Communications on Health Care Utilization and Health Behaviors: A Systematic Review." *Medical Care* 50:466–78.

Food and Drug Administration. 2011. "Guidance for Industry: Hypertension Indication: Drug Labeling for Cardiovascular Outcome Claims." (OMB Control No. 0910–0670).

Food and Drug Administration. 2014. "Guidance for Industry: Distributing Scientific and Medical Publications on Unapproved New Uses—Recommended Practices." Revised Draft Guidance. (February). http://www.fda.gov/downloads/Drugs/GuidanceComplianceRegulatoryInformation/Guidances/UCM387652.pdf.

Guallar, Eliseo et al. 2013. "Enough Is Enough: Stop Wasting Money on Vitamin and Mineral Supplements." *Annals of Internal Medicine* 159:850–51.

Hall, Ralph F. 2013. "1st Amendment Cases: Our Most Important Judicial Trends." Food and Drug Law Institute Advertising and Promotion for the Pharmaceutical, Medical Device, Biological, and Veterinary Medicine Industries (September 17). http://www.fdli.org/docs/ap2013-slides/top-20-combined-final.pdf?sfvrsn=0.

Klasmeier, Coleen and Martin H. Redish. 2011. "Off-Label Prescription Advertising, the FDA and the First Amendment: A Study in the Values of Commercial Speech Protection." *American Journal of Law and Medicine* 37:315–57.

Kolata, Gina. 2007. "How Fen-Phen, A Diet 'Miracle,' Rose and Fell." *New York Times* (September 23).

Kesselheim, Aaron S. 2011. "Off-Label Drug Use and Promotion: Balancing Public Health Goals and Commercial Speech." *American Journal of Law and Medicine* 37:225–27.

Kesselheim, Aaron S. et al. 2013. "FDA Regulation of Off-Label Drug Promotion Under Attack." *Journal of the American Medical Association* 309:445–46.

Krumholz, Harlan M. et al. 2007. "What Have We Learnt from Vioxx?" *British Medical Journal* 334:120.

Levi, Marcel et al. 2010. "Safety of Recombinant Activated Factor VII in Randomized Clinical Trials." *New England Journal of Medicine* 363:1791–97.

Manchanda, Puneet and Elisabeth Honka. 2005. "The Effects and Role of Direct-to-Physician Marketing in the Pharmaceutical Industry: An Integrative Review." *Yale Journal of Health Policy Law & Ethics* 5:785–808.

Moynihan, Ray. 2010. "Rosiglitazone, Marketing, and Medical Science." *British Medical Journal* 340:c1848.

Outterson, Kevin. 2014. "The Drug Quality and Security Act—Mind the Gaps." *New England Journal of Medicine* 370:97–99.

Pear, Robert. 2011. "Health Spending Rose in '09, but at Low Rate." *New York Times* (January 5).

Radley, David C. et al. 2006. "Off-Label Prescribing Among Office-Based Physicians." *Archives of Internal Medicine* 166:1021–23.

Ross, David B. 2007. "The FDA and the Case of Ketek." *New England Journal of Medicine* 356: 1601–3.

San Francisco Arts & Athletics, Inc. v. United States Olympic Committee, 483 U.S. 522 (1987).

Schiff, A. B. 1995. "Corporate Compliance and Voluntary Disclosure." *Health Care Law Newsletter* 10:8.

Schultz, William B. 1996. "Promotion of Unapproved Drugs and Medical Devices." *Testimony Before the S. Comm. on Labor and Human Resources.* http://www.fda.gov/NewsEvents/Testimony/ucm115098.htm.

Sell, David. 2013. "U.S. Won't Pursue Case of Drug Representative and 'Off-Label' Promotion." *Philadelphia Inquirer* (January 26). http://articles.philly.com/2013-01-26/business/36550335_1_drug-companies-fda-misbranded-drug.

Silverman, Ed. 2013. "Off-Label Marketing: Free Speech or Illegal Promotion?" *British Medical Journal* 346:f320.

Silverman, Stuart L. 2009. "From Randomized Controlled Trials to Observational Studies." *American Journal of Medicine* 122(2):114–20.

Sorrell v. IMS Health, 131 S. Ct. 2653 (2011).

Spurling, Geoffrey K. et al. 2010. "Information from Pharmaceutical Companies and the Quality, Quantity, and Cost of Physicians' Prescribing: A Systematic Review." *PLoS Medicine* 7(10):e1000352.

United States ex rel. Lauterbach v. Orphan Medical Inc., Jazz Pharmaceuticals Inc., Civil Action No. 05-CV-0387-SJF-KAM (February 17, 2006).

United States v. Caronia, 703 F.3d 149 (2d Cir. 2012).

Virginia State Board of Pharmacy v. Virginia Citizens Consumer Council, Inc., 425 U.S. 748. (1976).

Wadman, Meredith. 2004. "Spitzer Sues Drug Giant for Deceiving Doctors." *Nature* 429:589.

Walton, Surrey M. et al. 2008. "Prioritizing Future Research on Off-Label Prescribing: Results of a Quantitative Evaluation." *Pharmacotherapy* 28:1443–52.

Wang, Philip S. et al. 2005. "Risk of Death in Elderly Users of Conventional vs. Atypical Antipsychotic Medications." *New England Journal of Medicine* 353:2335-39.

Waxman, Henry A. 2003. "A History of Adverse Drug Experiences: Congress Had Ample Evidence to Support Restrictions on the Promotion of Prescription Drugs." *Food & Drug Law Journal* 58:299-312.

Wood, Susan F. 2005. "Women's Health and the FDA." *New England Journal of Medicine* 353:16505-2.

Zauderer v. Office of Disciplinary Counsel of Supreme Court of Ohio, 471 U.S. 626 (1985).

CHAPTER ELEVEN

The FDCA as the Test for Truth of Promotional Claims

CHRISTOPHER ROBERTSON

OTHER CHAPTERS in this volume discuss the statutory basis for the Food and Drug Administration (FDA) and its regulation of off-label prescribing in particular (see especially Mello and Kesselheim's Chapter 10). In this chapter, I would like to focus on the importance of the manufacturer's intent as the predicate for FDA regulation and the production of knowledge as the purpose for FDA regulation. These two points are essential for understanding the controversy over off-label promotion, and they reveal a way forward.

I. THE MANUFACTURER'S SPOKEN INTENT

Before enactment of the Federal Food, Drug, and Cosmetic Act (FDCA), health care consumption occurred in a regime of cheap talk, where "the manufacturer was sole judge of the therapeutic benefits he should claim for his product. If a claim was fraudulent, *and if the government could prove it*, the federal government could deal with the problem. However, a false or misleading therapeutic claim was acceptable under the law in the absence of fraud. In other words, the more ignorant the manufacturer, the more sweeping his claims for drug benefit could be" (Rankin 1965:32; emphasis added).

The FDCA was designed to change that situation, putting physicians and consumers on firmer epistemic footing. Now, the FDCA requires that, before bringing a new drug to market, the drugmaker must submit an application to FDA with evidence that proves that the chemical compound is safe and effective for a particular, intended use.

There is an interesting conceptual, almost metaphysical, point at the core of this regime: the company's intention that a chemical compound be used to treat a disease is what makes that compound a new "drug," which then creates the burden to prove to FDA that the drug works for that intended purpose (21 USC § 321(g)(1)(B)–(C); 21 CFR § 201.128). Without such a drugmaker's "objective intent," the thing is just a chemical—not a drug—and the FDCA does not apply at all. In other words, "[t]o put the matter in practical terms: it is because of the 'intended uses' principle that hardware stores are generally free to sell bottles of turpentine, but may not label those bottles, '*Hamlin's Wizard Oil: There is no Sore it will Not Heal, No Pain it will not Subdue*,'" as they did before the FDCA (*United States v. Caronia* 2012:171).

It bears emphasis that this "intended uses" principle is a premarket approval system, a regulatory mechanism distinct from holding manufacturers liable *ex post* for making false or misleading claims about their drugs. Nonetheless, the FDCA does that too (21 USC §§ 331, 343, 352; *United States v. Ninety-Five Barrels, More Or Less, Alleged Apple Cider* 1924:442). The Supreme Court has read the labeling requirements in the statute broadly to impose liability for false and misleading promotional efforts (*United States v. Kordel* 1948:351). In contrast, the intended-use principle creates *per se* criminal liability for introducing a drug into interstate commerce that has not been approved by the FDA for its intended uses. Once FDA approves a drug to enter the market for a particular use (an "indication"), physicians are free to prescribe drugs for other "off-label" uses. Off-label usage is often worthwhile and sometimes based on sound evidence of efficacy and safety. However, off-label prescribing is a major driver of health care costs and sometimes presents real risks to patients that are not offset by proven benefits.

The regulatory logic for off-label uses is the same as for the first medical use of a novel chemical compound. The manufacturer may not promote the drug for new uses because doing so would evince an intention that the drug be used as such, even while the drug has not been proven effective, approved, or labeled for such use. The drug would then be misbranded and the introduction of it into interstate commerce would be a federal crime (21 USC § 352(f)(1)). In particular, the label may be inadequate if "[s]tatements of all . . . uses for which such drug is intended" and "usual quantities [of dose] for each of the uses for which it is intended" are insufficiently specified (21 CFR § 201.5(a)–(b)).

As the Senate Report on the FDCA explains, "[t]he manufacturer of the article, through his representations in connection with its sale, can determine the use to which the article is to be put" (Committee on Commerce 1935:240). This regulatory evolution has been controversial (e.g., Conko 2011:159–61). Yet, what other than the manufacturer's own speech could be better evidence of its intent?

This pattern in the law—using intent as the predicate for regulation and then using speech as evidence of intent—is quite common and not peculiar to pharmaceutical regulation. As early as 1888, the Supreme Court affirmed a criminal conviction for someone who manufactured an "oleaginous substance" because he intended for it to be used as food (*Powell v. Pennsylvania* 1888:679). Likewise, an automobile is not subject to regulation by the Federal Aviation Administration (FAA) unless it is "intended to be used for flight in the air" (14 CFR § 1.1; defining "aircraft").

This is all contrary to the suggestion of Redish and Klasmeier's chapter in this volume. They argue that the intended-uses standard is unworkable, because manufacturers are routinely *aware* of off-label uses, potentially thus imposing criminal liability everywhere. Although the current specification of the "intended uses" principle in 21 CFR § 201.128 could be read to make mere knowledge of off-label uses sufficient enough to impose liability, it is not applied in that way, and the distinction between awareness and intent is too fundamental for the criminal law to gloss over as such. A turpentine manufacturer's mere knowledge that its product may be abused for sundry purposes is not necessarily an intent that it be so abused. Similarly, when someone notifies General Motors that they are trying to fly a car off of a cliff, that awareness alone does not instantly and retroactively convert General Motors into an airplane manufacturer, subject to FAA regulation. In a world where off-label usage is ubiquitous, the law should draw a reasonable line that relies on overt evidence of the manufacturer's intent, such as the manufacturer's own promotional speech. Mere knowledge is not enough. For similar reasons, the FDA has created safe harbors permitting manufacturers to respond to questions about off-label uses and to distribute peer-reviewed literature about off-label uses (Kesselheim and Avorn 2012:2203). These dispensations allow for a workable regulatory balance.

II. PRESUMPTIONS OF TRUTH

It is peculiar that such regulatory mechanisms turn on the defendant's own stated intent: its speech. Yet the Supreme Court has emphasized that "the First Amendment . . . does not prohibit the evidentiary use of speech to establish the elements of a crime or to prove motive or intent" (*Wisconsin v. Mitchell* 1993:489). Although scholars have made provocative arguments to the contrary (Redish and Downey 2013:698), it remains true and routine that a murder case may turn on recordings of the defendant's stating an intent to join the conspiracy. Indeed, a federal rule of evidence specifically contemplates such situations (Fed. R. Evid. 801(d)(2); stating that "an opposing party's statement" is excluded from hearsay and may be admissible).

Although "commercial speech" was once altogether unprotected by the First Amendment, it has evolved into a zone of "intermediate scrutiny." In other chapters in this volume, Mello and Kesselheim and Klasmeier and Redish take opposite sides over the question of whether FDA's proscription on off-label prescribing should meet that test. Other scholars have noted that the commercial speech doctrine has itself "evolved into a strict scrutiny test in all but name" (Piety 2012:4). Indeed, FDA suffered a stinging defeat in a 2002 case concerning a relatively obscure practice of pharmaceutical compounding (*Thompson v. W. States Med. Ctr.* 2002:357). And, a 2011 health care information case held that a state law prohibiting the sale of data for pharmaceutical marketing purposes was unconstitutional as viewpoint discrimination (*Sorrell v. IMS Health Inc.* 2011:2670–71). A more fundamental reconceptualization of this regulatory regime may be necessary.

Indeed, the FDCA suffered a severe blow in a December 2012 case in which the United States Court of Appeals for the Second Circuit invoked the First Amendment to reverse a pharmaceutical sales representative's criminal conviction for conspiracy to sell a misbranded drug (*United States v. Caronia* 2012:164). "Caronia argue[d] that he was convicted for his speech—for promoting an FDA-approved drug for off-label use" (*Caronia* 2012:152). The Second Circuit agreed with that conceptualization of the case and said that it would "avoid constitutional difficulties by adopting a limiting interpretation" of the FDCA (*Caronia* 2012:162). The court "construe[d] the misbranding provisions of the FDCA as not prohibiting and criminalizing the truthful off-label promotion of FDA-approved prescription drugs"—a landmark, unprecedented holding (*Caronia* 2012:168).

The *Caronia* court made no real effort to cabin its holding in any principled way. Are manufacturers now free to transport drugs in interstate commerce that are labeled for one use, or perhaps not labeled at all, while explicitly stating that the manufacturer intends other, unlabeled and unproven uses? If so, the crime of "misbranding" and the entire FDCA premarket approval regime begins to seem precarious.

It may help to thematize precisely what the *Caronia* court took for granted: the truthfulness of the manufacturer's promotional claims. Accordingly, in *Caronia*, the court adopted this framing: "the First Amendment does not permit the government to prohibit and criminalize a pharmaceutical manufacturer's *truthful and non-misleading* promotion of an FDA-approved drug to physicians for off-label use where such use is not itself illegal and others are permitted to engage in such speech" (*Caronia* 2012:160; emphasis added). The court recognized the importance of this point because "[o]f course, off-label promotion that is false or misleading is not entitled to First Amendment protection" (*Caronia* 2012:165 n.10; citing *Cent. Hudson Gas & Elec. Corp. v. Pub. Serv. Comm'n of N.Y.* 1980:566). Indeed, the Supreme Court has for decades emphasized the role of truthfulness in its First Amendment analyses of commercial-speech regulations (e.g., *Bolger v. Youngs Drug Prods. Corp.* 1983:69).

Strikingly, the issue was not litigated: "The government [did] not contend that off-label promotion is in and of itself false or misleading" (*Caronia* 2012:165 n.10). Thus, the *Caronia* court apparently presumed that all of Mr. Caronia's off-label promotional claims were true.

Similarly, in a prominent case concerning FDA's regulation of promotional claims for dietary supplements, another circuit acknowledged that the predicate for Constitutional scrutiny is that "[t]ruthful advertising related to lawful activities is entitled to the protections of the First Amendment" (*Pearson v. Shalala* 1999:655; quoting *In re*

R.M.J. 1982:203). While conceding that "evidence in support of [the promotional] claim is inconclusive" (and thus the court could not really know whether the Constitutional predicate was met), the court nonetheless held that the First Amendment prohibited FDA from proscribing the commercial speech (*Pearson* 1999:659). Here again, truth was simply presumed.

Scholars have routinely made this same sort of presumption of truthfulness to get their First Amendment arguments off the ground. In his critique of the FDA regime, Osborn asks: "[W]here the challenged off-label information is *truthful*, what is the public interest in forbidding it?" (2010:307). Likewise, in this volume, Klasmeier and Redish criticize the FDA's regulation in this area, and offer "four core postulates of free speech theory that are indisputably contravened by the ban on off-label promotion," the first of which is "Government may not attempt to manipulate lawful citizen behavior by means of the selective suppression of *truthful* expression advocating lawful activity" (234) Strikingly, several of the recent scholarly treatments of the constitutionality of off-label promotion highlight "truth" in their titles (Conko 2011; Noah 2011). It is notable that each of these critiques prominently turns on the concept of "truthfulness," and each simply presumes that the regulated promotional claims are in fact truthful.

With this presumption in hand, the courts and commentators can paint FDA as a perverse paternalist, trying to keep people from the truth. Long before taking the bench, Justice Kagan wrote, "First Amendment law . . . has as its primary, though unstated, object the discovery of improper governmental motives. The doctrine comprises a series of tools to flush out illicit motives and to invalidate actions infected with them" (Kagan 1996:414). Accordingly, in *Western States Medical Center*, the Supreme Court emphasized that it has long "rejected the notion that the Government has an interest in preventing the dissemination of *truthful* commercial information in order to prevent members of the public from making bad decisions with the information" (*Thompson v. W. States Med. Ctr.* 2002:374; emphasis added).

Thus, for the regulation of off-label promotion, the courts and commentators seem to be presuming the truth about off-label promotional claims and presuming that FDA is motivated by a paternalistic desire to keep that truth out of the hands of consumers. As it happens, neither of these presumptions is warranted.

III. THE FDCA AS AN INCENTIVE FOR KNOWLEDGE PRODUCTION

If a sales representative says, "I wish you would try this drug as a treatment for epilepsy," the literal meaning is just a report of his personal desire for a sale. That report may be true, but only in a trivial sense. Pragmatically, utterances like these implicitly make a claim that the drug would be safe and effective for that purpose. (Why else inject a substance into someone's body?) That is the core propositional content of the claim.

Ultimately, we do not know whether such a drug promotional claim is true or false. And that information would be very costly to secure—millions of dollars and years of research by dozens of investigators and thousands of patients taking the drug and/or placebo, at real risk of side effects and often at real opportunity cost, compared to other treatments the patients could have tried instead.

Since truth or falsity is not knowable a priori, it instead depends on investments by those who have incentives to make those investments, in legal and institutional contexts that define those incentives. No individual patient or individual physician could find it rational to undertake a large-scale research study to prove safety and efficacy of some company's patented drug (Elhauge 1996:1526). That costly task must fall to the one entity that has the economic interest in reaping the aggregate benefits of its drug: the drugmaker.

Courts and commentators have long understood that intellectual property has a similar epistemic purpose: to incentivize the production and disclosure of knowledge (e.g., *Sinclair & Carroll Co. v. Interchemical Corp.* 1945:330–31; Turner 1969:451). In particular, the patent laws are aimed at producing "useful" knowledge (35 USC § 101). In the pharmaceutical sector, that means the manufacturer identifying "a specific disease against which the claimed compounds are alleged to be effective" (*In re Brana* 1995:1565). Although the patent system does not require the rigorous proof of the FDCA, it does include a disclosure requirement, designed to contribute to the larger body of scientific knowledge.

By promoting new unproven uses for their drugs, manufacturers attempt to expropriate additional value from their legally enforced monopolies and their sunk costs of research and development. They replace research with marketing. Similarly, the Federal Trade Commission Act is designed to combat "unfair methods of competition," which includes a proscription on making unsubstantiated health claims (*Fed. Trade Comm'n v. Direct Mktg. Concepts, Inc.* 2010:302). The problem is that such unproven off-label uses can become substitutes, whose real quality is unobservable in any particular case, thereby reducing the demand for drugmakers to invest in producing and proving the efficacy of new drugs or new uses of old drugs.

The FDCA can thus be understood as creating the incentive for drugmakers to produce knowledge to support new uses of the products that enjoy (or enjoyed) government-enforced monopolies. This epistemic and economic motive, tying investments to market rewards, is much different than the paternalist one caricatured by the courts and commentators.

IV. WHY COURTS CAN ACKNOWLEDGE THEIR IGNORANCE

In this light, should the courts continue to presume that off-label promotional claims are true and grant them immunity under the First Amendment, even while the drugmaker has rationally declined to make investments in discerning that truth? Arguably, no.

It is striking to compare the FDCA to the courts' own routine use of the rules of evidence to regulate the speech of attorneys and witnesses. The courts emphatically do not presume truth of self-interested conclusory assertions about scientific questions; they deride them as mere "ipse dixit" and altogether ban them from being uttered in court (*Gen. Elec. Co. v. Joiner* 1997:146).

Imagine a litigant who proposed to put a drug salesman on the stand to testify that a drug was in fact safe and effective—given the salesman's lack of expertise and lack of supporting evidence, the judge would exercise her "gatekeeping" role to exclude such unreliable testimony (*Daubert v. Merrell Dow Pharm., Inc.* 1993:597; Fed. R. Evid. 702). Under this regime, the proponent of a claim bears the burden of proving that the proposition is reliable and thus admissible. The courts' orders in limine thus become prior restraints, carefully delineating what may and may not be said, ultimately under threat of jail for contempt of court.

There is some irony in the courts telling other regulators that the First Amendment bars their efforts to regulate speech while the courts blithely use speech regulations to achieve the courts' own aims in their own domain. Other scholars have noted, and passed over, this difficulty, saying that "no one would challenge seriously under the First Amendment" such judicial regulations of speech (Sullivan 1998:569). Indeed, research has failed to reveal a single case in which the rules of evidence have been challenged on First Amendment grounds.

Courts admittedly do presume truth in other contexts. In the particular setting of a criminal trial, the burden is generally upon the government to prove the elements of any crime, which is to say that there is a presumption of innocence. For example, the Ninth Circuit recently affirmed pharmaceutical company executive Scott Harkonen's conviction for wire fraud (*United States v. Harkonen* 2013:636). The jury found that Harkonen issued a press release making materially false statements claiming the efficacy of a drug. The Court overruled his First Amendment claims, holding that the wire fraud statute's elements required proof that the speech was false or fraudulent, a burden met by the government, thereby putting the speech outside the First Amendment.

Falsity is not an element, however, in off-label promotion cases under the misbranding statute. In *Caronia*, for example, the truth or falsity of Mr. Caronia's promotional claims was irrelevant to the charged crime, since the utterances need only show his intention to sell the drug for an unapproved purpose. Thus, the elements of the crime do not themselves impose a burden on the government to prove falsehood. There is no presumption of truthfulness. Whether the First Amendment should independently impose such a burden on the prosecutor is the remaining question.

It is also important to distinguish these cases from ones where the drugmaker's claim is concededly truthful but arguably misleading. If the government concedes truth of the claim itself, but asserts that it may engender misunderstandings by hearers, the government bears the burden of proving those effects (e.g., *Zauderer v. Office of Disciplinary Counsel of Supreme Court of Ohio* 1985:646). Once truth has been conceded, that approach is sensible, given the epistemic value of truth and our aversion to paternalism as a motivation for speech regulation. Here, however, the present question is distinct: Who should have the burden when the government makes no such concession about the safety and efficacy of a drug and the truthfulness is unknown? That question is open.

Typically, "the standard First Amendment rule [is] that the burden of proof as to constitutionally relevant facts must lie on the party who would stifle the speech, not the speaker herself" (Netanel 2013:1113). Similarly, in this volume, Redish and Klasmeier say that the burden "naturally" falls as such. Closer inspection of the doctrine, however, suggests that in these circumstances the burden could instead fall on the defendant-speaker.

One point of reference is defamation and libel. Under the common law, still existing in many states, "when a plaintiff proves publication of words that are defamatory per se [such as an allegation of sexual deviance], the elements of *falsity* and malice (or fault) are *presumed*, but may be rebutted by the defendant" (*Costello v. Hardy* 2004:140–41). In the 1986 case of *Philadelphia Newspapers, Inc. v. Hepps*, the Supreme Court held "at least where a *newspaper* publishes speech *of public concern*, a private-figure plaintiff cannot recover damages without also showing that the statements at issue are false" (1986:768–69). How this doctrine would apply to cases not involving newspapers, and involving commercial speech instead, remains unclear. In 2013, the Iowa Supreme Court concluded that unless the speech is a matter of public concern by media defendants (such as newspapers), then the First Amendment permits courts to place the burden of proving truth on the defendant-speaker (*Bierman v. Weier* 2013:448). In this light, it would similarly be permissible to put the burden on the drugmakers to prove the truth of their promotional claims as a predicate for any First Amendment claims they assert.

Beyond the domain of truth and falsity, there are several related questions of who bears the burdens in First Amendment litigation. For intellectual property cases under the Lanham Act, the defendant bears the burden to prove that the use of a valid mark is protected speech (e.g., *Parks v. LaFace Records* 2003:444). Similarly, the Supreme Court has said that to the extent that there is a First Amendment right to use copyrighted works, it is satisfied by the fair use exception, which functions as an affirmative defense (*Campbell v. Acuff-Rose Music, Inc.* 1994:590). On the other hand, in the child pornography case of *Ashcroft v. Free Speech Coalition*, the Supreme Court rejected the notion that an affirmative defense satisfied the First Amendment problem. Here, the burden of the affirmative defense was not truthfulness of the representation but rather a showing that the depicted persons were of legal age. The Supreme Court held that such a burden was simply infeasible, given that "the defendant is not the producer of the work, he may have no way of establishing the identity, or even the existence, of the actors" (2002:255). Arguably, in contrast, a drugmaker is in a much better position to prove the truth of its own representations about the efficacy of its own drugs.

The foregoing suffices to suggest that there is room within current constitutional doctrine to allow for a more realistic approach to truth and falsity in the regulation of off-label promotion. The fundamental insight is this: it would be rather strange if the Constitution required the courts to adopt a presumption about the safety and efficacy of any given chemical compound for treating any given disease, a leap of faith that Congress, FDA, physicians, and scientists would prudently refuse to take, and which the rules of evidence prohibit the courts themselves from taking, when that very question is put into issue. Instead, the Constitution allows courts candidly and prudently to concede their ignorance about safety and efficacy, until given a warranted basis for belief. Whether it *should* defer to the FDCA as the test for the truth depends on whether the courts understand its larger epistemic purpose.

V. THE FDCA AS THE TEST OF TRUTH

While not purporting to be a comprehensive treatment of the First Amendment issues implicated by the regulation of drug promotions, this chapter has developed a few themes. First, it has shown how the FDCA's regulatory regime, like so many other areas of the law, turns on the defendant's own speech revealing its intent, but this chapter has also shown how that regime is precarious under an expanding First Amendment doctrine. While most clear in the domain of off-label promotion, the precariousness seems to cut even deeper, to undermine the FDCA's New Drug Application requirement itself. This threat motivates the project to explore whether FDA regulation of promotion could be reconstructed in a constitutionally sufficient way.

Second, this chapter has shown that the notion of truthfulness is at the core of much scholarly and judicial analyses in this domain but that the notion is deployed presumptuously. In fact, the truth or falsity of the drugmaker's promotional claims is unknown, largely because the drugmaker has declined to invest in making such a proof. The FDCA is designed to incentivize the drugmakers to make that investment, a function that is undermined if courts presume the truth as a predicate for providing immunity under the First Amendment.

The argument here is focused on the particular circumstances of off-label promotion: a nonmedia defendant, engaging in self-interested commercial speech, about a product that it has declined the option of proving to FDA the efficacy and safety for the indication suggested, where talk is cheap since efficacy and safety cannot be known through accumulation of anecdotes alone, in a domain where the costs of falsehood can be life-or-death for patients that forgo other treatments and suffer side effects, and where the actual crime alleged regulates behavior (the introduction of a drug in interstate commerce), not speech in the first place. In other contexts, the First Amendment analysis may be quite different. But here, the First Amendment should protect truthful speech only where we can say with some warrant that the label applies.

Once shown in this light, it becomes clear that the current FDCA premarket approval regime *already* provides a constitutional framework for truthful promotion of drugs. The advantage of using this procedure, which Congress enacted, is that FDA has the institutional capacity to provide a robust and scientifically rigorous assessment of the drugmaker's promotional claims. Because the FDA approval regime provides an avenue for truthful speech, prosecution for off-label promotion would thus raise no constitutional problem.

If the courts nonetheless refuse to defer to the epistemic test established by the coequal branches, the courts still need not adopt judicial naiveté. If a drugmaker refuses or fails to prove a promotional claim to FDA, the courts should, at the very least, impose on drugmakers the burden of proving their claims true as an affirmative defense at a jury trial. Conceiving FDA as something of a safe harbor, this alternative framework gives drugmakers two chances to prove truth, without going so far as to presume whatever they say is true (*Washington Legal Foundation v. Henney* 2000:335–36). In some cases, the epistemic basis for the claim will be quite strong; there are instances in which Medicare and payers have done their own systematic reviews and found enough evidence to merit billions of dollars of reimbursements. Still, when the evidence is weak, the manufacturer will contemplate the difficulty of actually proving the truth of its off-label promotional claims in a court of law, controlled by the limits on the admissibility of expert testimony and scientific evidence. Thus, the opportunity to prove the affirmative defense of truthfulness may actually provide little succor to defendants who lack rigorous scientific support for their claims.

Without some sort of reconstruction of FDA regulation of off-label promotion—of the sort suggested here, or by Mello and Kesselheim in the prior chapter—the Supreme Court's expansive First Amendment doctrine threatens to return us to a pre-FDCA world where drugs are presumed to cure any disease that the drugmakers say they cure, at least until somebody proves them false. Especially in contexts where information is expensive, the speaker is in the best position to purchase that information, and the stakes are high, the First Amendment should not proceed on bald presumptions about the truth.

NOTE

The ideas in this chapter are developed in significantly more detail in the 2014 article "When Truth Cannot Be Presumed," *Boston University Law Review* 94:545. Disclosure: The author has represented litigants asserting First Amendment defenses (outside the context of food and drug regulation) and represented amicus curiae the *New England Journal of Medicine* et al. in the *Sorrell* case discussed herein.

REFERENCES

Ashcroft v. Free Speech Coalition, 535 U.S. 234 (2002).
Bierman v. Weier, 826 N.W.2d 436 (Iowa 2013).
Bolger v. Youngs Drug Prods. Corp., 463 U.S. 60 (1983).
Campbell v. Acuff-Rose Music, Inc., 510 U.S. 569 (1994).
Cent. Hudson Gas & Elec. Corp. v. Pub. Serv. Comm'n of N.Y., 447 U.S. 557 (1980).
Committee on Commerce, Food, Drugs, and Cosmetics, S. Rep. No. 74-361. 1935.
Conko, Gregory. 2011. "Hidden Truth: The Perils and Protection of Off-Label Drug and Medical Device Promotion." *Health Matrix: Journal of Law-Medicine* 21:149–87.
Costello v. Hardy, 864 So. 2d 129 (La. 2004).
Daubert v. Merrell Dow Pharm., Inc., 509 U.S. 579 (1993).

Elhauge, Einer. 1996. "The Limited Regulatory Potential of Medical Technology Assessment." *Virginia Law Review* 82:1525–1622.
Fed. Trade Comm'n v. Direct Mktg. Concepts, Inc., 569 F. Supp. 2d 285 (D. Mass. 2008).
Gen. Elec. Co. v. Joiner, 522 U.S. 136 (1997).
In re Brana, 51 F.3d 1560 (Fed. Cir. 1995).
In re R.M.J., 455 U.S. 191 (1982).
Kagan, Elena. 1996. "Private Speech, Public Purpose: The Role of Governmental Motive in First Amendment Doctrine." *University of Chicago Law Review* 63:413–517.
Kesselheim, Aaron S. and Jerry Avorn. 2012. "The Food and Drug Administration Has the Legal Basis to Restrict Promotion of Flawed Comparative Effectiveness Research." *Health Affairs* 31:2200.
Kordel v. United States, 335 U.S. 345 (1948).
Netanel, Neil Weinstock. 2013. "First Amendment Constraints on Copyright After *Golan v. Holder*." *UCLA Law Review* 60:1082–1128.
Noah, Lars. 2011. "Truth or Consequences?: Commercial Free Speech vs. Public Health Promotion (at the FDA)." *Health Matrix: Journal of Law-Medicine* 21:31–95.
Osborn, John E. 2010. "Can I Tell You the Truth? A Comparative Perspective on Regulating Off-Label Scientific and Medical Information." *Yale Journal of Health Policy, Law, and Ethics* 10:299–356.
Parks v. LaFace Records, 329 F.3d 437 (6th Cir. 2003).
Pearson v. Shalala, 164 F.3d 650 (D.C. Cir. 1999).
Philadelphia Newspapers Inc. v. Hepps, 475 U.S. 767 (1986).
Piety, Tamara R. 2012. "'A Necessary Cost of Freedom'? The Incoherence of *Sorrell v. IMS*." *Alabama Law Review* 64:1–54.
Powell v. Pennsylvania, 127 U.S. 678 (1888).
Rankin, W. B. 1965. "The Future Relationships of FDA and the Pharmaceutical Industry." *Food, Drug, Cosmetic Law Journal* 20:632–37.
Redish, Martin H. and Michael Downey. 2013. "Criminal Conspiracy as Free Expression." *Albany Law Review* 76:697–734.
Sinclair & Carroll Co. v. Interchemical Corp., 325 U.S. 327 (1945).
Sorrell v. IMS Health Inc., 131 S. Ct. 2653 (2011).
Sullivan, Kathleen M. 1998. "The Intersection of Free Speech and the Legal Profession: Constraints on Lawyers' First Amendment Rights." *Fordham Law Review* 67:569–88.
Thompson v. W. States Med. Ctr., 535 U.S. 357 (2002).
Turner, Donald F. 1969. "The Patent System and Competitive Policy." *New York University Law Review* 44:450–76.
U.S. v. Ninety-Five Barrels, More Or Less, Alleged Apple Cider, 265 U.S. 438 (1924).

United States v. Caronia, 703 F.3d 149 (2d Cir. 2012).
United States v. Harkonen, 510 F. App'x 633 (9th Cir. 2013).
Washington Legal Foundation v. Henney, 202 F.3d 331 (D.C. Cir. 2000).
Wisconsin v. Mitchell, 508 U.S. 476 (1993).
Zauderer v. Office of Disciplinary Counsel of Supreme Court of Ohio, 471 U.S. 636 (1985).

CHAPTER TWELVE

Why FDA's Ban on Off-Label Promotion Violates the First Amendment

A Study in the Values of Commercial Speech Protection

COLEEN KLASMEIER AND MARTIN H. REDISH

I. INTRODUCTION

In order to protect the nation from harmful or worthless drugs and medical devices, the Food and Drug Administration (FDA) is legislatively authorized to control the commercial distribution of new drugs and medical devices to those whose efficacy and safety have been reviewed and approved by the agency (21 USC §§ 355(a), 360(k), 360e (2006)). Drugs and devices are approved for specific medical purposes. In numerous instances, however, the medical profession has discovered that treatments approved for one purpose may also serve other valuable medical purposes. Indeed, on a number of occasions such "off-label" uses have proven to be essential to the successful treatment of some very serious illnesses.

In these off-label situations, FDA is faced with a dilemma. On the one hand, off-label use of prescription drugs and devices gives rise to a series of major problems for FDA. While the drugs and devices in question have been vetted and approved by FDA for their designated purpose, at no point has FDA reviewed the supporting scientific data to determine efficacy for the off-label purpose. It is therefore at least conceivable that if such off-label uses are permitted, the drug or device may in reality be worthless or even dangerous for its alternative use, yet doctors may be freely employing it for that purpose. Moreover, widespread off-label use of prescription drugs and devices conceivably undermines FDA's authority and deters manufacturers from seeking FDA approval for even widespread alternative uses. The fear, then, would be that FDA's initial approval could serve as a wedge to permit the industry's equivalent of the Wild West, where the rule of law was seen only rarely. It is true that the federal government could avoid this danger by categorically prohibiting all off-label uses, but this alternative is also not free from problems. As a result of such a ban, potentially valuable treatments—often for the most serious of diseases—could be effectively banned, as manufacturers often lack sufficient incentives to pursue costly, time-consuming, and uncertain supplementary approval. In any event, under the current regulatory framework, FDA asserts that it lacks legal authority to restrict the ability of doctors to prescribe drugs or devices for off-label uses.

FDA has sought to resolve this dilemma by effectively swimming halfway across a river, thereby pleasing no one and avoiding none of its quite legitimate concerns. It has resolved its dilemma by leaving off-label uses essentially unregulated but making a manufacturer's "promotion" of an off-label use categorically illegal. As a result, while the benefits of off-label use are roughly preserved, whatever dangers might accompany such uses are in no way avoided since off-label uses are still generally allowed. It is true, of course, that with no promotion off-label use *as a whole* may decrease, but there is no basis, ex ante, on which to assume that the off-label use likely to be deterred is predominately the harmful version rather than the beneficial version. At the same time, while it is true that the ban on off-label promotion by manufacturers prevents at least some false or misleading claims, because the ban is all-inclusive it simultaneously prevents manufacturers from disseminating accurate information (to which they have unique access) about valuable off-label uses of which doctors otherwise may well be unaware. To be sure, those other than the manufacturer are still permitted to discuss the relative merits of off-label uses, and even manufacturers are authorized to speak in certain narrowly defined circumstances about an off-label use. But FDA could not have stated more clearly that manufacturers are prohibited from advocating off-label uses in any way. This is so even where FDA in no way challenges the accuracy, truthfulness, or completeness of that promotion. As a result, in many instances doctors are likely to be deprived of valuable information about important off-label uses that are totally lawful and extremely beneficial to some very sick people.

It should not be difficult to see that FDA's ban on off-label promotion violates the First Amendment right of free expression. The ban, after all, prohibits truthful speech advocating lawful activity, the paradigm situation for finding suppression—even of purely commercial speech—to be unconstitutional. This does not mean that commercial speech must receive absolute protection or that FDA should be wholly deprived of authority to regulate drug and device promotion in the interests of public health. It means, simply, that governmental imposition of a categorical ban of truthful promotion of lawful commercial activity contravenes both established commercial speech doctrine and core notions of free speech theory.

In this chapter, we seek to accomplish two complementary goals. Initially, we seek to establish the unambiguous unconstitutionality of FDA's current prohibition on off-label promotion as measured by controlling Supreme Court decisions concerning the First Amendment's protection of commercial speech. Secondly, we seek to glean from this analysis far more significant insights about the core premises of the constitutional protection of free expression. We point to four basic postulates of American constitutional and political theory that, we assert, underlie the social contract between citizen and government implicit in a societal commitment to liberal democracy. These core premises, we believe, represent the normative foundations of the constitutional protection of not only *commercial* speech but indeed *all* speech. By demonstrating this important philosophical overlap, we demonstrate the inherent linkage between commercial speech protection and the broader premises of the constitutional guarantee of free expression.

II. FDA'S SUPPRESSION OF OFF-LABEL PROMOTION: DEVELOPMENT OF THE REGULATORY FRAMEWORK

A. Off-Label Use and Accepted Medical Practice

While the phrase "off-label use" may seem to imply dubious medical practice, the reality is quite the opposite. Two important points must be noted as we begin our analysis of the constitutionality of FDA's prohibition on a manufacturer's ability to promote off-label use: (1) the prescription and use of approved drugs and medical devices for purposes not set forth in their FDA-authorized labeling is legal (48 Fed. Reg. 26,733 (June 9, 1983); see also 21 CFR § 312.3(d)), and (2) not only are such practices legal, they are quite common and, indeed, often accepted medical practice (Chen 2009; see also DeMonaco 2006). The conclusion is therefore inescapable that both patients and prescribers would often be aided by the dissemination of information to the medical profession about these valuable off-label uses.

Despite the indisputable value of off-label uses, FDA has categorically banned manufacturers of drugs and devices from promoting such uses. This is so despite the obvious reality that manufacturers are in a unique position to provide valuable information about off-label uses to the medical profession. Manufacturers possess a combination of incentive and resources to disseminate information to practitioners about off-label uses. To be sure, instances may arise in which particular claims about off-label uses could be deemed false or misleading. The same could be said about the advertising of *any* product or service. But FDA's ban draws no such distinctions. Even totally truthful, valuable information, if provided in the form of "promotion," is banned. It is difficult to comprehend the logic of such a distorted system.

B. FDA's Rationale for the Suppression of Off-Label Promotion

Although the antecedents of FDA's current stance on off-label promotion arose in the context of off-label use of methotrexate more than four decades ago, FDA has generally not invoked the methotrexate example to justify its restrictive approach to manufacturer dissemination of off-label use information. Naturally, FDA has relied on selected statutory provisions to assert that it is compelled to commence enforcement action against any manufacturer engaged in off-label promotion, but its efforts at justification have not ended there.

FDA has bolstered its legal arguments against off-label promotion by reciting a "parade of horribles" that agency officials assert have resulted or will likely result as a consequence of such promotion. FDA's regulation of speech is not the ultimate objective of the regulatory scheme, although that is surely its immediate target. Instead, the purpose of the off-label promotion ban is to regulate indirectly, through enforcement aimed at manufacturers, what FDA has failed or refused to do directly—namely, the regulation of the physician's decision to prescribe or use a drug or medical device for a particular off-label use. As careful examination of the First Amendment protection of commercial speech shows, FDA's practice is clearly unconstitutional.

III. COMMERCIAL SPEECH DOCTRINE AND THE FDA'S PROHIBITION OF OFF-LABEL PROMOTION

A. The Existing Doctrinal Framework for Commercial Speech Protection

1. The Evolution of the Central Hudson *Test*

Long excluded completely from the First Amendment's scope, commercial speech first received substantial constitutional protection in the Supreme Court's 1976 decision in *Virginia State Board of Pharmacy v. Virginia Citizens Consumer Council, Inc.* (1976), where the Court invalidated the state's prohibition on advertising of prescription drug prices. However, for many years thereafter, the Court afforded commercial speech only "a limited measure of protection, commensurate with its subordinate position in the scale of First Amendment values" (*Ohralik v. Ohio State Bar Ass'n* 1978).

In its 1980 decision in *Central Hudson Gas & Electric Corp. v. Public Service Commission* (1980), the Court adopted a four-part test to determine whether commercial speech is protected. This test continues to control, though a number of Justices have argued for a more protective test, and in its modern application the test has taken a far more protective form than it did originally. Under that test, the first inquiry asks whether the speech in question promotes the sale of an unlawful product or service or is found to be false or misleading. If the answer to either question is yes, the speech is automatically unprotected. Assuming the speech in question has passed this first hurdle, the remaining three questions scrutinize the nature of and justification for the speech regulation. For the regulation to be upheld, it must satisfy all three of the test's remaining prongs.

Under the test's second prong, the government must demonstrate that its regulation of commercial speech serves a "substantial" governmental interest (ibid., 566). Once that test has been satisfied, the reviewing court "must determine whether the regulation directly advances the governmental interest asserted. . . ." (ibid.). The court will invalidate the regulation if it "only indirectly advance[s] the state interest involved" (ibid., 564). Moreover, the regulation must *materially* advance the state's interest. Government has the burden of establishing beyond mere speculation that its regulation does in fact do so (*Edenfield v. Fane* 1993). Even if this requirement is satisfied, the regulation must still be found to be "[no] more extensive than is necessary to serve [the substantial governmental] interest" (*Central Hudson Gas & Electric Corp. v. Public Service Commission* 1980).

Although in the early years of the test's application one might have been able to characterize the Court's protection of commercial speech as far below the level afforded more traditionally protected varieties, in recent years the Court has afforded commercial speech a level of protection that in many ways rivals full protection. It has consistently invalidated regulations of commercial speech for their failure to satisfy the third prong, the fourth prong, or a synthesis of the two (see, e.g., *Lorillard Tobacco Co. v. Reilly* 2001; *Greater New Orleans Broad. Ass'n v. United States* 1999; *44 Liquormart, Inc. v. Rhode Island* 1996; *Rubin v. Coors Brewing Co.* 1995; *City of Cincinnati v. Discovery Network, Inc.* 1993). Indeed, on occasion it has invalidated expressive restrictions even in the case of FDA's regulation of drug advertising (see, e.g., *Thompson v. W. States Med. Ctr.* 2002). The Court has made clear that the government may not justify its regulations of truthful commercial speech simply by invoking sweeping and unsupported assertions of justification.

2. The Move Toward a More Categorical Standard

On a number of occasions, members of the Court have advocated or adopted an approach to commercial speech protection that, in certain instances, turns on more categorically established factors than does the interest balancing approach of even the more protective version of the *Central Hudson* test. In a number of opinions, Justices have argued that when the speech sought to be regulated or suppressed is truthful and advocates lawful purchase or use, governmental restriction of that speech is properly deemed categorically unconstitutional.

The first decision to openly advocate such a position was the plurality opinion of Justice Stevens in *44 Liquormart v. Rhode Island* (1996). Justice Stevens reasoned that bans of truthful advertising for lawful products or services when those bans are designed to protect consumers from commercial harms "rarely protect consumers from such harms. Instead, such bans often serve only to obscure an 'underlying governmental policy' that could be implemented without regulating speech" (ibid., 502–3 (citation omitted)). Justice Thomas continues to adhere to a view similar to Justice Stevens's (see ibid., 518–28 (Thomas, J., concurring); *Thompson v. W. States Med. Ctr.* 2002 (Thomas, J., concurring); Such bans, Justice Stevens argued, "usually rest solely on the offensive assumption that the public will respond 'irrationally' to the truth. The First Amendment directs us to be especially skeptical of regulations that seek to keep people in the dark for what the government perceives to be their own good" (*44 Liquormart v. Rhode Island* 1996).

Although a majority of the Court has never formally adopted this categorical approach in place of the *Central Hudson* test, on more than one occasion a majority opinion has expressed a similar sentiment (*Edenfield v. Fane* 1993). Over the years, extension of categorical protection to truthful advertising for a lawful product has gained a substantial following. Indeed, at no point in recent years has a majority of the Court issued a holding on commercial speech protection inconsistent with the categorical approach.

B. Measuring FDA'S Prohibition of Off-Label Promotion Under Supreme Court Commercial Speech Doctrine

1. *The Justifications for the Off-Label Prohibition: An Overview*

There are three major constitutional justifications for FDA's categorical prohibition on off-label advertising. First, it is argued that the regulated activity is not "expression" at all but instead amounts to nonexpressive conduct. Therefore, the First Amendment is wholly inapplicable. Second, even if it were assumed that expression is being regulated, the government has a substantial interest in protecting the public against false or unsupported claims about off-label uses that could lead to unsafe or economically unjustified use, and absent formal FDA approval of the drug or device for that specific use, there can be no assurance of the accuracy of the manufacturer's off-label claims. Third, it is argued that the government has a substantial interest in inducing manufacturers to seek FDA approval for off-label uses, and prohibition on manufacturer promotion of off-label uses does just that.

None of the arguments justifies FDA's prohibition of off-label promotion under existing commercial speech doctrine. Equally important, these arguments reveal a fundamentally flawed understanding of core premises underlying the constitutional protection of free expression in general and the constitutional protection of commercial speech in particular. Their acceptance would therefore result in adoption of a foundationally pathological version of free speech theory.

2. The Off-Label Prohibition as Regulation of Nonexpressive Conduct

FDA has argued that the prohibition of off-label promotion is not speech regulation at all but rather the regulation of nonexpressive conduct. Rather than myopically focus on the promotion itself, the argument proceeds, it is necessary to focus on the broader commercial transaction (Brief for the Appellants at 28, *Wash. Legal Found. v. Henney* 2000). Though it is perfectly legal for a prescription drug or medical device to be prescribed or used off label, it is unlawful for a manufacturer of such a product to sell its product with the intent that it be used off label. Promoting an off-label use, then, is deemed conclusive proof that the manufacturer is violating the law. Under FDA's argument, the manufacturer's promotion of off-label use is deemed to be nothing more than evidence of the broader illegal behavior of intending off-label use. Any negative impact on expression is therefore merely incidental to the regulation of nonexpressive behavior—a form of regulation subjected to a far less stringent form of constitutional analysis (see *United States v. O'Brien* 1968). If this argument were to be accepted, then all of the questions surrounding the protection of commercial speech would of course be rendered irrelevant. However, the argument must be rejected; it is manipulative and disingenuous on its face and inconsistent with controlling Supreme Court precedent.

Initially, it is worth noting the bizarre incongruity of punishing the sale for the purpose of off-label use while simultaneously treating the off-label use itself as completely lawful. But more important is the fact that contrary to its assertion, FDA does *not* employ the fact of off-label promotion merely as *evidence* of the illegal act. If FDA were truly concerned with the manufacturer's nonexpressive act of sale with intent that the product be used off label, it would logically prohibit *all* sales of a drug widely used off label because *any* time the manufacturer sells its drug, it would do so with knowledge that it will be used for off-label purposes. The fact that a manufacturer fails to promote off-label use surely does not imply that it is unaware that its product will be used off label. Indeed, reimbursement for specified off-label uses of prescription drugs is well established (Kesselheim 2011).

It could perhaps be suggested that a distinction must be recognized between *awareness* and intent; the mere fact that a manufacturer is aware that its product is being used in a particular way does not necessarily mean that it intended that it be used in such a manner. But in this instance, the suggestion of such a dichotomy would be naïve. Countless amounts of money are made from off-label sales that were openly funded by insurance reimbursement. Yet the government makes no effort to reduce or restrict sales. Indeed, there is no indication that FDA has *ever* pursued a manufacturer for selling its drug with knowledge that it will be used for off-label purposes, absent off-label promotion, and the government in litigation has denied that mere knowledge is actionable. Where a manufacturer does not seek to promote off-label use, FDA makes no objection, *though there can be no doubt that the manufacturer is aware when it sells its product that it will be used off label*. A party is presumed to intend the natural consequences of its acts.

In any event, FDA's own actions have made clear that its goal is suppressing speech, not in regulating conduct. In *United States v. Caronia* (2012), the Second Circuit rejected FDA's argument that its goal was merely to use off-label promotion as evidence of intent to misbrand because it "is simply not true" (ibid., 161). The court pointed to numerous statements at trial by FDA that the harm was the promotion itself, leading it to conclude that "the government clearly prosecuted Caronia for his words—for his speech" (ibid.). Why, one might reasonably ask, if FDA is so clear that it is the misbranding, not the promotion, that it is seeking to punish, in *Caronia* would it have completely abandoned that argument and instead directly challenged the speech for its own sake?

More importantly, in its proposed "Guidance for Industry: Distributing Scientific and Medical Publications on Unapproved New Uses—Recommended Practices," FDA proposed to accept manufacturers' distribution of articles, texts, and similar materials describing off-label uses, including through company sales representatives, as long as certain restrictions were followed. But if it is the intent to misbrand that is key to FDA, how could it logically accept *any* off-label promotion, regardless of whether or not it complies with predetermined restrictions? Even if those restrictions are satisfied, the fact remains that the promotion is for an unapproved purpose—by FDA's own reasoning conclusively demonstrating the manufacturer's intent to distribute in an off-label manner. The proposed guidance document, then, should once and for all end FDA's disingenuous charade that what it is attacking is conduct rather than pure expression.

When the dust settles, then, it is clear that FDA's categorical prohibition of promotional speech concerning off-label use amounts to a classic suppression of commercial speech. What makes this prohibition so problematic, it should be recalled, is that the actual behavior being advocated is perfectly lawful; neither FDA nor Congress has prohibited off-label uses themselves. Absent a showing that the suppressed promotion is false or misleading, it is by no means clear that under current doctrine suppression is *ever* constitutional. At the very least, suppression of truthful promotion of lawful use must satisfy increasingly demanding constitutional standards. It is highly doubtful that current FDA practice can meet them.

3. The Ban on Off-Label Promotion as a Form of Consumer Protection

One conceivable justification for the ban on off-label promotion is the fear that self-interested manufacturers will mislead consumers into using prescription drugs or devices for harmful or ineffective purposes. A drug's use for the off-label purpose, after all, presumably has never been established through proper use of the scientific method. But the very fact that many off-label uses are widely accepted undermines this notion. There is no inherent reason why claims made on behalf of off-label uses are *inherently* false or misleading. Moreover, in many instances in which such a danger arises due to the incompleteness of the claim, required disclaimers are surely a less-invasive means of regulating speech than is suppression.

C. Applying Commercial Speech Doctrine to the Prohibition of Off-Label Promotion

1. Applying the Categorical Standard

The argument that off-label promotion might convince practitioners to prescribe drugs and devices that would be ineffective or harmful and that such promotion can therefore constitutionally be suppressed cannot survive scrutiny under *44 Liquormart*'s categorical standard. There is no doubt that use of prescription drugs and devices off label is perfectly legal, and there exists no basis, ex ante, to categorically assume falsity of the claims made in the promotion of off-label use. In effect, then, the government is saying that the medical profession cannot be trusted to make accurate judgments as to the advisability of off-label use on the basis of free and open truthful communication. Indeed, the affront to the First Amendment is even more invidious in this context than it is under the categorical standard of *44 Liquormart*. In that case, the government's refusal to trust the judgment of the common person was declared antithetical to the very foundations of the First Amendment. In the case of off-label promotion, the government refuses to trust even trained professionals.

2. Applying the Central Hudson *Test to the Prohibition of Off-Label Promotion*

a. First prong: Is the speech false or misleading? Even if we were to proceed on the assumption that the Court would apply the four-part *Central Hudson* test, rather than the categorical standard, there is little doubt that FDA's prohibition on off-label promotion would fail. It should be recalled that while in its original incarnation the test provided a level of constitutional protection far lower than that given more traditional forms of expression, recent case law should disabuse us of the notion that the *Central Hudson* test is undemanding.

For reasons already discussed, it would be incorrect to conclude that off-label promotion fails *Central Hudson*'s first prong on the grounds that it is inherently false or misleading. It is true that while prescription drugs and devices have necessarily passed rigorous FDA review of the scientific basis for the manufacturer's claim of efficacy, no such review has ever been conducted for claims made on behalf of off-label uses. But it surely does not follow that *all* claims made on behalf of off-label uses are *inherently* false or misleading. Surely it is possible to imagine promotion that accurately describes how a drug or device has ameliorated disease even when the product has never been formally approved for such use. Indeed, FDA has itself openly acknowledged the efficacy of certain off-label uses. At the very least, then, promotion claiming the same efficacy for a drug or device that FDA has already conceded is effective for the purpose in question would have to be deemed truthful. It is therefore absurd to postulate some sort of ex ante, categorical presumption of the falsity of all off-label claims. But once one concedes that *some* off-label promotion is truthful, the burden naturally falls on government—as it does in the case of all regulation of false advertising—to prove the falsity of a particular advertisement before it may constitutionally suppress it.

Of course, like any form of advertising protected by the commercial speech doctrine (or any fully protected advocacy, for that matter), off-label promotion is just that—promotion. For that reason, it is not likely to explore fully both sides of the issue. But the strategic selectivity of expression designed to promote the interests of the speaker in no other context automatically renders the expression false or misleading. To the extent that the failure of commercial speech to include negative information has the potential to mislead the listener, under established doctrine government may require the speaker to communicate that information (see *Zauderer v. Office of Disciplinary Counsel* 1985). In the case of off-label promotion, for example, at the very least it would make sense for government to require that the manufacturer indicate that FDA has never approved the particular use described in its promotion. But that does not mean that government may suppress the communication completely when the danger can be avoided by the provision of more information rather than less.

b. Second prong: What is the government's "substantial interest"? Assuming the regulated speech is found not to be false or misleading, a reviewing court will proceed to an examination of the three remaining elements in the *Central Hudson* test. The first of those remaining inquiries—whether the regulation furthers a substantial governmental interest—is usually not very difficult to satisfy. That is equally true in the context of the suppression of the promotion of off-label use, at least to the extent that interest is defined in the broadest manner. No one could deny that government has a substantial interest in preserving the populace's health and safety, and to the extent the complete suppression of the promotion of off-label use furthers that interest in a material way and goes no further than necessary, then the *Central Hudson* test would be satisfied. But neither of those final two requirements is met by FDA's prohibition.

c. Third prong: Does the regulation materially advance the government's substantial interest? If one assumes that FDA's "substantial interest" under the second prong is to promote the safety and welfare of the populace by reducing the dangers of off-label use, it is purely speculative whether a ban on off-label promotion will achieve that goal. The federal government has conceded that not only are many off-label uses not only not harmful, they may in fact be quite beneficial. It is presumably for this reason that the federal government provides for reimbursement of off-label uses in at least some cases. The question then becomes, should we want practitioners to be made fully aware of valuable off-label uses? The answer most assuredly must be yes; it would be unacceptable for ill patients to be deprived of valuable and lawful treatments for the simple reason that their doctors were unaware of their existence. Yet FDA's categorical ban on manufacturer promotion will likely have this impact. Indeed, regional variations in practice patterns and between community-based practitioners and those based in academic settings, and corresponding variations in patient outcomes, strongly suggest that this is the case. It is therefore impossible to assert that public health and safety will be advanced by the categorical prohibition of off-label promotion.

d. Fourth prong: Does the regulation go further than necessary? The fourth prong of *Central Hudson* requires that the regulation of commercial speech go no further than necessary to serve the government's substantial interest. While the Court noted early on that government does not necessarily have to employ the very least restrictive alternative in order to satisfy *Central Hudson*'s fourth prong (see *Board of Trs. v. Fox* 1989), in recent years it has significantly increased the burden that this requirement imposes on government, on more than one occasion invalidating the suppression of commercial speech because of its failure to satisfy its demands. The Court has invalidated commercial speech regulations under this prong either because alternative nonspeech means of achieving the government's goal were available or because the regulation swept too far, impinging upon protected speech that failed to give rise to the harm sought to be prevented (see, e.g., *44 Liquormart, Inc. v. Rhode Island* 1996). The off-label promotion prohibition violates both aspects of the fourth prong.

Initially, there can be no doubt that to the extent the government's goal is to prevent false or misleading promotion, the categorical prohibition contravenes the fourth prong. There is no way FDA can establish, ex ante, that promotion is either false or misleading merely because it promotes an off-label use. FDA's categorical ban subsumes truthful promotion of these uses, as well as any conceivably improper uses. Secondly, we have already demonstrated that viable means of controlling nonexpressive behavior are available to the government to prevent dangers to the public health arising from unjustified or dangerous off-label uses—simply prohibit some or all off-label uses. There can be absolutely no doubt of Congress's constitutional power to enact such a ban because it would be controlling the use of products that have traveled in interstate commerce. Indeed, the fact that the federal government refuses to employ its authority to prohibit off-label use renders rather hollow its argument that promotion of such uses threatens the public health.

Our argument underscores the inconsistency of the off-label promotion ban with *both* versions of *Central Hudson*'s fourth prong. On the one hand, the goals of the ban could be achieved by nonexpressive regulations. On the other hand, the ban sweeps much too far, including both expression that does give rise to the asserted danger and expression that does not do so.

IV. OFF-LABEL PROMOTION, COMMERCIAL SPEECH, AND THE FOUNDATIONS OF FREE EXPRESSION

A. Recognizing the Overlap Between Commercial Speech and the Traditional Categories of Protected Expression

By exploring the serious constitutional pathologies underlying the off-label promotion ban, we are able to grasp how suppression of commercial speech simultaneously undermines foundational precepts of American political and constitutional theory that underlie the First Amendment protection of *all* speech. In so doing, we are able to underscore the important theoretical overlap between the protection of commercial speech on the one hand and the protection of other more traditional categories of expression on the other.

If we are correct in this suggested overlap, we will be able to draw important implications for the First Amendment's currently accepted theoretical framework. A number of respected scholars have either categorically rejected extension of First Amendment protection to commercial speech (Baker 1976) or proposed significant restrictions on its protection (Post 2000), reasoning that the profit motivation of the speaker distinguishes it from more traditionally protected categories of expression. While in recent years the Supreme Court has extended a significant level of First Amendment protection to commercial speech, it has long pointed to what it considers the "common sense differences" between that type of speech and the more traditionally protected forms of expression (see *Virginia State Board of Pharmacy v. Virginia Citizens Consumer Council, Inc.* 1976). These conclusions have already been subjected to serious challenge. The constitutional flaws that inhere in FDA's suppression of off-label promotion, we believe, underscore the fallacies in the argument that commercial speech is less deserving of First Amendment protection.

B. The Underlying Postulates of the Theory of Free Expression

There are many different, and often conflicting, theories of free expression. There are, however, certain core postulates of political theory without which any system of free expression would be incoherent. At some level, dispute may well exist as to the exact number and content of these postulates. But there are at least four such postulates that are clearly contravened by the governmental suppression of off-label promotion. This is so even though most, or all, of the expression suppressed by FDA's prohibition is described as commercial speech.

The four core postulates of free speech theory that are indisputably contravened by the ban on off-label promotion are the following:

1. Government may not attempt to manipulate lawful citizen behavior by means of the selective suppression of truthful expression advocating lawful activity.
2. The self-motivated nature of expression does not automatically render it false or misleading, thereby removing it from the scope of constitutional protection.
3. Government has greater power to regulate conduct than it has to regulate expression.
4. Government may not hold fully protected expression hostage as a means of extortion.

When viewed in the context of noncommercial speech, presumably no educated observer could dispute the accuracy of these postulates or their centrality to any coherent system of free expression. Violation of any of these four postulates would correctly be deemed to undermine core notions of free speech theory. The social contract implicit in the relationship of citizen to government in a liberal democratic society would be undermined if government could disrespect the intellectual dignity of its citizens. Yet, as the prohibition on off-label promotion illustrates, the exact same pathologies may occur when government violates these core postulates in the context of commercial speech, just as much as when other types of expression are suppressed; the same lines between government and citizen have been improperly crossed in both contexts.

1. Postulate One: Government May Not Attempt to Manipulate Lawful Citizen Behavior by Means of the Selective Suppression of Truthful Expression Advocating Lawful Activity

In a democratic society, basic choices of policy are, ultimately, made by the populace—if only indirectly, through their elected representatives.

To be sure, in certain instances the Constitution imposes limitations on majority choices, but even in those cases a supermajority of the populace may alter constitutional commands where it deems them to be no longer acceptable. Government may, of course, prohibit certain behavior, but the inherent logic of democracy prevents it from prohibiting the populace from debating the merits of those prohibitions. Where government has left behavioral choices to the individual, the premises of liberal democracy prevent the government from attempting to influence those choices through selective suppression of one side of a debate.

The point can be made clearer by viewing it through the lens of traditional political debate. If the First Amendment means anything, it prohibits government from suppressing one side of a political debate because it fears that the public might be convinced to make "the wrong" choice. For example, had the government sought to suppress opposition to the Iraq war because it feared that the public might be convinced to end that war, the suppression would undoubtedly have been found to infringe the First Amendment. In a democratic society, government may seek to influence the choices of the populace not by means of selective suppression but rather by making its own contributions to that debate. Indeed, it is impossible to point to a single respected First Amendment scholar who would deem such a selective ban to be constitutional. The reason is that vesting such power in government would necessarily prove too much, for it would undermine the fundamental premise underlying the commitment to self-government in the first place—namely, the citizen's ability to make lawful choices on the basis of free and open debate.

Given that no one could seriously challenge the constitutional pathology of paternalistically driven selective suppression in political debate, it is puzzling that so many have no difficulty authorizing such paternalism when the subject of the speech is commercial rather than political. There is no way one may legitimately compartmentalize respect for citizens' ability to make lawful choices on the basis of free and open debate. Either a democratic society trusts citizens to make such choices or it does not—in which event it automatically transforms from a democratic society to an authoritarian one. The fact that a dictatorship is benevolent does not make it a democracy. Yet, if citizens are deemed incapable of making lawful choices on the basis of free debate in the commercial realm, why all of a sudden do we deem them capable of making such choices in the political realm? When viewed through the lens of the relationship between government and citizen, the fact that in the commercial realm the speaker is seeking to make a profit is wholly irrelevant; the pathology in terms of democratic theory is not so much the suppression of the speaker's right as it is the lack of respect for the citizen's ability to make lawful choices—a lack of respect that inheres in the government's selective suppression, regardless of the speaker's purposes.

Recognition of this fundamental element of the liberal democratic social contract between government and individual, then, leads to a rejection of the core distinction drawn between commercial and political speech. FDA's prohibition of off-label promotion, despite its concession that many off-label uses are extremely valuable and the fact that the promotion is not inherently false or misleading, serves effectively to underscore the point.

2. Postulate Two: The Self-Interested Nature of Expression Does Not Automatically Reduce the Level of Constitutional Protection

FDA, it should be recalled, prohibits *manufacturer promotion* of off-label use; it does not, however, prohibit others from discussing such use. FDA, then, has necessarily made the decision that the harm it seeks to prevent derives exclusively from manufacturer promotion. The assumption implicit in that dichotomy is that the self-interested nature of the promotion somehow renders the expression both less worthy and more dangerous. As previously noted, the logic of such a dichotomy, to the extent it were to justify categorical suppression, is inherently inconsistent with the Supreme Court's commitment to the protection of commercial speech in the first place; under the Court's definition, *all* commercial speech is manufacturer speech; yet under the Court's doctrine it is extended a significant level of constitutional protection. Thus, the categorical suppression of speech for no reason other than the fact that it is made by an economically self-interested party is inconsistent with accepted Supreme Court jurisprudence.

Beyond its doctrinal difficulties, it is interesting to explore the consistency of FDA's expressive discrimination against manufacturers from the broader perspectives of constitutional and political theory. In a democracy, it is generally understood that individuals will often act out of their own self-interest and will employ expression as a means of convincing others to take actions to advance that self-interest. The long history and current power of political interest groups are conclusive proof of that political reality. In the world of noncommerical speech, no one would seriously suggest that the self-interested nature of expression somehow reduces the level of First Amendment protection it receives. When the unemployed urge others to support the extension of unemployment benefits, no one would ever contemplate the possibility that their expression receive reduced protection merely because they possess a self-interested economic motivation for their expression.

One could easily come up with hundreds of similar examples. Yet in the case of off-label promotion, FDA appears to have made the wholly unsupported, ex ante assumption that manufacturer speech will be inherently distorted, and therefore suppressed, for no reason other than the self-interested economic nature of the speaker's expression. Once again, when viewed through the lens of accepted political practice and modern democratic theory, the categorical discrimination against sellers for no reason other than the fact of the self-interested nature of their expression is rendered wholly incoherent. Few examples underscore this theoretically flawed dichotomy better than does FDA's ban on off-label *promotion* but not on off-label *discussion*.

3. Postulate Three: The Government Has Greater Power to Regulate Conduct Than It Has to Regulate Speech

In the traditional world of free expression, no one would ever question that expression has considerably greater constitutional protection than most forms of conduct. For example, under certain circumstances, at least, one has the right to advocate unlawful conduct (see, e.g., *Brandenburg v. Ohio* 1969); yet one surely has no constitutional right to engage in that unlawful conduct. While some scholars have challenged the conclusion that speech is truly "special" (Schauer 1984), it is not all that difficult to grasp the key differences. Expression deals more directly with the unique ability to think and reason on multidimensional levels—qualities that are essential in a democratic society (see Redish 1982). Moreover, while of course expression can cause harm, as an ex ante matter it is reasonable to presume that the harm caused by expression will generally be less immediate and concrete than harm caused by physical conduct. For purposes of a constitutional protection in which broad judgments must be made, it is therefore reasonable to draw a categorical distinction between expression and conduct. By providing special protection to expression, this is exactly what the First Amendment does.

In the area of commercial speech, the distinction has not always been so clearly understood. For a time, the Court proceeded on the wholly misguided assumption that government's greater power to regulate conduct subsumed the "lesser" power to regulate expression advocating that conduct, even where the conduct is not prohibited (see *Posadas de P.R. Assocs. v. Tourism Co. of P.R.* 1986). This "logic," of course, completely ignores the fact that because the First Amendment extends greater protection to expression than to conduct, it is, in reality, the regulation of the expression that is the "greater" power. Indeed, acceptance of this logic outside of the commercial speech context (and there is no logical basis for confining it in such a manner) would overturn over eighty years of Supreme Court jurisprudence in the constitutional protection of unlawful advocacy (see Redish 1984). In his opinion announcing the judgment of the Court in the 1996 decision in *44 Liquormart v. Rhode Island* (1996), Justice Stevens categorically and vigorously rejected the specious logic of prior Supreme Court jurisprudence (ibid., 509). The Court's vigorous protection of commercial speech since that decision confirms that the Court as a whole has concurred in Justice Stevens's rejection of the "greater includes the lesser" logic.

FDA's ban on off-label promotion runs directly counter to this core First Amendment premise and in doing so highlights its importance. Whether the federal government in general or FDA in particular wishes to control or even prohibit off-label uses is an issue on which the First Amendment is wholly agnostic. This is so because such actions would regulate conduct, not expression. Yet while categorically prohibiting *promotion* of off-label use, the government has for the most part not restricted off-label use. Indeed, as we pointed out earlier, FDA has gone so far as to permit the use of drugs that have not been approved for any purpose at all. Instead of allowing the advocacy and restricting the conduct, FDA has prohibited the advocacy but permitted the conduct, leaving doctors to fend for themselves in finding out about potentially life-saving off-label uses. In following this policy, FDA has turned the First Amendment on its head, creating an Alice-in-Wonderland world in which speech receives less protection than the conduct it advocates. Such an inversion of constitutional values would surely not be permitted when traditionally protected noncommercial speech is involved; there is no basis in logic or experience to justify it in the context of commercial speech.

4. Postulate Four: Government May Not Hold Fully Protected Expression Hostage as a Means of Extortion

Recall that one of the conceivable justifications turns not at all on the constitutional unworthiness of the off-label promotion itself. It turns, rather, on the asserted need for FDA to be able to coerce drug manufacturers into making off-label uses on label. Doing so requires the manufacturers to conduct expensive and detailed scientific studies supporting the efficacy and safety of the off-label use, something they understandably may often not be inclined to do. By prohibiting their ability to promote off-label use, FDA is able to pressure them into undertaking the necessary steps to make the uses on label. In effect, the government is asserting that it is constitutionally authorized to suppress speech that it readily concedes, if only for purposes of argument, is on its four corners fully deserving of protection, in order to extort certain actions out of private entities that have absolutely no legal obligation to undertake such action.

It is hard to imagine government even attempting such outrageous actions when the speech in question is noncommercial, but if it did so there can be little doubt that its actions would be held unconstitutional. By much the same reasoning, government could presumably prohibit individuals from attending religious services unless individuals engaged in specified behavior, despite the existence of the First Amendment's Free Exercise Clause. One shudders to think what might result if government sought to achieve the same result by suspending citizens' Eighth Amendment right against cruel and unusual punishment.

Where citizen behavior on its face falls within the scope of a constitutional protection, it would be absurd to permit government to suppress that behavior in order to induce citizens to engage in behavior in which they have no legal obligation to engage. Yet that is exactly what

FDA is doing when it seeks to justify the suppression of off-label promotion on the grounds that it is seeking to induce manufacturers to make the uses on label. The surreal aspect of such extortive behavior is underscored by the fact that the federal government has full constitutional authority to directly require the manufacturers to transform the off-label uses into on-label form, *simply by directly prohibiting the prescription of approved drugs for off-label use* (U.S. Const. art. i, § 8, cl. 3 (Commerce Clause); U.S. Const. art. i, § 8, cl. 18 (Necessary and Proper Clause). While the court in *United States v. Caronia* (2008) reasoned that the federal government lacked such constitutional authority, such a conclusion preposterously ignores the last sixty years of Commerce Clause jurisprudence (see, e.g., *Gonzales v. Raich* (2005) (upholding federal power to criminalize sale of drugs traveling in interstate commerce); *Katzenbach v. McClung* (1964) (upholding Title II of 1964 Civil Rights Act, which prohibited discrimination in restaurants that sell food that traveled in interstate commerce)). Whether FDA itself, under the current regulatory framework, possesses such authority is wholly irrelevant to the constitutional analysis. If the federal government, as an entity, wishes to require that a prescribed drug be on label, there is no doubt of its constitutional authority to do so. The government therefore cannot be permitted to suppress expression as an indirect means of achieving its goal, when it has available a perfectly legitimate direct means of achieving the same end. Whether it would be simpler—or simply more expedient—for the government to achieve its goal indirectly by suppressing protected speech is irrelevant to the constitutional inquiry. The suppression of fully protected, potentially valuable expression is far too high a price to pay for governmental convenience.

V. CONCLUSION

In light of the Supreme Court's vigorous protection of commercial speech rights in recent years, FDA's continued suppression of off-label promotion is quite puzzling. There is no basis to support the assumption that all off-label uses are harmful. To the contrary, off-label uses are often extremely valuable, a fact that government officials have readily conceded. There is absolutely no basis to support the assumption that all manufacturer promotion of off-label uses is inherently false or misleading, and FDA makes no such assumption. Indeed, it is quite conceivable that manufacturer promotion could inform doctors of lifesaving or improving off-label uses of which they would not otherwise have been aware. Thus, we are dealing with a situation in which the suppressed speech truthfully advocates lawful action that might well maximize the welfare of all involved. Yet an agency of the federal government has categorically prohibited such expression.

Under current Supreme Court commercial speech doctrine, truthful commercial advertising for a lawful product or activity is extended substantial First Amendment protection. Government must show beyond the level of mere speculation that the suppression of such speech materially advances a substantial interest and goes no further than necessary to achieve that end. The Court has regularly invalidated regulations of commercial speech because of their failure to satisfy those requirements. Moreover, there is a strong indication in modern Supreme Court doctrine that *any* suppression of truthful advertising for a lawful activity designed to protect the listener from making a lawful choice is categorically unconstitutional.

When viewed on the more foundational levels of American political and constitutional theory, the ban on manufacturer promotion of off-label use becomes even more bizarre. The suppression of truthful commercial advertising promoting lawful activity does far more than contravene the First Amendment rights of the seller. Far more invidious to the foundations of the liberal democratic social contract is the government's lack of respect the ban demonstrates for the citizenry's ability to make well-informed, lawful choices on the basis of truthful advocacy.

For all of these reasons, the ban on the promotion of off-label use must—and, at some point in the near future, almost certainly will—be held to violate the First Amendment. It is unfortunate that those vested with governmental power have not themselves recognized this fact. But rest assured that the judiciary will soon educate them.

NOTE

This chapter is a modified version of the article "Off-Label Prescription Advertising, the FDA, and the First Amendment: A Study in the Values of Commercial Speech Protection," 37 *American Journal of Law & Medicine*

315 (2011), copyright ©2011 by American Society of Law, Medicine & Ethics, Boston University School of Law; reprinted with permission.

REFERENCES

44 Liquormart, Inc. v. Rhode Island, 517 U.S. 484, 506 (1996).
Baker, Edwin C. 1976. "Commercial Speech: A Problem in the Theory of Freedom." *Iowa Law Review* 62:1.
Board of Trs. v. Fox, 492 U.S. 469, 477 (1989).
Brandenburg v. Ohio, 395 U.S. 444 (1969).
Brief for the Appellants at 28, *Wash. Legal Found. v. Henney*, 202 F.3d 331 (D.C. Cir. 2000) (No. 99–5304).
Central Hudson Gas & Electric Corp. v. Public Service Commission, 447 U.S. 557 (1980).
Chen, Donna T. 2009. "U.S. Physician Knowledge of the FDA-Approved Indications and Evidence Base for Commonly Prescribed Drugs: Results of a National Survey." *Pharmacoepidemiology & Drug Safety* 18:1094.
City of Cincinnati v. Discovery Network, Inc., 507 U.S. 410 (1993).
DeMonaco, Harold J. et al. 2006. "The Major Role of Clinicians in the Discovery of Off-Label Drug Therapies." *Pharmacotherapy* 26(3):323.
Edenfield v. Fane, 507 U.S. 761, 770–71 (1993).
Gonzales v. Raich, 545 U.S. 1 (2005).
Greater New Orleans Broad. Ass'n v. United States, 527 U.S. 173, 176, 190 (1999).
Katzenbach v. McClung, 379 U.S. 294 (1964).
Kesselheim, Aaron S. et al. 2011. "Strategies and Practices in Off-Label Marketing of Pharmaceuticals: A Retrospective Analysis of Whistleblower Complaints." *PLoS Medicine* 8(4).
Lorillard Tobacco Co. v. Reilly, 533 U.S. 525, 565 (2001).
Ohralik v. Ohio State Bar Ass'n, 436 U.S. 447, 456 (1978).
Posadas de P.R. Assocs. v. Tourism Co. of P.R., 478 U.S. 328 (1986).
Post, Robert. 2000. "The Constitutional Status of Commercial Speech." *UCLA Law Review* 48:1.
Redish, Martin H. 1982. "The Value of Free Speech." *University of Pennsylvania Law Review* 130:591.
Redish, Martin H. 1984. "Freedom of Expression: A Critical Analysis." Lexis Law Pub (June) 173–212.
Rubin v. Coors Brewing Co., 514 U.S. 476 (1995).
Schauer, Frederick. 1984. "Must Speech Be Special?" *Northwestern University Law Review* 78:1284.
Thompson v. W. States Med. Ctr., 535 U.S. 357 (2002).

United States v. Caronia, 576 F. Supp. 2d 385, 401 (E.D.N.Y. 2008).
United States v. Caronia, 708 F.3d 149 (2d Cir. 2012).
United States v. O'Brien, 391 U.S. 367 (1968).
Virginia State Board of Pharmacy v. Virginia Citizens Consumer Council, Inc., 425 U.S. 748 (1976).
Zauderer v. Office of Disciplinary Counsel, 471 U.S. 626 (1985).

PART FOUR

Timing Is Everything
Balancing Access and Uncertainty

Introduction

W. NICHOLSON PRICE II

TRADITIONALLY, THE FDA process of regulating drugs—like that of other agencies around the world—has focused on evaluating the drug before permitting the drug to be marketed. This section reflects the growing recognition that the realities of modern drug development mean that a heavy focus on premarket approval is no longer sufficient. Premarket clinical trials are slow by the standards of bringing new drugs to market but relatively small and short term by the standards of fully evaluating long-term effects in large populations. Thus, they fail in opposite ways: the requirement of clinical trials creates delays and keeps patients from accessing the new drugs they see coming down the pipeline, but they also fail to prevent completely the occurrence of significant adverse effects that are either too rare or too slow to be observed in small and relatively short trials.

This section includes three chapters addressing the context, contours, and challenges of appropriately timing and managing the entry of drugs into the market. A central theme across all chapters is recognition that the current system is insufficient to address the problems mentioned above. If adverse events and effectiveness in diverse populations cannot be determined in premarket clinical trials, postmarket surveillance must supply this information. But current postmarket surveillance mechanisms, comprised of an alphabet soup of FDA programs, are inadequate: those applying to all drugs are weak, those applying to new drugs are only marginally less so, and all are poorly enforced. More and better postmarket surveillance is needed, especially in a lifestyle regime under which more drugs come to market faster.

Within this general theme of increased surveillance, one key distinction is how postmarket surveillance should impact the balance of regulation across a drug's lifespan. Gibson and Lemmens express concerns that postmarket surveillance not result in a loosening of the standards for premarket approval. Charo, on the other hand, notes that this may be an inevitable trade-off, given pressure for faster drug development and the limitations of the current clinical-trial regime.

Differences also exist regarding implementation, particularly whether FDA already has the necessary tools to bring about a more effective postmarket surveillance regime. Parasidis provocatively suggests that the existence of off-label use could be enough to give FDA a statutory hook to mandate active surveillance. Charo, Gibson, and Lemmens suggest that more-direct legislative action may be required, especially to implement the possibility of phased or conditional approval.

This section grapples further with the incentive questions raised by postmarket surveillance. Mechanisms like phased or conditional approval challenge the current financial landscape for drugs, where the majority of profits are earned in the earlier part of a drug's lifespan. Similarly, the entrenchment of a profitable status quo makes regulatory action threatening that status quo harder; withdrawing drugs once they have entered the market is much more difficult than preventing the same drug from entering the market in the first place.

Though it may not seem an obvious connection, all three chapters in this section note a connection between postmarket surveillance/regulation and the expanding limits placed on FDA's regulatory authority by First Amendment free-speech jurisprudence. As discussed elsewhere in this volume, the Second Circuit held in *United States v. Caronia* that the FDA's prosecution of a drug-company representative for promoting off-label uses of an approved drug violated the First Amendment. Such off-label prescription is itself legal, but FDA regulation of marketing regarding such off-label uses has been an important tool in limiting the impact of unapproved uses. This ruling—and reasoning like it—may make it difficult for FDA to integrate marketing restrictions into a program of postmarket surveillance.

Ironically, in a realm of greater postmarket surveillance, *Caronia* and its ilk may prove a mixed blessing for the drug industry. If FDA cannot impose marketing restrictions on the basis of a postsurveillance finding of ineffectiveness of adverse effects in a particular subpopulation or for a particular indication, it might need to resort to more drastic measures, such as withdrawing the drug from the market or requiring the institution of severe Risk Evaluation and Mitigation Strategies (REMS) to prevent certain uses entirely.

Several questions remain open about timing and postmarket surveillance, particularly regarding implementation. An advantage to the current systemic dichotomy between pre- and postmarket phases is its clarity; a drug is approved or not, and it has a specific set of label indications. The regulatory submission to seek drug approval is presented as a unified body of data to answer specific questions: safety and efficacy for specific populations and specific indications. FDA can then evaluate whether those questions have been adequately answered. As a larger volume of more equivocal and more granular information arises from postmarket surveillance, how should FDA address it? Should FDA generally seek to translate such information into direct action—label changes, new REMS, or drug withdrawals—or should it instead simply be passed on to consumers and doctors through new information portals?

If an information-sharing approach is preferred, how should that occur? Should raw reports from surveillance be made available, despite likely problems with data quality and the possibility of competitor gamesmanship? If the information is instead to be curated and collected before distribution, should FDA fill that role or should the drug sponsor? The sponsor has responsibility for the drug's safety and the greatest expertise—but that is true under the current regime, which has clearly inadequate postmarket surveillance. As Parasidis notes, limitations on state tort suits may limit the already-insufficient incentives for collecting and responding to postmarket surveillance information. Are there regulatory or tort incentives available to encourage more active management by sponsors, or will any such incentive invariably pale beyond the financial gains of keeping drugs broadly available on the market?

Translating postmarket surveillance information into regulatory action requires answering similar issues. Must sponsors suggest regulatory actions and changes, or should FDA be charged with initiating such actions? And should third parties have the right—or the responsibility—to intervene (perhaps through the mechanism of Citizen Petitions), which would depend on access to detailed surveillance information?

All of these questions, of course, are complicated by the presence of generic drugs. When issues of quality control may differ among manufacturers, how should surveillance information be segregated? Should generic companies have the same surveillance obligations as brand-name drug sponsors, or should their responsibilities be tempered by the different treatment of generic firms under federal preemption doctrine?

Even given broad acceptance that much greater postmarket surveillance is needed, these questions require answers. At the extremes, the proper response may be relatively clear: the Vioxx scandal clearly called for regulatory action, and individual reports of minor side effects would almost as clearly not require action. But there is a vast area in between, and how FDA reacts to those in-between cases will powerfully impact the landscape of drug regulation going forward.

Overall, postmarket surveillance promises to fill some of the problematic information gaps that result from the current evolving system of drug approval. But as the shortcomings of the optimistically implemented current surveillance mechanisms demonstrate, implementation of surveillance is difficult. Even assuming that information is accurately collected and analyzed, knowing what to do with that information presents its own set of significant challenges.

CHAPTER THIRTEEN

Speed Versus Safety in Drug Development

R. ALTA CHARO

AN IDEAL pharmaceutical development and regulatory system delivers new drugs or new indications for old drugs in a timely manner, with an assurance of both safety and effectiveness. It incentivizes innovation, particularly for those medical products that address conditions for which we need options that are more effective, less risky, and more affordable than existing offerings. And it is capable of self-correction when errors are made. Indeed, from a systems analysis viewpoint, it can sometimes be more important to build in the capacity to detect and correct inevitable errors than to build a system so chock full of protections that error minimization comes with overwhelming rigidity and stifling regulation.

It is this last observation that is perhaps at the center of the newer approaches now being discussed for drug development in the United States. A number of them focus on getting drugs out faster and correcting mistakes later. The Food and Drug Administration (FDA) already has a number of expedited programs. "Fast track" designation offers closer collaboration with the agency to ensure collection of appropriate data needed to support drug approval, including such things as the design of the proposed clinical trials and use of biomarkers. "Accelerated approval" allows marketing based on a surrogate marker that is believed to be indicative of a disease state and treatment effect but not demonstrative of a direct health gain to the patient. "Priority review" will have the agency acting on applications within six (rather than the standard ten) months. And the "breakthrough therapy" designation is a kind of "fast track" on steroids, with intensive agency guidance as early as Phase I in an effort to expeditiously move the drug through premarket testing. To make these as effective as possible, a final "Guidance for Industry Expedited Programs for Serious Conditions—Drugs and Biologics" was issued in May 2014 (FDA 2014).

For this to work, though, there are some parts of the system that are in need of reform but are not getting the attention they deserve. Specifically, marketing restrictions would help to identify mistakes early and correct them before too many people are harmed. And attention must be paid to the business model for drug development and to de-risking the process so that there will be less resistance to regulatory actions that limit or withdraw marketed drugs as needed.

I. SPEED AND SAFETY

Historically, drug regulation began with an emphasis entirely on error correction; there was either no premarket screening or, as came in 1938, there was premarket screening only for risks (FFDCA 1938). Postmarket realization that a drug was overly dangerous could be addressed by removing the drug from the market. But this error correction system was hampered by inadequate postmarket surveillance and many litigation barriers for regulators.

The turning point undoubtedly lies in the experience with thalidomide, which produced thousands of grossly disabled newborns in other countries. It was never approved in the United States, however, because one medical reviewer, Frances O. Kelsey, found the anecdotal evidence and short-term studies inadequate to confirm the safety of the drug, which was being prescribed to pregnant women to control extreme morning sickness (Bren 2001). As a result, FDA was lauded as a "gold standard" for drug regulatory agencies, and its cautious approach was applauded as the best balance between patient safety and patient needs.

The thalidomide experience led U.S. regulators to shift emphasis to the premarket period, with extensive requirements to demonstrate both safety and efficacy (Kefauver-Harris 1962). This resulted in vast improvements in the quality of the pharmaceuticals reaching market, but it made innovation slow and expensive. In response, incentives were created in the form of limitations on commercial speech that allowed the government to give innovators who submitted to regulation the opportunity to cash in on extended periods of exclusive marketing rights, separate from and in addition to patent protections. The restrictions and the recent attacks thereon are covered in depth in other chapters of this book.

Controlled clinical trials are at the core of evidence-based medicine. Without them, it can be impossible to distinguish between coincidence and connection, correlation and causation. There are notable examples of public health failures due to the decision to forego or truncate clinical trials. In one such case, a study of perioperative beta blockers was stopped early because while initial results in a small trial indicated reduced risk of heart attacks, a much larger study later showed a dramatic increase in the risk of strokes and death (Poldermans 1999). Another example would be the Women's Health Initiative (WHI) study of estrogen and progesterone in treating postmenopausal women. It was expected to confirm the widely held view that these hormone replacement therapies could reduce cardiovascular disease in older women, but in fact, large, long-term trials found quite the opposite (Majumdar 2004).

By the 1970s, pressure began to build for faster access to therapies for serious diseases, for example, from cancer patients who had few options in standard medicine and so were drawn to try ultimately ineffective alternatives such as laetrile, a substance found in apricot pits (Wilson 2013). This trend toward emphasizing speed rather than cautious trials for safety and effectiveness continued in the 1980s with the AIDS crisis and the emergence of an organized effort in the gay community to obtain access to investigational drugs in the face of an incurable disease without any approved therapies (Harden 2012). That this coincided with a political atmosphere emphasizing economic competitiveness and deregulation only made the call for loosening some of the "gold standard" regulatory restrictions that much more compelling.

Even into the 1990s, concerns about drug development seemed to focus more on lack of access to new (presumably better) drugs rather than fear that existing drugs on the market were proving to be less effective or more dangerous than anticipated. These concerns even extended to debates about expanding clinical trials to subject populations previously thought to be particularly vulnerable to injury or coercion (e.g., children, fertile and pregnant women, ethnic and racial minorities) on the theory that participation in research was no longer to be viewed as a form of exploitation but rather as an opportunity for early access to superior care (Charo 1993; Mastroianni 1998; Noah 2003; Coleman 2007).

And yet, as one moves into the current century, the pendulum begins to swing wildly between public outrage over newly discovered problems with common drugs such as Vioxx and Avandia and frustration with slow approvals to the point of constitutional challenges to the very basis of FDA regulation and its careful attention to phases of drug development (*Abigail Alliance* 2008). And outside of limited alternatives such as priority review vouchers (Robertson 2012), running alongside this debate is the repeated use of the same technique—extending market exclusivity to enhance potential profits and incentivize R&D—to promote development of products that meet special needs, for example, for orphan diseases or pediatric populations (FDA 2006).

Using exclusive market rights to enhance premarket control has led to extensive product development, but it has also created its own problems. Among them are lengthy development times; uncertainty about approvals; resistance to limiting or withdrawing drugs once marketed; and higher prices for consumers, due both to development costs and the lack of early competition from generics (Moors 2014).

The result, it seems, is that no one is happy with FDA. Industry spokespersons complain it is too picky, too temperamental, and generally too slow. The result, they say, is that drugs do not get developed or—if they do—are unnecessarily expensive (Kaitin 2010). Patients with diseases as yet imperfectly managed with existing pharmaceuticals, ranging from Alzheimer's to hepatitis C, also clamor for new options (Stix 2012; Herschler 2014). At the same time, the predictable, periodic failures of pharmaceuticals on the market—whether less effective than predicted or more dangerous than expected—lead to criticism from the media, watchdog groups, and Congress.

Against this backdrop is a growing appreciation for the inherent limitations of traditional premarket clinical trials, which for reasons of practicality, finance, and diminishing returns will often be neither long enough, nor large enough, nor demographically comprehensive enough to achieve the level of postmarket safety and effectiveness that the public expects (Kramer 2010; Singh 2012). This has only energized the calls for far less time in premarket testing, followed by enhanced postmarket surveillance to confirm effectiveness and detect problems (von Eschenbach 2012).

Shorter, smaller trials could facilitate faster patient access to innovative therapies and lower the financial hurdles that can deter development of new drugs. But, as many have noted, clinical trials are already inadequate with regard to duration and scale. Although powered to confirm efficacy in carefully selected populations—often characterized by an absence of comorbidities or complicating drug–drug interactions and narrow age ranges emphasizing young adulthood and early middle age—clinical trials are not necessarily good predictors of a new drug's effectiveness in the broader range of ages and medical conditions in clinical practice. As to safety, almost no trial can be powered to pick up the rare adverse event or a subtle increase in frequency of otherwise common events (Duijnhoven 2013). Because these shortcomings cannot be overcome without increasing the duration, scale, and cost of trials, some have said that we get too little at too high a price and have called for much of Phase III testing to be abandoned completely so we can at least answer the demand for getting drugs to market more efficiently (von Eschenbach 2012).

Using this approach, accelerated drug approval through use of breakthrough designations (Sherman 2013; Calabrese 2013), pharmacogenomics for targeted testing, biomarkers, surrogate endpoints, and abbreviated trials can be integrated with better postmarket surveillance and the use of Risk Evaluation and Mitigation Strategies (REMS), topics covered in great depth in other parts of this book. Indeed, FDA has already taken precisely these steps for a number of products. For example, the agency announced that it would look for changes in biomarkers rather than clinical endpoints for Alzheimer's disease, a slow-developing, degenerative condition (Kozauer 2013). But as noted above, some proposals argue that once proof of concept and the absence of major risks for a new drug have been confirmed, the remaining effort to prove efficacy and safety should move entirely to the postmarket phase (von Eschenbach 2012).

Without stronger mechanisms to take action based on postmarket data, however, it is unlikely that surveillance and FDA's REMS can satisfactorily offset any increased risks due to the abbreviated premarket testing. Experience with postmarket measures does not provide confidence in their effectiveness (Gibson 2015; Parasidis 2015). For years, postmarket (Phase IV) study commitments went unfulfilled, and even the new REMS have yet to demonstrate consistent use and effectiveness (HHS 2013). And while some studies suggest that the risks associated with the most innovative new drugs are not significantly greater than for all drugs generally (Mol 2013), others note that accelerated processes and deadlines lead to just-in-time decision making associated with higher subsequent rates of adverse events (Carpenter 2011).

While REMS are a beginning, offsetting increased risks due to abbreviated clinical trials probably also requires stronger marketing limitations. Particularly for drugs whose new mechanisms of action or new molecular entities pose the greatest uncertainties, it may be important to control their speed of diffusion, for example, by limiting their use to narrow populations, restricting off-label prescribing, and curbing the direct-to-consumer advertising known to drive patient demand and physician acquiescence (IOM 2006). The challenge here is to increase control over diffusion of a drug without running up against constitutional protection of freedom to engage in commercial speech and without straying beyond regulation of products and beginning to invade physician judgment in the practice of medicine. A system of "conditional approvals" akin to that now being used in Europe might help (EC 2006).

In addition, the postmarket measures that limit drug sales or even cause drugs to more frequently be removed from the market may make return on pharmaceutical investments less predictable and less profitable. The existing approach, which relies exclusively on extended periods of postmarket exclusivity to provide financial incentives for innovation, may need to be supplemented by more premarket incentives, such as the research grants, tax credits, awards, and prizes that have been proposed for antimicrobial development, another area in which postmarket returns are not sufficient to promote investment by companies (So 2011).

II. MARKETING LIMITATIONS

As the White House Office of Science and Technology stated, effective postmarket measures would be helped enormously by limiting marketing and prescribing of drugs in the first years after approval, at least for drugs that are the first in their class; the properties for such drugs are most likely the least tested and predictable (PCAST 2012). Such limitations would slow the diffusion of new drugs into clinical practice and provide time for postmarket studies to reveal more robust information about effectiveness and safety.

But limitations on the marketing and prescribing of recently approved drugs pose legal and strategic challenges. The explosive growth in direct-to-consumer advertising has fueled patient demand for new drugs, and FDA has little regulatory authority over physician prescribing. For companies, restrictions on commercial speech have been slowly eliminated through a series of federal court decisions, the most recent of which comes perilously close to calling the current restrictions on off-label marketing unconstitutional (*United States v. Caronia* 2012). The implications of this case are discussed in several chapters of this book.

FDA is not primarily a research agency. It offers the incentive of marketing exclusivity to companies that invest in proving the safety and efficacy of new drugs or new indications for existing drugs. It does this by deeming the promotion of an unapproved drug or indication a form of prohibited misbranding, on the theory that the drug cannot be considered safe and effective for its intended use, even if evidence not submitted to FDA would suggest otherwise. This has given FDA what might loosely be considered a monopoly on truth. Because FDA will not approve an identical drug or indication for a period of years, this gives companies that invest in proving safety and efficacy to FDA a period of market exclusivity that provides high-profit margins. These margins decrease sharply as soon as generics can be sold. As a result, any innovation that threatens to shorten the years of exclusivity or limit the scope of sales within those years makes uncertain a company's calculation of the return on the substantial investment required for clinical trials. Strategically, therefore, one can predict strenuous resistance to any effort to withdraw drugs, or to add new limits to the kinds of patients or indications for which they can be used, such as was the case for midodrine, an antihypotensive agent that FDA unsuccessfully attempted to withdraw for failure to meet postmarket study commitments (Mitka 2012).

The European Medicines Agency is in the early stages of implementing a system of conditional drug approval that might remedy two problems: getting drugs out more quickly but with greater regulatory control in the postmarket stage (Duijnhoven 2013). In 1993, the European Union introduced a system to offer approval and early market access for drugs that met the definition of exceptional circumstances (ECs). This included situations in which comprehensive safety and efficacy data was unattainable, for example, for an orphan disease. The sponsors would gain early approval but then commit to a variety of "specific obligations" for postmarket surveillance and management after obtaining marketing approval. Over time, the EC pathway came to be used not only for rare diseases but also for common diseases that still had unmet needs. In the latter situation, however, it was at least in theory possible to obtain comprehensive data.

As a result, the European Commission created two pathways. The first, like the original EC pathway, works for drugs for which it is not possible to provide the European Medicines Agency with additional data. The second, called "conditional approvals" (CAs), allows drugs to enter the market with less than the usual level of safety and efficacy data, if they have a good risk–benefit ratio demonstrated in the initial trials and it is expected that more data will be obtainable once marketed (Tsang 2005; Arnardottir 2011).

In some ways, this resembles the proposals made in the United States to dispense with Phase III trials for drugs that have good Phase II results. The "breakthrough therapies" designation, mentioned above, has been greeted enthusiastically by developers, and as of late 2013 had already been applied to more than thirty new products (Mullard 2013). FDA's draft guidance requires that applicants show "preliminary clinical evidence" that a drug offers "substantial improvement" over available therapies on at least one "clinically significant" endpoint in a "serious or life-threatening disease" (FDA 2013).

It also has parallels to the Limited Population Antibacterial Drug (LPAD) Pathway legislation supported by the Infectious Diseases Society of America. LPAD legislation would provide a faster approval mechanism for antibacterial drugs for patients with serious or life-threatening infections and few or no treatment options. LPAD drugs would be approved based upon smaller clinical trials, as these infections typically occur in too few patients to populate a large traditional trial. LPAD drugs would be clearly labeled and monitored to guide their appropriate use in the limited indicated population for whom the benefits have been shown to outweigh the risks.

As with the effort to address the need for new antimicrobials to address the serious and rapidly growing threat of resistance, the CA pathway in Europe is only for drugs intended to fulfill unmet needs, such as for serious diseases—orphan as well as common—for which there are no good treatments available or for which the new drug offers a substantial improvement over existing options. Thus, the CA pathway moves some drugs into the market when early indications suggest a positive benefit-to-risk ratio.

Importantly, however, CA does this without losing control over the medication, by incorporating sunset provisions. In other words, "conditionally" approved drugs lose their approval status and are removed automatically from the market if sponsors fail to meet their postmarket commitments for further trials. While it is too early to document a pattern of approval sunsets, initial research at least has shown that the frequency of adverse events and special communications to providers has not been significantly higher for drugs that went to market with this abbreviated clinical trial period (Boon 2010).

Conditional approval creates, in theory, a real assurance that the necessary information for drug evaluation will be obtained in a timely manner. It is always easier for regulatory agencies to do nothing than to do something. In the United States, withdrawal of a marketed drug takes positive action. Under conditional approval in the European Union, the mere failure to either submit postmarket data to support the initial release or to apply for a limited-time renewal of the conditional approval means that the drug can no longer be marketed.

III. ADVERTISING

Another means of balancing the need for safety with the desire for shorter development times lies in restricting direct-to-consumer advertising (DTCA) for the initial period following approval, at least for truly new drugs built on unfamiliar mechanisms and formulations. Given that there is not yet broad experience with these drugs, it can be argued that the safest course is to slow diffusion and allow physicians an unpressured atmosphere in which to play their appointed role as "learned intermediaries" who consider all aspects of a patient's condition before choosing the best medication. Under these circumstances, physicians might well choose an older, better-understood medication, at least until there was some reason to think a newer drug offered distinct benefits.

Drug companies tend to invest in DTCA for drugs that treat chronic conditions, have few side effects, have little or no generic competition, or that have recently experienced a change (new indication approved, expecting to switch to over-the-counter status, patent about to expire so competition is expected soon, etc.). One study indicates that by 2001, 25 percent of Americans had asked their doctor about a drug they had heard about or seen advertised. Later studies show this percentage to be between 10 and 35 percent (Atherly and Rubin 2009; Rosenthal 2003; Rosenthal 2002).

Restricting DTCA will not completely stop overly rapid diffusion of new drugs into population-wide use, as a formulary or payer's preference for a particular drug is more likely determining which drug in a drug class the doctor prescribes rather than the patient's request for a specific drug (Atherly and Rubin 2009). In the managed care setting, providers' prescription choices are often limited, and DTCA may do more to increase prescriptions for a type of drug than a specific brand. A study published by the National Bureau of Economic Research in January 2003 found that the effects of DTCA for prescription drugs are more significant and consistent for the drug class than for an individual drug (Rosenthal 2003). And only 5.5 percent of the time do doctors admit to prescribing a DTC-advertised drug when they think another drug is more effective (Atherly and Rubin 2009).

But even if the data are mixed depending upon the study, it does appear that the explosion of radio and television DTCA has affected consumer behavior, with patients now requesting brand-name drugs (i.e., newer drugs). To the extent that physicians comply, rather than prescribing older medications, there is the risk that newer drugs are being used broadly without the benefit of a slower introduction within a smaller population from whom new insights might be gained (Ventola 2011). This was, indeed, one of the ideas addressed in the 2006 IOM report (IOM 2006) and considered for FDAAA.

However, in the end, concerns that legislating such limits might run afoul of constitutional free speech protections led to abandoning the idea in favor of REMS and industry voluntary actions regarding DTCA. Given subsequent developments in the *Sorrell* and *Caronia* cases, the constitutional concerns seem justified.

IV. COMMERCIAL SPEECH JURISPRUDENCE AND ITS EFFECT ON POSTMARKET SAFETY MEASURES

The fundamental issue presented in *Caronia* concerns FDA's prohibition on off-label marketing, a prohibition based on a presumption that any marketing for an indication, dose, or population for which the drug was not tested is presumably a form of misbranding (21 USC § 321). But in *Caronia*, the court noted that off-label prescribing is perfectly lawful. In a substantial blow to FDA's enforcement theory, it held that FDA cannot prosecute off-label promotion of a drug if the off-label use is lawful and the promotion used information that was truthful and not misleading. The court narrowly avoided overturning the entire structure of the drug regulatory system by focusing on some very specific aspects of the case, but the dissent nonetheless cautioned that the decision may effectively prohibit all uses of off-label promotional speech to support a misbranding conviction under the Federal Food, Drug, and Cosmetic Act (FDCA). Indeed, all three judges on the *Caronia* panel acknowledged that the decision may have left the enforcement of the misbranding provisions of the FDCA on shaky ground.

Limiting FDA's ability to restrict off-label promotion has a number of effects that can undermine efforts to promote drug safety efforts, both before and after drug approval (Kesselheim and Mello 2015; Robertson 2015), and it might even have the effect of turning pharmacogenomics into a short-cut that allows narrower approval trials to nonetheless result in wide population use beyond the label. It could mean that drugs will be used in large populations and at doses or for indications for which there has been no independent governmental arbiter of safety and effectiveness.

In many cases off-label use is perfectly justifiable. But the combination of pharmacogenomics and accelerated approval pathways coupled with a retreat from the ability to limit off-label marketing means that companies would no longer have much R&D incentive at all—not even an exclusive market for supplemental indications—and investment in exploring these supplemental uses will likely decrease, with no obvious alternative player inclined to spend the time and effort to confirm the value of these drugs for these new uses (Stafford 2012).

V. CONDITIONAL APPROVAL

In the United States, a conditional approval system akin to that used in Europe might avoid some of the tension between marketing restrictions and commercial speech. The *Caronia* decision is grounded in an analysis of off-label marketing of an *approved* drug. Conditional approval—which would be given when less than the usual amount of data is available—is fundamentally different. In essence, a conditionally approved drug or indication would still be considered investigational, and the protections for commercial speech provided in *Caronia* might not apply, thus paving the way for restrictions on DTCA labeled indications as well as on all promotion of off-label indications.

But a conditional approval system may still fail to address the need for a predictable and reasonable return on investment for companies. It is uncertain that a conditionally approved drug could be sold for profit; many insurers may assume that such a drug is experimental and therefore not pay for it. Even if these issues were addressed—and they should be, through coverage with evidence development or other means—some promising new drugs may lack financial backing. Other factors would create substantial uncertainty, including conditional approval itself, the prospect of strong postmarket measures, and the threat of liability. Thus, it makes sense to consider other types of financial incentives for industry.

The European Union's "Innovative Medicines Initiative" is one example. Its "Joint Technology Initiatives" bring together companies, universities, public laboratories, patient groups, and regulators, with cofunding from industry and the EU budget, to the tune of several billions of euros. They establish their own research agenda and selectively fund projects based on open competition, citing successes in developing new or improved tools and models for diseases as disparate as schizophrenia and diabetes (European Commission 2013).

This approach has been advocated for antibiotics; market exclusivity is a poor incentive if new antibiotics are only used as a last resort against multidrug resistant organisms. The Infectious Diseases Society of America and others have claimed that the antimicrobial market fails to adequately compensate manufacturers for the investment necessary to develop new antimicrobials because they will likely be stockpiled or used judiciously and not be blockbusters that generate large returns (IDSA 2014). In such a situation, a lengthy period of market exclusivity—known as a "pull" mechanism—is not sufficient to offset development costs. Additional pull mechanisms might help, such as offering prizes or awards for reaching specified development markers. In addition, "push" mechanisms could be used, for example, by reducing the obstacles to research or offering grants and tax breaks (Pogge 2005; Outterson 2014).

Similarly, in order to offset the return-on-investment uncertainties created by moving to a conditional approval system that slows the rate at which a new drug penetrates the market and increases the risk of limitations on its uses or even the loss of the approval to market, a new set of push-and-pull mechanisms could de-risk development and maintain the incentives to innovate. In general, a system of grants, rewards, prizes, and tax credits could provide a supplemental source of revenue for research and development.

VI. FINAL COMMENTS

As long as speed and safety are seen as opposing forces in drug development and approval, progress will be halting. But enhanced safety measures can generate the confidence needed to permit earlier release of drugs. Those enhanced safety measures should include measures to slow the uptake of those new drugs and new indications for which uncertainties are the greatest. To do that, one must address the constitutional protections for commercial speech, include an action-forcing mechanism that ensures continued surveillance and data collection, and provide an economic incentive to developers other than the prospect of uninterrupted and unconstrained market exclusivity for decades to come. With all these moving pieces, there might well be a way to achieve a system of pharmaceutical development and regulation that is both faster and safer.

NOTE

The author gratefully acknowledges the assistance of Jennifer Carter, University of Wisconsin Law, 2014.

REFERENCES

Abigail Alliance for Better Access to Developmental Drugs v. von Eschenbach, 495 F.3d 695 (D.C. Cir), cert denied 128 S. Ct. 1069 (January 14, 2008).
Arnardottir, Arna, Flora M. Haaijer-Ruskamp, Sabine M. J. Straus, Hans-Georg Eichler, Pieter A. de Graeff1, and Peter G. M. Mol. 2011. "Additional Safety Risk to Exceptionally Approved Drugs in Europe?" *British Journal of Clinical Pharmacology* 72(3):490–99.
Atherly and Rubin. 2009. "The Cost Effectiveness of Direct-to-Consumer Advertising for Prescription Drugs." (May). http://prescriptiondrugs.procon.org/sourcefiles/Atherly_DTC_Profits.pdf.
Boon, W. P. C., E. H. M. Moors, A. Meijer, and H. Schellekens. 2010. "Conditional Approval and Approval Under Exceptional Circumstances as Regulatory Instruments for Stimulating Responsible Drug Innovation in Europe." *Clinical Pharmacology & Therapeutics* 88(6):848–53.
Bren, Linda. 2001. "Frances Oldham Kelsey: FDA Medical Reviewer Leaves Her Mark on History." *FDA Consumer* 35, 2 (March/ April):27.
Calabrese, D. 2013. "Breakthrough Therapy: FDA's New Designation for Accelerating Approval of Promising Pipeline Therapies." *Formulary* 48(8):275–76.
Carpenter, Daniel, Jacqueline Chattopadhyay, Susan Moffitt, and Clayton Nall. 2011. "The Complications of Controlling Agency Time Discretion: FDA Review Deadlines and Postmarket Drug Safety." *American Journal of Political Science* 56:98–114.
Charo, R. Alta. 1993. "Protecting Us to Death: Women, Pregnancy, and Clinical Research Trials." *St. Louis University Law Journal* 38 (Fall):135–67.
Coleman, Doriane Lambelet. 2007. "The Legal Ethics of Pediatric Research." *Duke Law Journal* 57 (December):517–624.
Duijnhoven R. G., S. M. Straus, J. M. Raine, A. de Boer, A. W. Hoes, and M. L. De Bruin. 2013. "Number of Patients Studied Prior to Approval of New Medicines: A Database Analysis." *PLoS Medicine* 10(3):e1001407. doi:10.1371/journal.pmed.1001407.
European Union Commission Regulation (EC) No. 507/2006 (2).
European Union Commission. 2013. "Innovative Medicines Initiative 2: Europe's Fast Track to Better Medicines." See http://www.imi.europa.eu/ (accessed June 10, 2014).

Federal Food, Drug, and Cosmetic Act (FFDCA), Pub. L. No. 75-717, 52 Stat. 1040 (1938).

FDA Draft Guidance. 2013. "Guidance for Industry Expedited Programs for Serious Conditions—Drugs and Biologics."

Gibson, Shannon and Trudo Lemmens. 2015. "Overcoming 'Pre-Market Syndrome': Promoting Better Post-Market Surveillance in an Evolving Drug Development Context." In *FDA in the Twenty-First Century: The Challenges of Regulating Drugs and New Technologies*, ed. Holly Fernandez Lynch and I. Glenn Cohen. Columbia University Press (2015).

Harden, Victoria. 2012. *AIDS at 30: A History*. Dulles, VA: Potomac Books.

Health and Human Services Office of Inspector General. 2011. "FDA Lacks Comprehensive Data to Determine Whether Risk Evaluation and Mitigation Strategies Improve Drug Safety." https://oig.hhs.gov/oei/reports/oei-04-11-00510.pdf (accessed March 30, 2013).

Herschler, Ben. 2014. "WHO Joins Clamor to Make New Hepatitis C Pills Affordable." *Reuters* (April 9). http://www.reuters.com/article/2014/04/09/us-hepatitis-idUSBREA38ORK20140409 (accessed April 17, 2014).

Infectious Diseases Society of America (IDSA). "Antibiotic Development: The 10 x '20 Initiative Bringing New Antibiotics to Patients Who Need Them." http://www.idsociety.org/10x20/#sthash.xvCqYUWy.dpufhttp://www.idsociety.org/10x20/ (accessed April 12, 2014).

Institute of Medicine (IOM). 2006. *The Future of Drug Safety: Promoting and Protecting the Health of the Public*. National Academy Press.

Kaitin, K. I. 2010. "Deconstructing the Drug Development Process: The New Face of Innovation." *Clinical Pharmacology and Therapeutics* 87(3):356–61.

Kefauver-Harris Amendment to the FFDCA, Pub. L. No. 87-781, 76 Stat. 780 (1962).

Kesselheim and Mello, Aaron. 2015. Prospects for Regulation of Off-Label Drug Promotion in an Era of Expanding Commercial Speech Protection, in *FDA in the Twenty-First Century: The Challenges of Regulating Drugs and New Technologies*, I. Glenn Cohen and Holly Fernandez Lynch, eds., Columbia University Press (2015).

Kozauer, N. and R. Katz. 2013. "Regulatory Innovation and Drug Development for Early-Stage Alzheimer's Disease." *New England Journal of Medicine* 368(13):1169–71.

Kramer, Daniel B, Elias Mallis, Bram Zuckerman, Barbara Zimmerman, and William Maisel. 2010. "Premarket Clinical Evaluation of Novel Cardiovascular Devices: Quality Analysis of Premarket Clinical Studies Submitted to the Food and Drug Administration 2000–2007." *American Journal of Therapeutics* 17(1):2–7.

Majumdar, Sumit, Elizabeth A. Almasi, and Randall S. Stafford. 2004. "Promotion and Prescribing of Hormone Therapy After Report of Harm by the

Women's Health Initiative." *Journal of the American Medical Association* 292:16 (October):1983–84.

Mastroianni, Anna. 1998. "HIV, Women, and Access to Clinical Trials: Tort Liability and Lessons from DES." *Duke Journal of Gender Law and Policy* 5 (Spring):167–72.

Mitka, M. 2012. "Trials to Address Efficacy of Midodrine 18 Years After It Gains FDA Approval." *Journal of the American Medical Association* 307(11):1124–27.

Mol, Peter, Arna Arnardottir, Domenico Motola, et al. 2013. "Post-Approval Safety Issues with Innovative Drugs: A European Cohort Study." *Drug Safety* 36(11):1105.

Moors, Ellen, Adam F. Cohen, and Huub Schellekens. 2014. "Towards a Sustainable System of Drug Development." *Drug Discovery Today* 19(11):1711–20.

Mullard, Asher. 2013. "Learning from the 2012–2013 Class of Breakthrough Therapies." *Nature Reviews Drug Discovery* 12:891–93.

Noah, Barbara. 2003. "The Participation of Underrepresented Minorities in Clinical Research." *American Journal of Law and Medicine* 29:221–22.

Outterson, Kevin. 2014. "New Business Models for Sustainable Antibiotics." *Centre on Global Health Security: Working Group Papers* (visited April 17, 2014).

Parasidis, Efthimios. 2015. "FDA's Public Health Imperative: An Increased Role for Active Postmarket Analysis". In *FDA in the Twenty-First Century: The Challenges of Regulating Drugs and New Technologies*, I. Glenn Cohen and Holly Fernandez Lynch, eds., Columbia University Press (forthcoming 2015).

Pogge T. W. 2005. Human Rights and Global Health: A Research Program. *Metaphilosophy* 36(1/2):1–28.

Poldermans, Dan et al. 1999. "The Effect of Bisoprolol on Perioperative Mortality and Myocardial Infarction in High-Risk Patients Undergoing Vascular Surgery." *New England Journal of Medicine* 341 (December):1789–94.

President's Council of Advisors on Science and Technology (PCAST). 2012. "Report to the President on Propelling Innovation in Drug Discovery, Development, and Evaluation." http://www.whitehouse.gov/administration/eop/ostp/pcast/docsreports (accessed April 12, 2014).

Robertson A. S., R. Stefanakis, D. Joseph, and M. Moree. 2012. "The Impact of the US Priority Review Voucher on Private-Sector Investment in Global Health Research and Development." *PLoS Neglected Tropical Diseases*. 6(8):1750.

Rosenthal, Meredith, Ernst R. Berndt, Julie M. Donohue, Richard G. Frank, and Arnold M. Epstein. 2002. "Promotion of Prescription Drugs to Consumers." *New England Journal of Medicine* 346:498–505.

Robertson, Christopher. 2015. "FDA's Authority to Regulate Marketing Under an Expanding First Amendment," in *FDA in the Twenty-First Century: The Challenges of Regulating Drugs and New Technologies*, I. Glenn Cohen and Holly Fernandez Lynch, eds., Columbia University Press (forthcoming 2015).

Rosenthal, M. et al. 2003. "Demand Effects of Recent Changes in Prescription Drug Promotion." In *Frontiers in Health Policy Research, Volume 6*, ed. David M. Cutler and Alan M. Garber, 1–26. Cambridge, MA: MIT Press. http://www.nber.org/chapters/c9862.pdf.

Sherman, Rachel E., Jun Li, Stephanie Shapley, Melissa Robb, and Janet Woodcock. 2013. "Expediting Drug Development—FDA's New 'Breakthrough Therapy' Designation." *New England Journal of Medicine* 369(20):1877–80.

Singh, Sonal and Yoon K. Loke. 2012. "Drug Safety Assessment in Clinical Trials: Methodological Challenges and Opportunities." *Trials* 13:138.

So A.D., N. Gupta, S. K. Brahmachari et al. 2011. "Towards New Business Models for R&D for Novel Antibiotics." *Drug Resistance Updates* 14(2):88–94.

Sorrell v. IMS Health Inc., 131 S. Ct. 2653 (2011).

Stafford, R. S. 2012. Off-Label Use of Drugs and Medical Devices: A Review of Policy Implications. *Clinical Pharmacology & Therapeutics* 91:920–25.

Stix, Gary. 2012. "Patients Clamor for Cancer Drug That Shows Promise for Alzheimer's in Mice." http://blogs.scientificamerican.com/observations/2012/02/13/patients-clamor-for-alzheimers-drug-that-shows-promise-in-mice/ (accessed April 17, 2014). doi:10.1038/nbt0905–1050.

Tsang, Lincoln. 2005. "Overhauling Oversight—European Drug Legislation." *Nature Biotechnology* 23:1050–53.

United States v. Caronia, 703 F.3d 149 (2d Cir. 2012).

U.S. Food and Drug Administration (FDA). 2014. Guidance for Industry Expedited Programs for Serious Conditions—Drugs and Biologics. OMB Control No. 0910-0765. (May).

U.S. Food and Drug Administration (FDA). 2006. "Promoting Safe and Effective Drugs for 100 Years." *FDA Consumer Magazine* (The Centennial Edition/January-February).

Ventola, C. L. 2011. "Direct-to-Consumer Pharmaceutical Advertising: Therapeutic or Toxic?" *Pharmacy & Therapeutics* 36(10):669–74, 681–84.

von Eschenbach, Andrew. 2012. "Medical Innovation: How the U.S. Can Retain Its Lead." *Wall Street Journal* §A:19 (February 14).

Wilson, Benjamin. "The Rise and Fall of Laetrile." http://www.quackwatch.org/01QuackeryRelatedTopics/Cancer/laetrile.html (accessed April 11, 2013).

CHAPTER FOURTEEN

Overcoming "Premarket Syndrome"

Promoting Better Postmarket Surveillance in an Evolving Drug-Development Context

SHANNON GIBSON AND TRUDO LEMMENS

I. ADDRESSING PREMARKET SYNDROME

The drug-approval system in most advanced economies focuses predominantly on a review of safety and efficacy data from various premarket clinical trials. These trials are largely organized and controlled by the pharmaceutical industry—an industry with a vested and significant financial interest in demonstrating the safety and efficacy of the products tested. Once approved by regulatory authorities, prescribers and consumers often uncritically presume that drugs are reasonably safe and effective. Subsequent postmarket clinical trials are the exception rather than the rule.

The regulatory fixation on premarket activities, which we refer to as "premarket syndrome," contributes to a range of problems that can negatively impact both patient health and the health care system. Premarket clinical trials are typically conducted under controlled conditions that generally do not reflect how a drug will be used in the real world (Ahmad 2003) and rarely assess whether a drug is actually more effective than existing therapies (Flood and

Dyke 2012). These trials are often of short duration, so rare or longer-term side effects may remain hidden for many years, by which point the drug is already widely prescribed (Lasser et al. 2002). Moreover, drugs are frequently prescribed "off label" to patients and disease groups never assessed in clinical trials (Wiktorowicz, Lexchin, and Moscou 2012), as is discussed in greater depths in other chapters in this book.

Two recent trends—one in drug development and one in drug regulation—are reinforcing the importance of moving drug evaluation beyond the premarket stage. First, in recent years the pharmaceutical industry has been showing a growing interest in developing drugs for niche markets where the symptoms of premarket syndrome can be particularly acute. The narrow population base for these therapies often inherently limits the amount of safety and efficacy data available to support their approval—a fact that heightens the importance of assessing the benefits and risks of these drugs through the ongoing collection and analysis of postmarket data.

Second, as the shortcomings of the current focus on premarket activities become increasingly apparent, drug regulatory systems in various jurisdictions have proposed reforms that move away from the "artificial dichotomy" of the pre- versus postmarket stages (Eichler et al. 2012). In the United States, a 2007 report by the Institute of Medicine (IOM) recommended that the Food and Drug Administration (FDA) adopt a "life-cycle" approach to drug approval where the benefits and risks of drugs are monitored not simply based on premarket evidence but on the entire body of evidence that is collected throughout the life cycle of the drug (Committee on the Assessment of the US Drug Safety System 2007). The report endorsed such measures as aggressive assessments of drug effects throughout the product life cycle; an overhaul of adverse events reporting; greater public–private funding of postmarket studies; and authority for FDA to demand postmarketing reports. In February 2013, FDA released a draft benefit–risk framework to implement the life-cycle approach (FDA 2013a).

In this chapter, we explore both the promises and risks associated with rising interest in niche market development and, concurrently, how reform efforts toward the life-cycle approach and more postmarket evidence generation are both a response to and a driving force behind the shift toward niche markets. We argue that despite the promise of the life-cycle approach in combatting premarket syndrome, significant questions remain about whether regulatory authorities are prepared to address the attendant challenges that accompany the shift toward the life-cycle approach, particularly the concern that increasing the focus on postmarket monitoring and evaluation may lead to a softening of regulatory control at market entry.

A. Premarket Syndrome in Niche Markets

Pharmaceutical developers have traditionally shied away from smaller drug markets due to the revenue limitations presented by the reduced patient base (Woodcock 2007). However, in recent years, the pharmaceutical industry has turned more attention toward high-value niche therapies for the treatment of smaller patient populations. The orphan drug market, the main focus of Greenwood's chapter in this volume, has expanded rapidly in recent years, largely due to legislative initiatives—most significantly the Orphan Drug Act of 1984, Pub. L. No. 97-414, 96 Stat. 2049 (ODA)—that have incentivized drug development related to orphan diseases (Coté, Xu, and Pariser 2010). The ODA offers a special designation to drugs that treat a rare disease—defined as conditions affecting less than 1 in 200,000 people in the United States—and provides a number of incentives including tax benefits on clinical trials, fast-track approval, grants, and seven years of market exclusivity. FDA now reports that almost 200 orphan drugs enter development each year, and approximately one-third of new drug approvals are for the treatment of rare diseases (Rockoff 2013).

A second trend that is contributing to burgeoning interest in niche markets is the advent of pharmacogenomics, the study of the influence that genetic factors have on drug response. Pharmaceutical products are now being developed in combination with companion diagnostic tests that can stratify patient populations based on genetic predisposition to respond to drug therapies (Collier 2011)—thereby dividing more common diseases into rarer disease genotypes. The number of pharmacogenomic products on the U.S. market has increased steadily over the past decade from thirteen prominent examples of personalized medicine in 2006 to seventy-two in 2011 (Personalized Medicine Coalition 2011). Likely the two best-known examples are the breast cancer drug Herceptin (trastuzumab) and the leukemia drug Gleevec (imatinib), both of which have already achieved billions of dollars in annual sales (Keeling 2007). Further, a growing number of products in clinical development rely on biomarkers, presaging their increasing importance in drug development.

Premarket syndrome may be particularly acute in niche markets since these therapies present a number of evidentiary challenges that increase the imperative to continue to monitor drug use during the postmarket phase. First, clinical trials for niche markets often prove difficult to organize as a result of the inherently limited number of patients with the disease (Boon and Moors 2008) and "encounter a disadvantage compared with more widely used drugs, as large-scale clinical trial data are usually unavailable" (Owen et al. 2008:236). Consequently, drugs for niche markets may be approved on the basis of safety and efficacy data that are less robust than the data produced in larger-scale clinical trials. A second factor is enrichment strategies, the "prospective use of any patient characteristic . . . to select patients for study to obtain a study population in which detection of a drug effect . . . is more likely than it would be in an unselected population (FDA 2012a:2). Genetic biomarkers represent a powerful new tool that researchers can use to refine the study population. Yet while enrichment increases the power of a study to detect a clinical effect in *some* populations, questions remain around whether these results are generalizable to other populations, as well as around the level of data needed to establish selection criteria for enrichment. Moreover, as noted by Ioannidis, "the large majority of proposed genetic associations (including pharmacogenetics) made in the past . . . have not been replicated with larger-scale evidence and stringent statistical criteria" (2013:413). FDA notes that when enrichment strategies are employed, "post-market commitments or requirements may be requested to better define the full extent of a drug's effect" (FDA 2012a:32). Overall, a consequence of both smaller clinical trials and enrichment strategies, to the extent that they are justifiable, is a heightened need to assess the benefits and risks of these drugs through the ongoing collection and analysis of data after market entry.

II. THE LIFE-CYCLE APPROACH TO DRUG APPROVAL

Hamburg and Sharfstein, the commissioner and former deputy commissioner of FDA, respectively, once noted that "it has been said that the FDA has just two speeds of approval—too fast and too slow" (2009:2494). In making approval decisions, regulators in all jurisdictions face the fundamental dilemma of balancing the need for robust evidence on safety and efficacy with demands for timely access to promising new therapies. Decision makers face an apparent catch-22, as they are torn between demands from those who insist that patients facing life-threatening conditions need earlier access to treatments and those that fear drugs are often approved too soon on the basis of small clinical trials that do not accurately reflect drug use in the real world (Hamburg and Sharfstein 2009). Industry interests in getting products to market as fast as possible often intermingle with or hide behind the rhetoric of the need to provide access to life-saving medicines, and industry-funded advocacy groups may be used as a front in lobbying efforts. As we discuss in the following sections, while the objectives of reform efforts toward the life-cycle approach appear reasonable, it is important to remember that the devil is often in the details of implementation.

A. Expedited Access to Promising New Therapies

Over the past few decades, growing patient demands for early access to promising new therapies and for greater autonomy in drug treatment decisions have pushed regulatory authorities to adopt more flexible approval programs to speed the development and availability of drugs that treat serious diseases. In 1992, FDA introduced two new programs: the Priority Review designation, which shortened the review time for drugs that treat serious conditions and represent a significant improvement in safety or effectiveness; and the Accelerated Approval program, which allowed drugs for serious conditions that meet unmet medical need to be approved faster based on surrogate endpoints. Subsequently in 1997, FDA introduced Fast Track approval, which offers applicants more frequent interactions with the

FDA review team and allows for "rolling reviews" of applications (FDA 2013).

It is no coincidence that demands for expedited review are increasing alongside the growing interest in niche markets. Advances in fields such as pharmacogenomics and legislative initiatives such as orphan drug policies have made previously neglected rare disease markets much more attractive to industry. Patients are being presented with seemingly promising new treatment options and, consequently, demands for earlier access to these therapies are escalating. Moreover, since many niche market therapies treat serious or life-threatening diseases with few or no alternative treatment options, pressure on regulators to speed access to these therapies can be particularly intense.

Growing interest in niche markets is pushing drug regulators to adapt the drug approval system to accommodate the evidentiary challenges introduced by limited patient populations. Most recently, the FDA Safety and Innovation Act of 2012 (FDASIA) introduced a new "breakthrough therapy" designation and expanded the scope of the existing accelerated approval process to allow approval on the basis of both surrogate and intermediate clinical endpoints (FDA 2013). The new breakthrough designation is available to investigational drugs for the treatment of serious or life-threatening conditions where preliminary clinical evidence indicates the drug offers substantial improvement over existing therapies on at least one clinically significant endpoint. Sherman and colleagues (2013) note that the impetus for the new breakthrough designation can be traced to emerging trends in drug development and discovery, particularly targeted therapies aimed at subgroups of patients within broader disease categories who are expected to experience a much larger treatment effect, even in early trials.

While not directly included in the FDA's 2013 draft benefit–risk framework implementing the life-cycle approach, the expansion of FDA's expedited review programs should be considered along with the life-cycle approach as part of a broader movement that is increasing the focus on postmarket evidence generation and assessment. The accelerated approval process, in particular, lowers the threshold for initial approval on the condition of more postmarket commitments by the manufacturer, thereby pushing assessment beyond the premarket phase. The convergence of early approval mechanisms and the life-cycle approach is similarly evident in reforms that have been proposed by the European Medicines Agency (EMA), which move away from the artificial dichotomy of the pre- versus postmarket phases. In particular, the "staggered"-approval model endorsed by the EMA allows earlier market entry for well-defined or restricted populations of "good responders," followed by a broadening of the population postapproval as more "real-life" evidence becomes available (European Medicines Agency 2010). This staggered approval model combines both a more flexible approach to market entry (as under expedited approval programs such as accelerated approval) and more postmarket evidence generation and assessment (as under the life-cycle approach).

FDA claims that it has "been vigilant in assuring that reducing the time necessary for drug development has not compromised the safety and effectiveness of drugs for patients with serious conditions" (FDA 2013). Nonetheless, studies have shown that where drugs are expedited through the approval process, they are more likely to encounter problems during the postmarket phase, likely due to the faster review missing serious safety issues (Lexchin 2012a). Moreover, there are widespread concerns about approving drugs on the basis of surrogate markers or intermediate outcomes (versus long-term clinical endpoints) that may not be very informative in actual clinical practice (Ioannidis 2013). For example, a study by Berlin (2009) found that where oncology drugs are approved under the accelerated approval or priority review channels offered by FDA, drug labeling is revised significantly more frequently than the labeling for traditional drug products. Moore and Furberg (2014) note that, increasingly, innovative drugs are being approved more rapidly based on small clinical trials in narrower patient populations on the condition of expanded requirements for postmarket testing. Indeed, of the twenty drugs studied by Moore and Furberg that received expedited review, FDA required postmarket studies for nineteen of them. The authors note that this shift has made it more challenging to balance the risks and benefits of new drugs, which highlights the need for rigorous control on the rationale for expedited approval.

Another important concern is that expedited review programs could potentially be used by the pharmaceutical industry to get a foot in the regulatory door. Once a drug receives expedited approval, even if only for a very specific indication or patient group, industry may soon request label changes to expand the market. Moreover, even without explicit label change, the market for these drugs may also expand through off-label use; as earlier chapters in this volume suggest, recent case law may mean that off-label promotion becomes more routine and the rules against it less likely to be enforced. After market entry, withdrawing the drug based on subsequent information is likely to be met with strong resistance, particularly from patients who, rightly or wrongly, believe that they are benefiting from the drug. Industry may push regulators to keep the drug on the market through a variety of lobbying and pressure tactics, including the use of industry-supported patient advocacy groups (Hughes and Williams-Jones 2013). To the extent that it is appropriate to approve promising therapies based on a more limited level of evidence, there is a clear need to assess the benefits and risks of these drugs through the ongoing collection and analysis of data after market entry. However, as discussed in more detail below, there are serious questions about whether existing and proposed postmarket regulatory measures are properly equipped to counterbalance the risk posed by earlier market entry.

B. An Incremental Approach to Market Access

The dominance of the blockbuster model of drug development in recent decades has significantly impacted the nature of postmarket risk and, consequently, the risks associated with premarket syndrome. In particular, blockbuster drugs are so widely prescribed, and increasingly for prevention or the long-term treatment of chronic conditions, that even a minor change in relative risk can lead to a significant number of adverse drug reactions (Wiktorowicz, Lexchin, and Moscou 2012). Moreover, the vigorous promotion of blockbuster drugs by sponsoring pharmaceutical companies often leads to rapid market uptake that far outpaces the ability of the scientific community to develop the evidence to support such wide-scale use. The highly publicized market withdrawal of Vioxx demonstrates the risk that important safety data may only become apparent after many people start using the medication—the now-infamous drug has been associated with major adverse events, including myocardial infarctions and strokes, in tens of thousands of patients (Topol 2004).

To combat the risks posed by the rapid promotion and uptake of new drugs, the life-cycle approach involves "the pursuit and active management of emerging knowledge about the benefit-risk balance as drugs become more widely used by larger numbers of increasingly diverse patients" (Psaty, Meslin, and Breckenridge 2012:2491). Incorporating restrictions on the use of new products can help control the diffusion of drugs into the market and limit the number of patients who are exposed to these drugs before adequate data has been collected on their safety and efficacy (Ray and Stein 2006; Eichler 2012). Subsequently, access to the drug can be expanded or restricted as more evidence is generated on appropriate use. Controlling the diffusion of new drugs is particularly important when the drug is approved on the basis of a reduced evidence profile, such as based on a surrogate or intermediate clinical endpoints under the accelerated approval process. To increase the chances that only approved patients receive the drug, Eichler and colleagues argue that "[a]ppropriate targeting to the label population will need to be a high priority" and "at least after the initial approval, systemic restrictions and monitoring of prescribing may be required to prevent off-label use" (Eichler et al. 2012:429).

In recent years, as discussed in the Parasidis chapter in this volume, FDA has moved toward more proactive risk management for drug products. Perhaps most significantly, the FDA Amendments Act of 2007 (FDAAA) gave FDA the authority to require a manufacturer to develop Risk Evaluation and Mitigation Strategies (REMS) for certain products with "exceptional circumstances." REMS can mandate that patients meet safe-use conditions when further measures beyond drug labeling are needed to ensure that the drug's benefits outweigh its risks. For example, REMS may place restrictions on the distribution of drugs, including requiring advanced certification for prescribers and pharmacists, permitting dispensing only at authorized pharmacies, or mandating enrollment in patient registries. REMS may be imposed either before or after a drug is approved and may be associated with a single drug or a class of drugs (FDA 2012).

Pharmacogenomic drug products are well suited to an incremental approach to access because diagnostic testing could be used to identify those patients who are likely responders and therefore the best candidates to receive initial access to the drug. Indeed, REMS may specifically require the use of biomarker screening and pharmacogenomic tests before a drug is prescribed (Evans 2013). Ideally, if pharmacogenomic data can be gathered through improved pharmacovigilance systems and postmarket studies, this may allow for increasingly responsive tailoring of prescribing guidelines and the incremental expansion and tightening of approved drug indications based on evolving postmarket evidence. Indeed, FDA has acknowledged that if "new science enables us to determine that the adverse events are restricted to a small, identifiable segment of the population, public health could be improved by making the drug available to others who could benefit without undue risk" (FDA 2007:4). However, defining the appropriate target population in actual clinical practice is itself an uncertain process. In particular, Ioannidis argues that "[t]he lack of a systematic approach to the pharmacogenetic evidence and the inconsistent use of pharmacogenetic information in Food and Drug Administration labeling may create confusion in clinical practice" (Ioannidis 2013:415). Clinicians may have difficulty assessing what a label really means and how to interpret a genetic test result and its association with the effect of a drug, thus perhaps unnecessarily excluding patients who could benefit from therapy.

The effectiveness of REMS will ultimately depend on regulators having the authority, resources, and political will to police and enforce the prescribed measures. Unfortunately, a recent evaluation by the Office of the Inspector General (OIG) released in February 2013 found that FDA lacks the comprehensive data necessary to determine whether REMS are actually improving drug safety. The OIG reviewed forty-nine REMS and found that half did not contain the information requested by FDA, ten were not submitted within given time frames, and only seven achieved all of their objectives—all of which raise significant concern about the effectiveness of REMS. The report concludes that "[i]f FDA does not have comprehensive data to monitor the performance of REMS, it cannot ensure that the public is provided maximum protection from a drug's known or potential risks" (OIG 2013:22). The OIG also recommended that FDA be given the legislative authority to take enforcement actions when drug companies do not submit all information requested in assessment plans, which the agency currently lacks. The OIG analysis also hints at a broader issue in drug regulation: the gap between well-intended and seemingly reasonable regulatory adjustments to new developments and the reality of drug regulation on the ground—a reality characterized by limited public resources and a resulting power imbalance between industry and regulatory agencies.

C. A Stronger Focus on Postmarket Evidence Generation

Another consequence of premarket syndrome is that the focus on premarket activities means that regulators often have limited measures available to ensure continued compliance with regulations once a drug has received market approval. Earlier studies showed that although drug manufacturers are often expected to conduct further studies after a drug enters the market, they often fail to do so (Avorn 2007). However, a turning point has arguably been reached in recent years as FDA expanded regulatory requirements for postmarket commitments. Most notably, the FDAAA expanded FDA's authority to require manufacturers to conduct studies after market approval if new safety information comes to light (Evans 2010). FDA can even mandate that these studies be completed on a specific timetable. In a 2012 report on the fulfillment of postmarket requirements, FDA reported that of the 675 postmarket studies for new drug applications that had been mandated as of September 2011, 87 percent were on schedule (FDA 2012b). Nonetheless, in May 2012, the IOM issued a report on postmarket safety in which it urged FDA to be more aggressive in proactively dealing with the safety concerns that emerge following the entry of a drug into the market (Kuehn 2012).

Particularly for drugs approved based on a reduced evidence profile, the expansion of postmarket surveillance is integral to counterbalancing increased uncertainty around safety and efficacy at market entry. However, the impact of more postmarket studies and monitoring will be limited if insufficient attention is paid to how such evidence is generated. Developing the infrastructure necessary for the comprehensive, timely, and accurate collection of data on drug use in actual clinical practice remains a significant challenge (Evans 2010), as is discussed in the chapter by Parasidis. Moreover, a fundamental problem with the existing regulatory system is the pharmaceutical industry's control over the majority of clinical trials, which studies have shown may lead to manipulation of trial results through carefully crafted research design, choices made in the context of statistical analyses, exclusion of negative findings, over-inclusion of positive findings, and even outright misrepresentation (Lexchin 2012b). These issues will only be overcome through more significant changes to pharmaceutical knowledge governance (Lemmens and Telfer 2012; Lemmens 2013).

Enhanced transparency measures are fundamental to improving the reliability of pharmaceutical knowledge production (Lemmens 2013)—not only during premarket development but also after market entry. Industry may also be driven to manipulate or obfuscate unfavorable study results in the postmarket phase if such studies demonstrate, for example, the inferiority of a drug to other therapies or the inappropriateness of a drug for widely prescribed off-label uses. Transparency is important not only for reasons of public accountability but also because problems with drug safety and efficacy have sometimes been detected by independent scientific experts rather than by regulatory agencies. FDA has already taken important steps toward improving transparency, perhaps most notably the introduction of strict registration- and results-reporting requirements for all Phase II–IV trials under the FDAAA, with associated penalties for noncompliance. Some regulatory agencies, such as the EMA, have gone further and impose transparency for all clinical trial reports. Kimmelman and Anderson (2012) have even argued for the registration of preclinical research, which would be particularly relevant for pharmacogenomic drug development where the rationale for biomarker selection is based on preclinical data. However, transparency is only a first step, and reform efforts must also be directed at deeper issues in data production that stem from industry control over the design, conduct, and reporting of clinical trials.

D. Downstream Impact on Funding Decisions

The movement toward niche markets and the life-cycle approach will have important downstream impacts on how drug funding decisions are made, particularly since both trends may give rise to significant pressures to fund drugs for general distribution on the basis of a reduced evidence profile. Obtaining funding coverage is an increasingly important step in the uptake of new drugs, particularly for high-cost niche market therapies that few patients can afford to pay for out of pocket. Consequently, there is increasing interest in coverage with evidence development (CED), a funding arrangement where population-level payment or reimbursement is tied to prospective data collection in an attempt to gather more evidence and reduce decision uncertainty around funding coverage (Garrison Jr. et al. 2013). The increasing number of expensive niche market therapies with uncertain evidence profiles is placing pressure on funding authorities to enter into innovative risk-sharing approaches with drug sponsors (Owen et al. 2008).

CED may prove to be a natural complement to the life-cycle approach: both systems acknowledge that drug approval decisions should not be strictly binary but rather should be managed incrementally and continuously reassessed. Both recognize the uncertainty around premarket data and contribute toward ongoing evidence generation in the postmarket phase in an effort to deal with this uncertainty. Both facilitate a more incremental approach to the diffusion of new drug products into the market by limiting or attaching conditions to approval or coverage to prevent the drug from being widely prescribed before sufficient data on safety and efficacy have been gathered. Finally, both mandate more prospective data collection in an attempt to reduce decision uncertainty and better inform approval and coverage decisions.

Nonetheless, CED is also subject to many of the same challenges that may hinder the life-cycle approach, namely uncertainty around where to set the evidentiary bar for initial approval and how to ensure effective evidence generation and compliance with conditions of coverage once approval is granted. Discontinuing funding can be politically very difficult, even in light of new evidence of increased risks or questionable efficacy (Bishop and Lexchin 2013). For these reasons, CED must also be approached with caution—arguably even more so given that funding authorities bear the financial consequences of their approval decisions in a way that regulatory authorities do not. In this vein, there is increasing criticism of the often exorbitant prices demanded for many niche market therapies and of the industry's traditional justification that these prices reflect the need to recoup the high cost of development across a smaller population base. However, major questions are now arising about the research and development costs claimed by industry and, by extension, the prices demanded for many new drugs. Light and Warburton (2011), for example, point to various inflationary tactics used by industry to pump up estimated research and development costs and thus they reach a much lower estimate of average drug development costs than the billion-dollar-plus figures routinely claimed by industry sources—figures that are largely based on confidential industry data. Again, the importance of transparency and better oversight mechanisms also becomes apparent in the funding context.

III. CONCLUSION

Reform efforts toward the life-cycle approach are helping to reduce the signs and symptoms of premarket syndrome that have long plagued the drug regulatory system. Concurrently, the wave of new niche market therapies for serious and life-threatening diseases, coupled with rising demands for earlier access to these "promising" new therapies, has led regulators to expand exceptional access programs. Both of these trends are contributing to the increasing focus on the postmarket phase. While there are many positive aspects to this trend, such as more proactive risk management and more postmarket studies based on real-world use, significant questions remain about whether there has been concurrent lowering of the bar to market entry. Any arguments in favor of expediting the approval of drugs must be carefully weighed against the potential dangers that may arise from earlier approval. Given the widely recognized problems associated with the reliability of evidence in existing regulatory processes, transparency and data access, while not in themselves a complete solution, should be a priority of reform efforts. Ultimately, while expanding postmarket evidence generation is an important element in counterbalancing evidentiary uncertainty on drug safety and efficacy at market entry, these measures should be a complement to, not a substitute for, more rigorous premarket assessment.

NOTE

The authors would like to thank the organizers and attendees of the Petrie-Flom Center conference "The FDA in the 21st Century" at Harvard Law School on May 3–4, 2013, and of the conference "Old Markets, New Markets: Health Law After the 2012 Act" at the University of Sheffield School of Law on June 27, 2013, for useful comments and discussions. We are particularly grateful to I. Glenn Cohen, Tamara Hervey, Donald Light, Joel Lexchin, Julian Cockbain, and Sigrid Sterckx for comments on earlier versions of this and a related paper; to Kelly Tai for diligent research and work on the references; and to Brenda Robson for her editing comments. This chapter builds on the article: S. G. Gibson and T. Lemmens. 2014. "Niche Markets and Evidence Assessment in Transition: A Critical Review of Proposed Drug Reforms," *Medical Law Review* 22(2): 200–20. Research for this paper was supported by a Genome Canada grant on Ethical and Legal Issues of Cancer Initiating Stem Cell Research.

REFERENCES

Ahmad, S. R. 2003. "Adverse Drug Event Monitoring at the Food and Drug Administration." *Journal of General Internal Medicine* 18(1):57–60.

Avorn, J. 2007. "Paying for Drug Approvals—Who's Using Whom?" *New England Journal of Medicine* 356:1697–1700.

Berlin, R. J. 2009. "Examination of the Relationship Between Oncology Drug Labeling Revision Frequency and FDA Product Categorization." *American Journal of Public Health* 99(9):1693–98.

Bishop, D. and J. Lexchin. 2013. "Politics and Its Intersection with Coverage with Evidence Development: a Qualitative Analysis from Expert Interviews." *BMC Health Services Research* 13(88):1–10.

Boon, W. and E. Moors. 2008. "Exploring Emerging Technologies Using Metaphors—A Study of Orphan Drugs and Pharmacogenomics." *Social Science & Medicine* 66(9):1915–27.

Collier, R. 2011. "Bye, Bye Blockbusters, Hello Niche Busters." *Canadian Medical Association Journal* 183(11):E697–98.

Committee on the Assessment of the U.S. Drug Safety System. 2007. *The Future of Drug Safety: Promoting and Protecting the Health of the Public.* Consensus Report, Washington, D.C.: National Academies Press.

Coté, T. R., K. Xu, and A. R. Pariser. 2010. "Accelerating Ophan Drug Development." *Nature Reviews Drug Discovery* 9:901–2.

Eichler, H.-G., K. Oye, L. G. Baird, E. Abadie, J. Brown, C. L. Drum, J. Ferguson, et al. 2012. "Adaptive Licensing: Taking the Next Step in the Evolution of Drug Approval." *Clinical Pharmacology & Therapeutics* 91(3):426–37.

European Medicines Agency. 2010. "Road Map to 2015." (December 16). http://www.ema.europa.eu/docs/en_GB/document_library/Report/2011/01/WC500101373.pdf (accessed February 5, 2014).

Evans, B. 2013. "Legal Trends Driving the Clinical Translation of Pharmacogenomics." In *Principles of Pharmacogenetics and Pharmacogenomics*, ed. David Flockhart, David B. Goldstein, and Russ B. Altman, 81–94. New York: Cambridge University Press.

Evans, B. 2010. "Seven Pillars of a New Evidentiary Paradigm: The Food, Drug and Cosmetic Act Enters the Genomic Era." *Notre Dame Law Review* 85(2):419–524.

Flood, C. M. and P. Dyke. 2012. "The Data Divide: Managing the Misalignment in Canada's Evidentiary Requirements for Drug Regulation and Funding." *UBC Law Review* 45(2):283–328.

Garrison Jr., L. P., A. Towse, A. Briggs, G. de Pouvourville, J. Grueger, P. E. Mohr, J. L. Severens, P. Siviero, and M. Sleeper. 2013. "Performance-Based Risk-Sharing Arrangements—Good Practices for Design, Implementation and Evaluation: A Report of the ISPOR Good Practices for Performance-Based Risk-Sharing Arrangements Task Force." *Value Health* 16:703–19.

Hamburg, M. A. and J. M. Sharfstein. 2009. "The FDA as a Public Health Agency." *New England Journal of Medicine* 360:2493–95.

Hughes, D. and B. Williams-Jones. 2013. "Coalition Priorité Cancer and the Pharmaceutical Industry in Québec: Conflicts of Interest Influence in the Reimbursement of Expensive Cancer Drugs?" *Health Policy* 9(1):52–64.

Ioannidis, J. P. A. 2013. "To Replicate or Not to Replicate: The Case of Pharmacogenetic Studies: Have Pharmacogenomics Failed, or Do They Just Need Larger-Scale Evidence and More Replication?" *Circular: Cardiovascular Genetics* 6:413–18.

Keeling, P. 2007. "Personalized Medicine: The Absence of 'Model-Changing' Financial Incentives." *Future Medicine* 4(1):73–81.

Kimmelman, J. and J. A. Anderson. 2012. "Should Preclinical Studies Be Registered?" *Nature Biotechnology* 30:488–89.

Kuehn, B. M. 2012. "IOM Urges FDA to Be More Aggressive in Monitoring Safety of Approved Drugs." *Journal of American Medical Association* 307(23):2475–76.

Lasser, K. E., P. D. Allen, S. J. Woodhandler, D. U. Himmelstein, S. M. Wolfe, and D. H. Bor. 2002. "Timing of New Black Box Warnings and Withdrawals for Prescription Medications." *Journal of American Medical Association* 287(17):2215–20.

Lemmens, T. 2013. "Pharmaceutical Knowledge Governance: A Human Rights Perspective." *Journal of Law, Medicine and Ethics* 41(1):163–84.

Lemmens, T. and C. Telfer. 2012. "Access to Information and the Right to Health: The Human Rights Case for Clinical Trials Transparency." *American Journal of Law and Medicine* 38:63–112.

Lexchin, J. 2012a. "New Drugs and Safety: What Happened to New Active Substances Approved Between 1985 and 2010?" *Archives of Internal Medicine* 172(21):1680–81.

———. 2012b. "Those Who Have the Gold Make the Evidence: How the Pharmaceutical Industry Biases the Outcomes of Clinical Trials of Medications." *Science and Engineering Ethics* 18(2):247–61.

Light, D. W. and R. Warburton. 2011. "Demythologizing the High Costs of Pharmaceutical Research." *BioSocieties* 6:34–50.

Moore, T. J. and C. D. Furberg. 2014. "Development Times, Clinical Testing, Postmarket Follow-Up, and Safety Risks for the New Drugs Approved by the US Food and Drug Administration: The Class of 2008." *Journal of American Medical Association* 174(1):90–95.

Office of Inspector General. 2013. "Report (OEI-04-11-00510): FDA Lacks Comprehensive Data to Determine Whether Risk Evaluation and Mitigation Strategies Improve Drug Safety." *Office of Inspector General*. February 12. https://oig.hhs.gov/oei/reports/oei-04-11-00510.pdf (accessed February 5, 2014).

Owen, A. J., J. Spinks, A. Meehan, T. Robb, M. Hardy, D. Kwasha, J. Wlodarczyk, and C. Reid. 2008. "A New Model to Evaluate the Long-Term Cost Effectiveness of Orphan and Highly Specialised Drugs Following Listing on the Australian Pharmaceutical Benefits Scheme: The Bosentan Patient Registry." *Journal of Medical Economics* 11(2):235–43.

Personalized Medicine Coalition. 2011. "Personalized Medicine by the Numbers." *Personalized Medicine Coalition* (October). http://www.personalizedmedicinecoalition.org/sites/default/files/files/PM_by_the_Numbers.pdf (accessed January 30, 2013).

Psaty, B. M., E. M. Meslin, and A. Breckenridge. 2012. "A Lifecycle Approach to the Evaluation of FDA Approval Methods and Regulatory Actions." *Journal of American Medical Association* 307(23):2491–92.

Ray, W. A. and M. Stein. 2006. "Reform of Drug Regulation—Beyond an Independent Drug-Safety Board." *New England Journal of Medicine* 354(2):194–201.

Rockoff, J. D. 2013. "Drug Makers See Profit Potential in Rare Diseases." *Wall Street Journal* (January 30). http://online.wsj.com/article/SB10001424127887323926104578273900197322758.html.

Sherman, R. E., J. Li, S. Shapley, and M. Robb. 2013. "Expediting Drug Development—The FDA's New 'Breakthrough Therapy' Designation." *New England Journal of Medicine* 369(20):1877–80.

Topol, E. J. 2004. "Failing the Public Health—Rofecoxib, Merck, and the FDA." *New England Journal of Medicine* 351:1707–9.

U.S. Food and Drug Administration (FDA). 2013. "Fast Track, Breakthrough Therapy, Accelerated Approval and Priority Review." http://www.fda.gov/forconsumers/byaudience/forpatientadvocates/speedingaccesstoimportantnewtherapies/ucm128291.htm (accessed January 30, 2014).

———. 2013a. "Structured Approach to Benefit-Risk Assessment in Drug Regulatory Decision-Making." http://www.fda.gov/downloads/forindustry/userfees/prescriptiondruguserfee/ucm329758.pdf (accessed February 10, 2014).

———. 2012. "A Brief Overview of Risk Evaluation & Mitigation Strategies." http://www.fda.gov/downloads/aboutfda/transparency/basics/ucm328784.pdf (accessed February 5, 2014).

———. 2012a. "Enrichment Strategies for Clinical Trials to Support Approval of Human Drugs and Biological Products (Draft Guidance for Industry)." http://www.fda.gov/downloads/Drugs/GuidanceComplianceRegulatoryInformation/Guidances/UCM332181.pdf (accessed January 30, 2014).

———. 2012b. *Report on the Performance of Drug and Biologics Firms in Conducting Postmarketing Requirements and Commitments.* Federal Register Online. http://www.gpo.gov/fdsys/pkg/FR-2012-03-06/html/2012-5302.htm (accessed March 19, 2014).

———. 2007. "FDA's Response to the Institute of Medicine's 2006 Report." http://www.fda.gov/downloads/drugs/drugsafety/postmarketdrugsafetyinformationforpatientsandproviders/ucm171627.pdf (accessed January 30, 2014).

Wiktorowicz, M., J. Lexchin, and K. Moscou. 2012. "Pharmacovigilance in Europe and North America: Divergent Approaches." *Social Science and Medicine* 75(1):165–70.

Woodcock, J. 2007. "The Prospects for 'Personalized Medicine' in Drug Development and Drug Therapy." *Clinical Pharmacology & Therapeutics* 81(2):164–69.

CHAPTER FIFTEEN

FDA's Public Health Imperative

An Increased Role for Active Postmarket Analysis

EFTHIMIOS PARASIDIS

THE FUNDAMENTAL purpose of the U.S. Food and Drug Administration (FDA) is to promote and protect the public health (Hamburg 2010). An integral component of the agency's public health mission is the creation and maintenance of a robust framework for regulating medical products. Experts at FDA and throughout the health care industry have long heralded a "life-cycle" approach to regulation, whereby safety and efficacy are examined both pre- and postmarket (IOM 2007a). Despite the documented need for life-cycle analysis, FDA has focused its regulatory efforts on premarket review and has exerted comparably little energy on postmarket analysis.

This imbalance is largely the result of statutory provisions that have been ingrained in FDA's fabric for decades—not only has Congress underfunded and overburdened FDA, it has incentivized speed in the premarket review process and has enacted legislation that, to varying degrees, has limited the agency's ability to mandate postmarket studies (Parasidis 2011). This dynamic has directly impacted the reliability of risk–benefit profiles for marketed products. Patients, providers, and payors are affected significantly, as each is deprived of evidence that could prove integral to the decision of which treatment option to pursue, or whether a particular course of treatment is medically necessary. The risks borne by patients are exacerbated by preemption laws, which limit or preclude legal remedies in the event of adverse events, and an expanding commercial speech doctrine (discussed in part 3 of this volume), which hinders FDA's ability to regulate products once initial approval is granted.

In this chapter, I contend that the future of FDA as a public health agency is largely dependent on how well the agency is able to account for the current and evolving legal and political landscapes. In the context of medical products, this entails leveraging the agency's mandate and resources to address the limitations of premarket review and expand the instances in which postmarket analysis is required. Specifically, I argue that, in all cases where it has the authority to do so, FDA should require that sponsors conduct active postmarket analysis for the life of their products. As will be discussed, active postmarket analysis encompasses thorough, timely, and continuous monitoring of risks and benefits in real-world patients. Such monitoring includes utilization of health information technology (IT), observational studies, biomedical informatics, and, where appropriate, postmarket clinical trials. While FDA should assist in framing postmarket obligations and reviewing the results, sponsors must be held accountable for conducting and completing their postmarket studies.

I. THE ROLE OF POSTMARKET ANALYSIS IN THE DRUG REVIEW PROCESS

Since premarket clinical trials are limited in duration and scope, the risk–benefit profile that is derived from premarket studies is often inaccurate or incomplete. For example, premarket trials cannot capture latent adverse events nor can premarket studies sufficiently reveal adverse events that occur at low rates. In addition, premarket studies typically do not include patients with comorbidities; as such, the data do not always reflect how "real-world" patients react to treatment. Notwithstanding the ability of postmarket studies to mitigate the shortcomings of the premarket review process, decades of political and economic pressures have forced FDA to dedicate the majority of its resources to the premarket approval process (Parasidis 2011).

In situations where FDA has required postmarket studies, enforcement has been lax, and many sponsors have failed to complete their postmarket obligations (Kessler and Vladeck 2008).

In this section, I provide an overview of the framework under which FDA must work to fulfill its public health mission in the context of medical products. I begin by briefly summarizing FDA's passive framework for postmarket surveillance. I then discuss the symbiotic relationship between regulation and state tort claims, and explain how limitations on the availability of tort remedies have exacerbated the risks caused by FDA's passive postmarket surveillance regime. Although I will focus my discussion in this chapter on the regulation of pharmaceuticals, active postmarket analysis should be a component of all medical products, including medical devices and vaccines.

A. FDA's Passive Framework for Postmarket Surveillance

An influential report published by the Institute of Medicine (IOM) found that FDA has not historically had "adequate resources or procedures for translating preapproval safety signals into effective postmarketing studies, for monitoring and ascertaining the safety of new marketed drugs, for responding promptly to the safety problems that are discovered after marketing approval, and for quickly and effectively communicating appropriate risk information to the public" (IOM 2007a). Following the report—which was motivated in part by the public health catastrophe surrounding Vioxx, which caused thousands of deaths and tens of thousands of serious adverse events—Congress enacted the FDA Amendments Act of 2007 (FDAAA). Among its provisions, FDAAA requires that FDA create an active postmarket surveillance system.

In response to the legislative mandate, FDA established the Sentinel System. The Sentinel System is a nationwide electronic reporting system for monitoring medical products, which will have access to a wealth of health information that can be applied to study questions related to safety and efficacy. The Sentinel System supplements FDA's various postmarket surveillance schemes. For example, the FDA Adverse Event Reporting System (FAERS) has been FDA's primary source for managing and monitoring adverse events (Keuhn 2013). Under FAERS, formerly known as AERS, drug sponsors have an obligation to report known adverse health events to FDA in a timely manner. Although FAERS provides a conduit through which adverse events can be reported to FDA, the Federal Food, Drug, and Cosmetic Act (FDCA) does not contain a provision that requires drug sponsors to monitor their products for adverse events.

A supplement to FAERS is MedWatch, where health professionals and patients may voluntarily report adverse events. Reports submitted through MedWatch are comparatively minimal, accounting for less than 5 percent of all reported adverse events. As with drug sponsors, physicians and patients do not have an affirmative duty to seek out information related to adverse events. Furthermore, unlike drug sponsors, who must report what they know, doctors and patients are not obligated to report known adverse events to FDA. Coupled with the Sentinel System, FAERS, and MedWatch, the FDA's Document Archiving, Reporting and Regulatory Tracking System (DARRTS) facilitates tracking of postmarket safety issues. DARRTS allows FDA to manage activities related to evaluations of safety issues, such as due dates and postmarket safety reports (IOM 2007a). FDA also maintains an adverse event database for vaccines (VAERS) and medical devices (MAUDE).

In addition to mandating that FDA adopt an active postmarket surveillance system, FDAAA grants FDA the ability to require that drug sponsors complete a Risk Evaluation and Mitigation Strategy (REMS). The goal of the REMS program is to detect adverse events as quickly as possible in order to provide FDA with timely and accurate information that may be utilized to determine whether proactive measures should be taken, such as changing a drug's label or removing the drug from the market. FDA can require a REMS before or after a drug is approved, and a REMS can be required for a specific drug or an entire drug class. While FDA can require a REMS as a condition of approval, if the agency seeks to require a REMS after a drug is approved, the agency must demonstrate that new safety information has arisen (Gilhooley 2008). New safety information is defined as "a serious risk associated with use of the drug which FDA has become aware since the drug was approved, since a REMS was required, or since the last assessment of the REMS" (FDA 2014).

Despite FDA's extensive postmarket surveillance tools, studies estimate that FDA's postmarket network captures less than one percent of serious adverse reactions. This is largely due to the fact that sponsors do not have an affirmative duty to monitor their products for adverse events and that physicians do not have a duty to report known adverse events. Importantly, reported information is of limited clinical relevance because, from an epidemiological standpoint, "the FDA does not know how many people are using the drug and does not have adequate information about those who are"; as a result, FDA has "difficulty determining the incidence of a given adverse reaction" and whether it affects a particular subpopulation of patients (Kessler and Vladeck 2008). Taken together, FDA's postmarket surveillance framework has not consistently provided the agency with information sufficient to make intelligent decisions as to whether new safety risks should be communicated to the public.

B. Why Limitations on State Tort Claims Increase the Need for Active Postmarket Analysis

Since FDA does not have accurate information on postmarket risks, the agency is deprived of information that could help it make important decisions related to safety and efficacy of marketed products. This translates to an increased risk for patients and increased uncertainty on the part of providers and payors as to whether a particular treatment is medically necessary. Preemption of state tort claims—which refers to instances where patients injured from medical products are legally precluded from seeking remedies through the courts, a topic discussed in greater detail in other chapters of this volume—exacerbates the risks that stem from the regulatory regime.

In the context of FDA-regulated medical products, Congress has enacted targeted measures that immunize certain enterprises such as vaccine and generic drug manufacturers (Parasidis 2011). Preemption does not apply to manufacturers of brand-name drugs, nor does it apply to medical devices approved through the 510k process (contrast with medical devices approved through the PMA process, where preemption applies) (ibid.). Notably, in instances where preemption applies, a sponsor is not under a general obligation to research a marketed product for information related to safety and efficacy. In addition to federal immunities, a number of states have taken an active role in further limiting industry's liability exposure. These reforms include caps on noneconomic and punitive damages, a refusal to permit strict liability claims, and recognition of a "regulatory compliance defense" (Tobias 2008).

Michigan's regulatory compliance defense is paradigmatic—in Michigan, an FDA-approved drug that contains an FDA-approved label is neither defective nor unreasonably dangerous as a matter of law. In Utah, conformity with FDA regulations establishes reasonable care or nondefectiveness as a matter of law, while Texas views compliance with FDA regulations as strong and substantial evidence that a product is not defective. In New Jersey, a label approved by FDA constitutes adequate warning as a matter of law. Various states have rejected strict liability claims for defective design and have solely permitted claims based in negligence. Several states shield pharmaceutical companies from punitive damages unless an injured plaintiff can show that the company fraudulently obtained FDA approval. As with federal preemption of state tort claims, there is no state reform measure that predicates limited liability on active postmarket analysis for safety and efficacy (Parasidis 2011).

Although fear of lawsuits may incentivize drug sponsors to disclose adequate information regarding safety and efficacy, civil litigation is not the optimal means of regulating the life-science industry. Civil suits are not the ideal vehicle for setting the standard of care or determining adequate product warnings. Lawsuits involve retrospective debate of medical information that, often times, was unavailable to the provider or patient at the time medical treatment was provided. Indeed, due to variations in civil judgments, a sponsor may be subject to different standards in different states.

Notably, tort law is not an efficient means of transferring costs from tortfeasor to victim (Goldberg and Zipursky 2010). While a civil judgment may provide monetary compensation and serve to vindicate wrongful conduct, studies have shown that approximately fifty cents of every dollar recovered through tort claims accommodate administrative costs and attorney fees (Polinsky and Shavell 2010). More importantly, when a product kills or severely injures a patient, monetary damages through tort law provide limited relief to patients and their families.

While tort claims may not be the best vehicle for addressing information asymmetries, the availability of tort claims serves as an incentive for industry to be diligent and forthcoming with risk-related information. This incentive is diminished when preemptions laws preclude injured patients from having their day in court.

II. COMBINING REGULATORY AUTHORITY AND HEALTH INFORMATION TECHNOLOGY TO FACILITATE ACTIVE POSTMARKET ANALYSIS

To help fulfill its public health mission, FDA must capitalize on innovations in health IT and leverage its regulatory authority to create a comprehensive framework for active postmarket analysis. This framework can be structured around the Sentinel System, which has access to electronic health care data for more than one hundred million patients. Advancements in electronic medical record (EMR) systems and biomedical informatics provide important avenues through which sponsors, FDA, and researchers can access robust data sets. Along with providing near real-time medical information, there is broad consensus that the adoption of EMR systems and utilization of health IT in the provision of medical care will significantly improve the efficiency and effectiveness of health care (Jha et al. 2009).

Since enactment of FDAAA, FDA has been exploring new methods and tools that will enable it to capitalize on existing large postmarket databases. For example, the agency has been collaborating with the Centers for Medicare and Medicaid Services (CMS) and the Department of Health and Human Services (HHS) on a postmarket surveillance project that utilizes Medicare and Medicaid data. Meanwhile, FDA has stated that it "is aggressively recruiting more epidemiologists, statisticians, medical officers, safety evaluators, statistical programmers, [and] data managers" (FDA 2009). These experts will be called upon to timely and effectively access and analyze new safety data. Through collaborative efforts, the agency is looking to leverage existing databases and emerging health information technologies to target science at every stage of a drug's life-cycle.

These substantial additions to the postmarket arsenal of FDA set the stage for a transition from passive monitoring of adverse events to active postmarket requirements. As numerous reports have discussed, passive surveillance has serious limitations, which include "underreporting, biased reporting, and difficulties in attributing an adverse event to a specific drug" (IOM 2007b). Replacing the passive system of postmarket surveillance with an active postmarket framework is an intelligent next step toward a stronger and more effective drug safety system.

Under existing regulations, FDA can mandate postmarket studies as a condition of approval or when a new safety signal arises. Accordingly, FDA's first step should be to mandate active postmarket analysis as a matter of course for the life of all newly approved products. Each sponsor should be responsible for fulfilling requirements outlined in a postmarket surveillance plan and REMS. For each product, the drug sponsor would be responsible for providing postmarket update reports to FDA and the public on a regular basis. If postmarket research reveals that the rate or severity of an adverse effect exceeds that which is identified on the label (whether for the general population or a specific subpopulation), or other information relevant to the risk–benefit profile of the underlying product, the sponsor should be responsible for filing a report with FDA and publicly disclosing this new information.

The public disclosure component could take many forms. For example, FDA could create a website that allows users to search by medical product. FDA's recent collaborations with Drugs.com, WedMD, and Medscape are particularly promising avenues through which the public disclosure requirement could be fulfilled. A product's page could contain the current label, a complete history of submitted postmarket reports, and FDA's analysis of submitted reports. The product page should also contain summary information that identifies ongoing postmarket obligations. All summary information should use language that the general public could understand, with links to more detailed scientific discussion that is geared toward providers.

Active postmarket analysis is particularly important for new molecular entities, since such drugs contain molecules that have never been approved for use in humans. Similarly, active postmarket analysis is indispensable for drugs approved on surrogate endpoints, since the drug sponsor was not obligated to demonstrate an actual clinical benefit during the premarket review process. These concerns are not merely theoretical. According to one recent study, between 2005 and 2011, nearly half of new molecular entity approvals were granted solely on the basis of surrogate endpoints (Downing et al. 2014).

While mandating active postmarket analysis for all new drug approvals will help produce real-world risk–benefit information and mitigate information asymmetries, it will only capture a small percentage of drugs that are on the market. The key to a robust active postmarket system is ensuring that *all* marketed products are being continually monitored and evaluated. Current guidelines grant FDA the authority to mandate postmarket studies or REMS for approved products only where a new safety signal arises. The FDCA defines "new safety signal" to include serious risks that were previously unknown to the agency. The scope of this definition is unclear, and FDA has the discretion to define serious risks in a number of ways, either taking an expansive or limited view of the term.

Given the public health concerns raised by the existing pre- and postmarket framework, FDA should err on the side of patient safety and take an expansive approach to identifying the instances in which it has the authority to require a REMS or postmarket research for approved products. For example, a serious risk can include an instance where a drug is used for off-label purposes. Since off-label uses are not evaluated by FDA and drug labels do not provide guidance for off-label uses, the fact that patients are using a marketed product for an unapproved use raises a new safety signal. Selective publication that disproportionally favors positive findings is well documented, and the underreporting of negative results raises concerns for patients, providers, and payors, each of whom is left largely in the dark when it comes to determining whether a particular treatment is reasonable or medically necessary (Light 2010).

Under these circumstances, FDA could require active postmarket analysis as a condition of allowing the drug to remain on the market. There is nothing in the FDCA that limits the type of postmarket research that FDA could require. Rather, the FDCA sets the trigger that grants the agency the discretion to mandate postmarket studies. Although arguably FDA would have the authority to pull a drug from the market if adverse events from off-label uses occur, this is a drastic remedy that may preclude legitimate uses of an approved product. A more sensible alternative, and one that truly promotes the agency's public health mission, is using regulatory tools to bring forth meaningful information on each product's risks and benefits.

The Second Circuit's decision in *United States v. Caronia*, which opens the door to increased off-label promotion by drug sponsors, provides added justification for determining that off-label use constitutes a new safety signal for purposes of triggering FDA's ability to mandate active postmarket analysis (2012:149). An integral component of FDA's regulatory authority centers on the ability of the agency to monitor and limit drug marketing and advertising. As Kesselheim and Mello discuss in their chapter, *Caronia* casts doubt over the agency's ability to enforce these regulations. Though the *Caronia* decision is only binding within the Second Circuit—which encompasses New York, Connecticut, and Vermont—other courts may adopt the reasoning of *Caronia* should a similar first amendment challenge to an FDCA violation arise. Equally as important, the *Caronia* decision raises serious questions as to the government's ability to prosecute off-label promotion under federal and state qui tam laws.

III. PROJECTED COSTS OF ACTIVE POSTMARKET ANALYSIS

FDA projects that a postmarket framework focused on active analysis will require funding and resources similar to those of premarket review (FDA 2009). In current dollars, this would set the cost at approximately $500 million per year. This is a negligible amount, representing 0.01 percent of the annual federal budget. Further, when one considers that an active postmarket framework will directly and significantly benefit all Americans, the cost per capita—$1.56 per person per year—is miniscule.

Of course, government expenditures are not the full picture. In addition to government costs associated with an active postmarket surveillance system, each drug sponsor will be responsible for funding the research related to its postmarket obligations. Although total costs per product will vary, a typical observational study that relies on primary- and secondary-data costs between $100,000 and $250,000 for small studies of less than two years of duration, and $1.5 million to $3 million for larger and lengthier studies (Holve and Pittman 2009). Though not insignificant amounts, these figures represent a fraction of the total cost for drug development. For instance, industry figures place the cost of developing one drug at approximately $1.2 billion (Adams and Brantner 2010). Thus, as with the projected federal expenditures, the projected postmarket research budgets are negligible when placed into context.

At one extreme, these costs may be passed on entirely to patients; at the other, industry would fully absorb the costs and decrease its return on investment. While industry's incentive may be to pass the costs to consumers, public and private payors have other levers, such as negotiated rates, which could help balance the impact of the costs of an active postmarket system. To the extent current practices outside of the United States serve as a guide, Americans may be able to receive more postmarket analysis and cheaper drug prices. For example, although both Japan and the European Union maintain a more active postmarket system than that of the United States, drug prices in these regions are significantly lower than those in the United States. This is a direct result of nations negotiating drug prices with the drug sponsor. Unlike most industrialized nations, the United States generally does not negotiate drug prices with drug sponsors, though some large insurance companies and public payors do so. Legislation currently prohibits CMS—which is the largest payor, public or private, in the United States—from negotiating lower drug prices, although this prohibition has often been called into question.

As a result, Americans pay a significant premium for their drugs. In fact, prescription drug prices in the United States are among the highest in the world, and per capita prescription drug spending in the United States is substantially more than in all other nations (Kanavos et al. 2013). Even within the United States, the un-negotiated costs for individuals are more than double the negotiated costs. For example, within the public payor system in the United States, the Department of Veterans Affairs (VA) negotiates drug prices with pharmaceutical companies. These negotiations result in a 58 percent discount in drug prices for VA patients when compared to CMS patients (Steinberg and Bailey 2007).

At the macro level, the pharmaceutical industry is one of the most profitable in the world, earning over $900 billion in 2010, one-third of which came from sales in the United States. The industry had an average return on investment of 13.3% in 2010, which is 64 percent higher than the average return on investment across all industries. Between 2004 and 2008, the pharmaceutical industry's average return on investment was 19 percent, which places the industry among the highest of all industries. As Standard & Poor's concludes, this "lofty ratio" is "a function of the industry's . . . high profit margins" (Saftlas 2011). The operating profit margins of pharmaceutical companies averaged 32 percent in 2008, nearly double that for corporations in the S&P 500 index. "Net earnings as a percentage of sales averaged [approximately] 16% between 2004 and 2008," which is 150 percent more than the average of companies in the S&P 500 index. Notably, growth in the pharmaceutical industry is projected to increase at a margin twice that of the global market (ibid.).

In the United States, the pharmaceutical industry also enjoys significant government subsidies and tax incentives. Importantly, much of the initial pharmaceutical research in the United States is funded by American taxpayers through entities such as the National Institutes of Health. Once initial research demonstrates practical promise, industry will typically take over the development of the product. Balancing incentives to innovate with public health and rising health care costs is no simple task; that said, the pharmaceutical industry is a lucrative business in which there is room to incorporate active postmarket analysis as a component of marketed products.

IV. CONCLUSION

Because FDA's lackluster postmarket framework fails to capture the majority of adverse events, important information on safety and efficacy escapes the eyes of regulators, physicians, patients, and payors. While calls for increased postmarket analysis date back at least to the 1960s, FDA has not taken meaningful steps toward including active postmarket analysis as a regulatory requirement for all marketed products. Although an active postmarket framework comes at a cost, the projected savings are substantial. In the United States, it is estimated that adverse drug events cause more than 100,000 deaths and 2,000,000 serious injuries annually. The cost of treating patients with adverse drug events is approximately $1.6 to $5 billion per year, half of which may be preventable (Bond and Raehl 2006). This figure does not include the costs to patients and their families, or any resulting litigation. The figures also do not include the costs of lost productivity.

All stakeholders in the health care industry play an integral role in ensuring that maximum benefits flow from an active postmarket surveillance system. Patients must be mindful to fully disclose their medical histories and any adverse health events, while providers must ensure that their diagnosis and treatment regimens are carefully documented in the patients' records. Health IT experts are responsible for accurately transcribing patient records into searchable EMRs, creating a query system that captures pertinent medical information from the EMRs, and producing aggregated and de-identified data that are reflective of the underlying patient population. Sponsors must frame queries to capture relevant data and then interpret the data in an honest, transparent, and scientifically sound manner. Regulators must guide sponsors in determining which queries are appropriate and interpret results in a manner that furthers the public health. Independent researchers work to supplement the inquiries and analyses of sponsors and regulators.

The Sentinel System has the ability to address the adverse event gap in the United States, but only if FDA combines the potential of Sentinel data with its regulatory authority and establishes an active duty on the part of drug sponsors to seek out and report information related to postmarket safety and efficacy. More accurate information on safety and efficacy will result in fewer injuries and deaths due to adverse health events. Better information will also reduce health care costs by prescreening high-risk patients before treatment. Not only will the patient be spared the potential side effects, payors will not incur costs associated with use of the product.

Industry stands to reduce costs as well. The costs of litigating civil claims are immense, with settlement funds reaching into the billions of dollars. Individual claims against drug companies, where the adverse drug event caused a permanent disability, averaged $4.3 million per patient. Claims for death and other serious adverse events also resulted in judgments or settlements in the millions per patient (ibid.). Taken together, the avoidance of these financial and reputational costs, coupled with the health benefits to patients, far outweighs the costs of active postmarket analysis.

While FDA's ability to mandate postmarket obligations is somewhat limited, the agency is supplied with an arsenal of regulatory tools that, if properly utilized, could go a long way toward bringing important data on risks and benefits to the forefront. By capitalizing on biomedical informatics techniques and the wealth of information available through the Sentinel System, FDA is poised to reinvent its regulatory agenda and assert its regulatory power in ways that will bring meaningful information to stakeholders throughout the health care industry.

In the context of FDA regulation of medical products, the agency must work to balance access and uncertainty, and ensure that providers, patients, and payors have the best available evidence of safety and efficacy. Americans pay more for their medications and receive less from their regulators in terms of postmarket review and analysis than citizens of other nations. Promoting and protecting the public health requires that FDA modernize its approach to regulation by incorporating active postmarket analysis for all medical products.

NOTE

This chapter is based, in part, on E. Parasidis. 2011. "Patients Over Politics: Addressing Legislative Failure in the Regulation of Medical Products," *Wisconsin Law Review* 5: 929–1002.

REFERENCES

Adams, C. P. and V. V. Brantner. 2010. "Spending on New Drug Development." *Health Economics* 19(2):130–41.

Bond, C. A. and C. L. Raehl. 2006. "Adverse Drug Reactions in United States Hospitals." *Pharmacotherapy* 26:601–8.

Downing, N.S. et al. 2014. "Clinical Trial Evidence Supporting FDA Approval of Novel Therapeutic Agents, 2005–2012." *Journal of the American Medical Association* 311(4):368–77.

Gilhooley, M. 2008. "Addressing Potential Drug Risks: The Limits of Testing, Risk Signals, Preemption, and the Drug Reform Legislation." *South Carolina Law Review* 59:347.

Goldberg, J. C. P. and B. C. Zipursky. 2010. "The Easy Case for Product Liability Law: A Response to Professors Polinsky and Shavell." *Harvard Law Review* 123:1919.

Hamburg, M. A. 2010. "Innovation, Regulation, and the FDA." *New England Journal of Medicine* 363:2228–32.

Holve, E. and P. Pittman. 2009. "A First Look at the Volume and Cost of Comparative Effectiveness Research in the United States." *AcademyHealth*.

Institute of Medicine (IOM). 2007a. "The Future of Drug Safety: Promoting and Protecting the Health of the Public."

———. 2007b. "Challenges for the FDA: The Future of Drug Safety."

Jha, A. K. et al. 2009. "Use of Electronic Health Records in U.S. Hospitals." *New England Journal of Medicine* 360:1628–38.

Kanavos, P. et al. 2013. "Higher U.S. Branded Drug Prices and Spending Compared to Other Countries May Stem Partly from Quick Uptake of New Drugs." *Health Affairs* 32(4):753–61.

Kessler, D. A. and D. C. Vladeck 2008. "A Critical Examination of the FDA's Efforts to Preempt Failure-to-Warn Claims." *Georgetown Law Journal* 96:461–95.

Keuhn, B. M. 2013. "Scientists Mine Web Search Data to Identify Epidemics and Adverse Events." *Journal of the American Medical Association* 309(18):1883–84.

Light, D. W. 2010. *The Risks of Prescription Drugs*. New York: Columbia University Press.

Parasidis, E. 2011. "Patients Over Politics: Addressing Legislative Failure in the Regulation of Medical Products." *Wisconsin Law Review* 2011(5):929–1002.

Polinsky, A. M. and S. Shavell. 2010. "The Uneasy Case for Product Liability." *Harvard Law Review* 123:1437.

Saftlas, H. 2011. "Healthcare: Pharmaceuticals." *Standard & Poor's Industry Surveys*.

Steinberg, M. and K. Bailey. 2007. "No Bargain: Medicare Drug Plans Deliver High Prices." *Families USA*.

Tobias, C. 2008. "FDA Regulatory Compliance Reconsidered." *Cornell Law Review* 93:1003–38.

United States v. Caronia, 703 F.3d 149 (2d Cir. 2012).

U.S. Food and Drug Administration (FDA). 2014. "A Brief Overview of Risk Evaluation & Mitigation Strategies (REMS)." Accessed May 2, 2014. http://www.fda.gov/downloads/AboutFDA/Transparency/Basics/UCM328784.pdf.

U.S. Food and Drug Administration (FDA). 2009. "Changing the Future of Drug Safety: FDA Initiatives to Strengthen and Transform the Drug Safety System." Accessed May 2, 2014.

PART FIVE

Old and New Issues in Drug Regulation

Introduction

R. ALTA CHARO

We have an opportunity for everyone in the world to have access to all the world's information. This has never before been possible. Why is ubiquitous information so profound? It's a tremendous equalizer. Information is power.
—GOOGLE EXECUTIVE CHIEF ERIC SCHMIDT, 2009 COMMENCEMENT ADDRESS AT THE UNIVERSITY OF PENNSYLVANIA

DRUG SAFETY communication, obstacles to manufacturing innovation, and the DESI experience. Three topics that appear quite distinct. But threading through them are observations about the control of information, the importance of disclosure, and the power to pronounce the truth.

The real regulatory power of the U.S. Food and Drug Administration (FDA) lies in its power to pronounce the truth. Its enabling statutes prohibit marketing adulterated or misbranded drugs and require them to undergo a premarket review by FDA. As of 1938, any drug that has not been proven to FDA to be safe is presumed to be unfit for its intended purpose and therefore is misbranded and not marketable. As of 1962, FDA must be persuaded of the drug's effectiveness as well. That the sponsor or any third-party evaluator might conclude that the drug in fact is safe and effective, and entirely fit for its intended use, is not sufficient. This must be proven as well and declared by FDA before marketing is permitted. In that sense, FDA owns the power to pronounce on the truth of the drug's fitness.

Carpenter, Greene, and Moffitt's chapter, "The Drug Efficacy Study Initiative and Its Manifold Legacies," describes the mid-twentieth-century regulatory land grab for control of truth, when drugs were now required to prove efficacy to FDA, even if they had long been on the market. While sponsors and consumers might have formed their own judgments about the effectiveness of these drugs, FDA now had the exclusive authority to make that determination, which it did through the Drug Efficacy Study Initiative (DESI). FDA was not required to prove the "truth" to a third party in a drug-by-drug adjudicatory proceeding but could make sweeping categorical determinations through rulemaking and publication in the *Federal Register*. Thus, the power to declare the truth of a claim of efficacy shifted from manufacture to government.

Moving forward in time to the end of the twentieth century, Geoffrey Levitt's chapter, "Drug Safety Communication: The Evolving Environment," describes the decentralization of drug safety information gathering and reporting and asks how this will affect control of truthful information. Pointing to the experience with Vioxx, he describes the decline of trust in the sponsor–regulatory dyad and the move to more transparency about clinical trial results or even patient-level data. But while multiple third-party analyses may yield new insights that were missed or omitted by the sponsor, they undermine FDA's power to declare what is "true" and risk creating confusion for the public and the practitioners.

Nicholson Price's chapter, "Innovation Policy Failures in the Manufacturing of Drugs," looks not at truth itself so much as at the effects of disclosing it. For pharmaceuticals, the intellectual property incentives are sufficient to overcome the barriers to market entry. For manufacturing, however, the disclosure requirements associated with patents discourage innovation because it is easy to "invent around" a process and hard to identify or prosecute copycats. The resulting preference for trade secrets helps the pioneer but slows innovation by others, who must now develop processes wholly from scratch. With patents, disclosure rules mean control of information shifts from inventor to the public. With manufacturing process trade secrets, however, control of information remains entirely in the hands of the manufacturer, to the detriment of the public's interest in newer, better methods.

It is often said that we live in the information age. But perhaps it is more to the point to say that we live in the age of grassroots information, decentralized information, and ubiquitous information. It calls to mind the revolution occasioned by the printing press, which allowed the spread of learning beyond the narrow confines of monasteries and castles, ultimately helping to empower both the laity and the middle class.

In America, disclosure and transparency are the watchwords, and distrust of authority and expertise is the new clarion call. Whether in the growing legal tolerance for off-label promotion or the movement to disclose research data, there is a culture of crowd-sourcing truth. The 1962 Kefauver-Harris Amendments and the DESI experience may well represent the last major regulatory successes at owning pharmaceutical truth. And with innovation a primary goal of contemporary economic policy, the trade secrets of drug manufacturers may well represent a dying form of commercial ownership of information. But with transparency, disclosure, and democratization of truth comes the challenge of interpretation and responsible use of information.

> Information paints no picture, sings no song, and writes no poem.
> —R. F. GEORGY, *NOTES FROM THE CAFÉ* (2014)

CHAPTER SIXTEEN

The Drug Efficacy Study and Its Manifold Legacies

DANIEL CARPENTER, JEREMY GREENE,
AND SUSAN MOFFITT

STANDARDIZATION OF products is a ubiquitous and ineluctable characteristic of modern capitalism. This activity—the development of and adherence to common guidelines for the production of goods or delivery of services—might be undertaken by private actors, the state, or social organizations, or even by combinations of these. In function, standardization deposits criteria of judgment for commodities that permit consumers, producers, and other audiences to perform differentiation and commensuration among products and services. By enabling differentiation, standardization assists consumers in making quality or risk distinctions among products, distinctions that can support optimization (quality comparisons that, conditional on price differences, may be sufficient for choice), status rankings, and assessments of risk. By enabling commensuration (developing a common metric of performance), standardization can assist producers and consumers alike in "showing up" to the same marketplace, one in which products are similar enough on other characteristics as to enable the functioning of an informative price system and indeed of innovation (Greene 2011; Greene and Kesselheim 2011).

Yet standardization does much more than ascribe difference and similarity to products. It also serves as a basis for the exclusion of some products from a market. In other words, standardization functions not only within markets but also across them. It can help to define boundaries between "commodity spaces," some of which legitimately qualify (socially, legally, culturally) as a market, others of which do not. Standardization in this sense is linked to the regulation of marketplaces, as regulation often relies upon standardization. In the licensing of attorneys, accountants, physicians, and other professionals, the state and public-private associations decide which kinds of practices and credentials qualify for "membership" in the set of producers offering services.[1] Yet licensing entities perform this work by relying explicitly on standards of "accountancy" and legal and medical practice (and "malpractice").

In the realm of pharmaceuticals, standardization and screening are linked through the power of the state to limit production and market entry to legitimized and approved compounds. Students of FDA commonly trace these functions to two pieces of legislation. The Federal Food, Drug, and Cosmetic Act of 1938 created a premarket approval clearance requirement based upon a safety standard, leaving to the federal government the authority to determine market entry. Then in 1962, in the midst of the global health tragedy of thalidomide, Congress passed the Kefauver-Harris Amendments, which together converted the preclearance requirement to a positive approval standard and added an efficacy criterion to the safety hurdle of the 1938 legislation. Although FDA had been using efficacy judgments as dispositive standards in drug approvals since the 1950s, the Kefauver-Harris Amendments shifted the continuum of burden of proof toward drug companies and greatly regimented and standardized the process of drug development (Carpenter 2010; Carpenter and Sin 2007; Jackson 1970; Marks 1996).

Yet in significant ways, it was another institutional shift—in the Drug Efficacy Study Initiative (DESI)—that decisively reshaped modern drug regulation. In re-crafting drug regulation, it also refashioned modern pharmaceutical science, modern administrative procedure, and even the structure of the biopharmaceutical industry. DESI remade all these domains by participating in their standardization. DESI was a combination of the Drug Efficacy Study (DES) and its implementation. It was brought about by a congressional mandate in 1962, when "efficacy" considerations were officially added to the safety considerations under which FDA had been reviewing new drugs. Noting that from 1938 to 1962 FDA had been reviewing drugs officially on "safety" considerations only, Congress required FDA to review all drugs approved under the 1938–1962 "safety regime" and assess them for efficacy.[2] Those that failed the exercise would be withdrawn. A large-scale "efficacy" review of more than 4,000 medications ensued.

This joint exercise in standardization and screening "unmade" a set of previously stable drug markets and brought into being a set of markets that had not previously existed. To be sure, DESI did many other things. It brought far-reaching changes in law, pharmacology, and politics. Legally, it established "summary judgment" as an administrative technique whereby federal agencies could issue broad rulings on a range of products—withdrawing them or ending their economic life as commodities—without having to take each and every case to an administrative law judge (Carpenter 2010:356). Methodologically, it was in DESI more than in the post-thalidomide intensity of 1962 and 1963 that the randomized, double-blind, placebo-controlled trial was established, as the problem of evaluating "old drugs" became a template for revising the standards for evaluating "new drugs." Politically, DESI brought a rupture between some of the older-line pharmaceutical companies like Abbott, Parke-David, and Upjohn, and the newly academic industrials like Merck and Pfizer (Carpenter 2010).

Our purpose in reviewing the development of the Drug Efficacy Study and its implementation is threefold: (1) to demonstrate how the process of standardization emerged in FDA drug regulation and contributed to FDA's power; (2) to highlight the ways in which DESI and FDA's process of standardization evoke long-standing debate on the impact of product regulation; and (3) to suggest potential durable legacies DESI may have on pharmaceutical markets and public health.

I. REGULATING IN RETROSPECT

The Drug Efficacy Study was a by-product of a congressional attempt to regulate a population of "grandfathered" drugs. The Federal Food, Drug, and Cosmetic Act of 1938 prohibited the introduction into interstate commerce of any "new drug" whose safety had not been established by examination of the "Secretary," namely the administrative apparatus of FDA. Thousands of new drug applications were submitted to the agency in the period from 1938 to 1962, and thousands were approved. Although it was widely acknowledged that efficacy considerations were also being used by FDA reviewers in the Bureau of Medicine (later called the Bureau of Drugs), there is little doubt that safety was both the principal consideration and the variable most observable in tests of animal and clinical pharmacology.

As a result, when Congress passed the Kefauver-Harris Drug Amendments of 1962 and added "effectiveness" officially to the criteria of new drug review, a large number of drugs for which no efficacy review had ever officially been conducted remained on the market. In the rulemaking that followed a year later—the Investigational New Drug rules of 1963—FDA elaborated an architecture of phased clinical experiment that would restructure clinical experiment and R&D worldwide in the ensuing decades. Yet it was not until the late 1960s that FDA would finally turn systematically to meet its statutory charge to examine the pre-1962 approved drugs. Commissioner James Goddard, knowing that his agency was burdened by the launch of a much more comprehensive new drug review system, decided that his agency could not address the problem using its existing administrative resources alone and contracted with the National Academy of Sciences and the National Research Council to conduct the review.

The NAS-NRC effort resulted in the convening of a Policy Advisory Committee of 27 members and 21 separate panels of therapeutically specific expertise. The executive director was Duke C. Trexler of the National Academy of Sciences. The taxonomy of expertise and review reflected much of the prevailing understanding of categorization in therapeutics at the time (see Table 16.1; Trexler 1966:12).

Steps to establish the panels began in 1966.[3] In his March 31, 1966 memorandum to the director of the NAS-NRC Division of Medical Sciences, Dr. Keith Cannan, FDA Commissioner Dr. James L. Goddard sought NAS-NRC help to assess the efficacy of drugs approved between 1938 and 1962 (Rettig, Earley, and Merrill 1992:49–50). In his memorandum to Dr. Cannan (1966), Commissioner Goddard noted the scope and likely downstream importance of the DESI task. He wrote, "Although this is a one time task requiring evaluation of material somewhat different from that now obtained in current drug approval procedures, its long range significance exceeds that of all other drug activity pursued by the Food and Drug Administration."

TABLE 16.1
Desi Panel Organization by Therapeutic Category

1. Allergy	12. Gastroenterologic (2 panels)
2. Anesthesia	13. Hematologic
3. Anti-Infective (5 panels)	14. Metabolic
4. Anti-Neoplastic	15. Neurologic
5. Anti-Parasitic	16. Ophthalmologic
6. Cardiovascular	17. Psychiatric
7. Dentistry	18. Relief of Pain
8. Dermatologic (2 panels)	19. Reproductive System
9. Diagnostic	20. Respiratory Disturbances
10. Endocrine Disturbances	21. Rheumatic Diseases
11. Fluid and Electrolyte Balance	

FDA's public justification for contracting with the NAS-NRC was twofold: capacity and legitimacy. In its April 1966 press release, FDA stated that it sought NAS-NRC reviews because "The FDA itself does not have sufficient medical personnel to carry out a project of this scope" and because the NAS-NRC would be able to "tap the top medical and other scientific talent of the Nation" (FDA 1966). The task required reviewing and determining which drugs should be removed from the market based on inadequate evidence of efficacy. The scope and significance of the effort depended both on developing sufficient manpower and on securing legitimacy to make difficult and binding regulatory decisions. In his March 31, 1966 memo to Dr. Cannan, Commissioner Goddard continued, "Recommendations from the most expert sources are essential if this Administration is to suppress flagrant claims, eliminate worthless products and at the same time protect the physician's therapeutic resources" (Goddard 1966).

FDA officially contracted with the NAS-NRC in June 1966, and the NAS-NRC convened its Policy Advisory Committee in July 1966 to establish procedures for the efficacy review process. The Policy Advisory Committee subsequently established 30 review committees (see Table 16.2) comprised of six members each and assigned drugs to committees for review (Rettig, Earley, and Merrill 1992:49–52). The NAS-NRC, not FDA, selected panel members and organized panels roughly along the basis of therapeutic class. Based on evidence from trials submitted by firms, medical literature, and "the experience and judgment of panel members," panels assessed and reported drugs as effective, probably effective, possibly effective, or ineffective (Shorter 2008).[4]

The subsequent NAS-NRC review process occurred behind tightly closed doors. The committees' reviews were kept confidential, meetings were not publicly announced, and meeting locations were not publicly revealed (Bryan and Stern 1970:15).[5] This degree of secrecy was consistent with FDA practice at the time—deliberations

TABLE 16.2
Chairmen of Thirty Desi Panels, June 1968

Allergy: Dr. Bram Rose	Diagnostic Agents: Dr. Gilbert M. Mudge
Anesthesia: Dr. Emmanuel Papper	
Anti-Infective I: Dr. Heinz Eichenwald	Endocrine Disturbances: Dr. George Thorn
Anti-Infective II: Dr. William L. Hewitt	Fluid-Electrolyte Balance: Dr. Maurice Strauss
Anti-Infective III: Dr. William M. M. Kirby	Gastroenterology I: Dr. John T. Sessions
Anti-Infective IV: Dr. Calvin M. Kunin	Gastroenterology II: Dr. Henry J. Tumen
Anti-Infective V: Dr. William B. Tucker	Hematology: Dr. William H. Crosby
	Metabolic: Dr. Don H. Nelson
Anti-Neoplastic: Dr. Emil Frei III	Neurological: Dr. Melvis Yahr
Anti-Parasitic: Dr. Harold Brown	Ophthalmology: Dr. Irving H. Leopold
Cardiovascular I: Dr. Edward Freis	Psychiatric: Dr. Daniel X. Freedman
Cardiovascular II: Dr. Sol Sherry	Relief of Pain: Dr. Louis C. Lasagna
Dentistry: Dr. Edward Zegarelli	Reproductive System: Dr. Albert Segaloff
Dermatology I: Dr. Harvey Blank	
Dermatology II: Dr. Adolph Rostenberg, Jr.	Respiratory Disturbances: Dr. Sol Katz
	Rheumatic: Dr. Charles Ragan
Dermatology III: Dr. John A. Haserick	Surgery: Dr. George D. Zuidema

were kept private until regulatory decisions were issued, partly, FDA argued, to prevent market speculation and to encourage uninhibited deliberation.[6] Commissioner Goddard recalled that panel chairmen were publicly announced but not panel members in part to "protect panel members from potential pressure from industry" (Goddard 1969).[7] FDA also justified keeping the NAS-NRC reports confidential until the agency published orders in the *Federal Register* by arguing that releasing the reports earlier could induce panic, especially since FDA did not always adhere to NAS-NRC recommendations.[8] Once FDA released information and orders on drug efficacy, the reports from the NAS-NRC also became publicly available (Edwards 1970).

The scope of the panels' work was stunning both in terms of the vast swath of drugs reviewed and the outcome of the panels' reviews. By the close of the decade, the NAS-NRC panels produced 2,824 reports on 4,349 different drug formulations from 237 firms. Commenting on the scope of DESI panels' work, FDA Commissioner Charles C. Edwards noted, "At least 10,000 man hours were spent in formal decision-making" (Edwards 1970). Approximately two-thirds of the submitted drugs were evaluated by one NAS-NRC panel. The remaining one-third of the drugs were reviewed by multiple panels, ranging from two to fifteen different panels depending in part on whether the drugs were single-entity or combination drugs and in part on the number and nature of the efficacy claims made about the drugs (Edwards 1970). Estimates suggest that each panel reviewed, on average, 150 drugs (Rettig, Earley, and Merrill 1992:51). Commissioner Edwards noted that these reviews covered 80 percent "of the drugs in use" in 1970, that 10 percent of the reviews resulted in a panel determination that a drug was "ineffective," and that many more were deemed only "possibly effective" (Edwards 1970).[9]

Each panel had six members plus a chair (see Table 16.2), with the idea that the members would be chosen as representative of the expertise then existing in particular domains of clinical pharmacology and treatment. Each approved new drug application (NDA) was reviewed by the appropriate panel, and the NAS-NRC kept a separate database of its review proceedings in which the NDAs were assigned "log numbers." While the panels and FDA itself were criticized for proceeding too slowly, the Academy issued quarterly and annual reports on the drug-by-drug review.

The panels and the DES organization as a whole decided to assign pre-1962-approved NDAs into one of four categories,[10] arrayed in publications and below in decreasing order of their perceived value and effectiveness:

Category A: Effective
Category B: Probably Effective
Category C: Possibly Effective
Category D: Ineffective

These categories, invented on the fly by an organization of experts who convened behind closed doors with little publicity, would be implemented as public policy. Most of the initial attention around the DES categories, when the listings were first announced in the *Federal Register*, focused on those drugs assigned to Category D—and specifically scheduled for removal from the marketplace.

II. UNMAKING THE MARKETS FOR SAFE DRUGS

FDA's subsequent decisions to declare some drugs ineffective and to order them off the market were essentially binding. FDA brought the force of rulemaking to its withdrawal decisions by publishing orders in the *Federal Register* instead of using a formal adjudication hearing for each drug (Carpenter 2010:348).[11] Receiving a designation of ineffective meant withdrawing the product from the market, with some classes of drugs incurring the lion's share of the withdrawals. Some estimates suggest, for instance, that between 45 and 50 percent of psychopharmacologic drugs commonly used at the time were deemed ineffective and removed from the market (Shorter 2008). Fixed-combination antibiotics, such as Panalba, comprised a second class of drugs heavily represented in the category of drugs deemed "ineffective." Firms such as Upjohn sued FDA for having received an "ineffective" designation and for the drug's subsequent removal from the market, challenging the legality of the DESI process and ultimate decision. Critics alleged that the process, by forgoing a full administrative hearing for each drug adjudged and removed, denied firms their due process rights and disregarded clinicians' expert judgment (Carpenter 2010:329–30).[12] However, FDA's decisions to order drugs from the market prevailed, which we discuss in more detail later.[13]

Table 16.3 summarizes the withdrawals by selected categories of the Anatomical Therapeutic Chemical (ATC) Classification system used by the World Health Organization. The table helps to clarify the clear concentration of withdrawals in selected therapeutic categories, namely antibiotics, central nervous system drugs (where mental health drugs

TABLE 16.3
Partial Summary of Desi Withdrawals by ATC Therapeutic Code

ATC Code (Level 1 and Level 2)	Categorization	DESI Withdrawals Initiated, 1968 and After
A (Alimentary Tract and Metabolism)		
A01	Stomatological Preparations	17
A04	Antiemetics and Antinauseants	0
A06	Laxatives	0
A10	Drugs Used in Diabetes	0
A11 or A12	Vitamins or Mineral Supplements	18
A15	Appetite Stimulants	0
C (Cardiovascular System)		
C01	Cardiac Therapy	7
C02	Antihypertensives	1
C07	Beta Blocking Agents	0
D (Dermatologicals)		
D03	Preparations for Treatment of Wounds and Ulcers	13
D07	Corticosteroids, Dermatological Preparations	2
D10	Anti-Acne Preparations	0
J (Antiinfectives for Systemic Use)		
J01	Antibacterials for Systemic Use	127
J02	Antimyotics for Systemic Use	7
J07	Vaccines	0

TABLE 16.3
(Continued)

ATC Code (Level 1 and Level 2)	Categorization	DESI Withdrawals Initiated, 1968 and After
L (Antineoplastic and Immunomodulating Agents)		
L01	Antineoplastic Agents	0
L03	Immunostimulants	0
M (Musculo-Skeletal System)		
M01	Antiinflammatory and Anti-Rheumatic Products	12
M03	Muscle Relaxants	21
N (Nervous System)		
N05	Psycholeptics	22
N06	Psychoanaleptics	25
R (Respiratory Agents)		
R01 or R02	Nasal or Throat Preparations	30
R03	Drugs for Obstructive Airway Diseases	10
R05	Cough and Cold Preparations	3
V (Various)		
V09	Diagnostic Radiopharmaceuticals	2
V10	Therapeutic Radiopharmaceuticals	0

such as tranquilizers and reserpine-based therapies are classified), vitamin and mineral supplements on the prescription market, and respiratory agents. Yet those pharmacologic classes spared from DESI's enforcement are also notable, namely in various categories of oncology, beta-blockers in cardiology, many dermatologic therapies outside of wound and ulcer treatment, and antidiabetic and laxative drugs.

The stark differences across therapeutic categories bespeak both the variable perceptions of therapeutic efficacy in these cases, as well as the variable development of standards from which the NAS-NRC panels could draw in making efficacy determinations. The NAS-NRC panels did not conduct any new research but rather summarized samples of previous studies. Hence both the *amount* of research that had been conducted and the *character* of that research—in particular, whether it satisfied the "adequate and well-controlled trials" requirement of the 1962 law—were pivotal in shaping whether or not drugs were removed from legitimized pharmacological and medical practice in the United States. In related research, we have examined the possibility that these drug removals were associated with reduced mortality in the associated diseases, in part by removing subpar therapies from these areas, in part by creating "market space" (a sort of regulatory Schumpeterian creative destruction) for better therapies to emerge (Carpenter et al. unpublished). In other words, removing therapies from the market that failed to meet efficacy standards (perhaps destroying some firms' market share) may have created room for alternative, innovative therapies to take their place.[14]

The ultimate legal authority for removing drugs from the marketplace was not in dispute—FDA had the authority to declare invalid the new drug application of any drug and any company. Yet the central legal debate concerned the administrative efficacy with which the administration could do so. The Upjohn Company, headquartered in Kalamazoo, Michigan, in particular contested the process by which drugs (and its fixed-combination antibiotic Panalba) were removed by summary judgment and publication of removal lists in the *Federal Register*, demanding instead that each and every drug withdrawn proceed through a separate administrative hearing. Still, the removals proceeded apace in the 1970s, especially after the U.S. Supreme Court legitimized FDA's summary judgment procedure in a critical set of cases called the *Hynson* quartet: *Weinberger v. Hynson, Westcott & Dunning; USV Pharmaceutical Corp v. Weinberger; Weinberger v. Bentex Pharmaceuticals, Inc.*; and *Ciba Corp v. Weinberger*. In *Weinberger v. Hynson, Westcott & Dunning* (1973), the court found that firms' right to a hearing was conditional on firms' submission of "substantial evidence" of effectiveness. In terms of FDA standards established by 1970, substantial evidence of effectiveness meant producing two well-controlled studies that supported a drug's efficacy claim (Carpenter 2010:356). Justice William O. Douglas's opinion, writing for the Court in the *Hynson* case, affirmed that the 1962 amendments and subsequent FDA rules "express well-established principles of scientific investigation, in their reduction of the 'substantial evidence' standard to detailed guidelines for the protection of the public." In other words, the "well-controlled studies" standard FDA promulgated in 1970 was deemed "appropriate." The *Hynson* case marked the Court's affirmation of FDA's capacity to make efficacy judgments about drugs. It also paved the way for FDA's subsequent removal of over-the-counter drugs, 300 of which FDA found "ineffective" (Carpenter 2010:356–57).

III. MAKING NEW MARKETS FOR OLD DRUGS

For those drugs found ineffective and then removed from the market, the DESI program was the first step in a bitterly contested dismantling of vast segments of the pharmaceutical market. For those drugs found to be effective, however, DESI added robust academic and federal claims of therapeutic value to which any subsequent manufacturer could refer when marketing their own generically named product.[15] This was especially relevant as several pre-1962 DESI drugs were already off patent, and many more were going off patent each year.[16]

Several members of the Drug Efficacy Study were concerned that a DES Category A rating of "effective" applied to a drug *by generic name* might inappropriately be read as a federal certification that all versions of the same chemical compound, whether produced by brand-name manufacturers or small "fly-by-night" houses, were the same. Several of the members of the NAS-NRC committee, most notably the pharmacologist Alfred Gilman, had significant concerns that the DES judgments really applied only to the brand-name version of the drug evaluated and not to every pill, capsule, or table that might contain the same active ingredient.

The problem of generic equivalence, Duke Trexler recalled later (1969), was a question that "continually intruded into the work of the panels." When the DES final report was issued in 1969, it was accompanied by a set of comments solicited from the nearly two hundred panel members regarding their reflections on the problem of therapeutic equivalence of generic and brand-name drugs. There was a great deal of concern, but little consensus. As the DES reviews of effective drugs were completed and phased into regulatory implementation, the members of the review panels continued to voice discomfort that DES standards of efficacy would be attached to generically named drug products (Bryan and Stern 1970).[17]

As Gilman had predicted, after DESI, FDA found itself in an unusual bind regarding *effective* drugs. To require all new copies of off-patent drugs to submit an extensive new drug application (NDA) merely to market a drug that had been approved and used for at least seventeen years seemed both economically wasteful and potentially unethical. Instead, FDA proposed in 1969 that manufacturers of newly off-patent drugs need only file an *abbreviated* new drug application (ANDA) that showed proof of chemical equivalence according to standard pharmacopeial compendia. In other words, manufacturers needed only to show that the active chemical was present at the dose specified without any contaminants. This regulatory action simultaneously lowered barriers to market entry for potential generic manufacturers and opened up a clearly sanctioned set of DESI-effective drugs as a new market space for would-be generic manufacturers.[18]

Take for example, the small company Bolar, which, by the late 1960s, was trying to transform itself into a legitimate manufacturer of generic drugs. By 1969, an intensified FDA inspection described Bolar as a generic drug manufacturer whose product line consisted nearly entirely of amphetamines and vitamin tablets and a smattering of minor tranquilizers ("Report of Intensified Inspection, 4 November 1968" 1969). Only two of their drug products—both versions of the vasodilator pentaerythritol tetranitrate—had even begun the process of formal FDA approval through the ANDA process. For much of the late 1960s, FDA waged a slow campaign against Bolar products as adulterated and misbranded goods (Warner 1968).

It was doubly ironic then that when FDA and NAS-NRC released the first list of ineffective drugs published by the DESI project, one of the drugs found ineffective was pentaerythritol tetranitrate. Faced with the prospect of losing permission to market the only drug it was selling with legitimate FDA approval, Bolar's CEO, Robert Shulman, wrote a frustrated plea to the head of the Bureau on Drugs, Henry Simmons. Would FDA please let Schulman and Bolar know which drugs the DESI review had deemed "effective"—and therefore a viable market—so that Bolar might focus its efforts at regulatory approval in more productive channels? "We are a small manufacturer of generic drugs. In view of the recent NAS-NRC studies and the

Food and Drug Administration actions based on these studies, we are very interested in re-evaluating our product line. Accordingly, we are asking for a little guidance along these lines. Would it be possible to obtain a list of drugs that have been evaluated and into which category (effective, ineffective, etc.) that they fall [sic]. If we knew this information, we would be able to delete the ineffective items and replace them with effective drugs" (Shulman 1970).

The evolution of Bolar's correspondence with FDA over the 1960s documents the means by which new generic drug firms could be recruited into new regulatory regimes—especially when led by the carrot of the new markets potentiated by the promulgation of federal standards of drug efficacy. The standardization of efficacy performed by ANDA and the NAS-NRC review was a double-edged sword—the same action that cut away profitable but DESI-ineffective items such as pentaerythritol tetranitrate could also stabilize markets for generic production of DESI-effective drugs. Further correspondence between Bolar and FDA documents the transformation of this firm from a gray-market producer of illegitimate copies to an increasingly ethical (if still occasionally intransigent) regulated producer of legitimate copies.[19] By 1973, FDA inspectors could note with some satisfaction that "as a result of the NAS/NRC studies they have discontinued manufacturing all amphetamine products which formerly constituted the bulk of their business" and Bolar had shifted to a product line made up predominantly of "pending or approved ANDAs filed as the result of DESI announcements."[20]

IV. CONCLUSIONS: DESI AS AN EXERCISE IN MANIFOLD POWER WITH POTENTIAL LONG-TERM IMPACT AND AS A RESEARCH TOOL

Several features of the DESI exercise merit discussion here. For one, the DES implementation linked the gatekeeping power of the regulatory state with conceptual power of the regulatory state.[21] While there was appreciable contestation over differentiations among drugs that remained on the market,[22] it was the possibility that a drug would be removed from its profit-making role in the therapeutic marketplace that occasioned the greatest anxiety among American and international drug companies. So, too, the decision to delegate the decision making, review, and processing of the thousands of pre-1962 approvals to the National Academy of Sciences marked a critical late-twentieth-century precedent: the practice of FDA delegating to an advisory committee or to the National Academies the advisory task, but not the ultimate disposition, of potentially controversial tasks (Moffitt 2010). These legacies of DESI persist over a half-century after the NAS-NRC panels convened.

In addition, the DES's development of methods for effectiveness ratings relied upon developments in clinical pharmacology. The deployment of these methods, and the panels' judgment that well-controlled or randomized clinical studies were the best demonstration of effectiveness, marked both an important moment in statutory interpretation (the "well-controlled studies" plank of the 1962 amendments) as well as crucial standard-setting activities. The development of criteria for effectiveness judgments gave rise to newly cemented conceptions of experimental control, based upon reflections of how to separate treatment from control groups in clinical experiments—not only in the legitimation of randomization but also in the emergence of new conceptions of balanced treatment and control groups in both randomized and nonrandomized studies. In addition, the DES mounted a square attack on "therapeutic equivalence" as a concept separable from chemical equivalence and as a necessary component of any demonstration of effectiveness (Carpenter and Tobbell 2011).

For yet another feature, DESI provides unique opportunities to assess the long-term market and public health impacts of standardization. The scope of the National Academy of Sciences-National Research Council panels' work, for one, was vast. Some estimates suggest that DESI "affected . . . the indications and the ingredients for 75 percent of the products on the marketplace at the time" (Crout 1998). The closed process that the panels followed and that FDA maintained enabled recommendations to manifest as an abrupt shock to the market, not something that firms could anticipate and adjust to preemptively. The recommendations the panels offered and the decisions FDA subsequently promulgated were ultimately binding.

What was the market impact of DESI? There is much to be gained by further empirical study of DESI and its consequences. Theoretically, DESI evokes competing predictions. From the standpoint of libertarian economic theory, the forced removal of a product is a supply restriction that should generate at least three consequences. First, equilibrium price in the market from which products are removed should rise. Second, because in an unregulated marketplace consumers (patients) will have been optimally matched to heterogeneous products, consumer value (and perhaps social welfare generally) should decline. Finally, because withdrawal and standards of entry dampen incentives to producers to invest in specific markets, innovation might also be retarded.[23] These are of course posed as the simplest propositions from an account of efficient market operation, yet we note that all three of them have been stated in various ways in the economics, law and economics, and policy literature, as characterizations of FDA's program of efficacy-based drug regulation (Epstein 2006; Gieringer 1985; Peltzman 1973, 1974).

Another set of arguments, based upon "screening" and "approval regulation" theories, suggests that when there are clear standards of efficacy, market removals may increase social welfare. First, behavioral dynamics in the marketplace for new therapeutics (placebo effects, inefficiencies in prescribing, poor physician and patient learning about quality, etc.) may lead patients to be matched suboptimally to therapies. If some of these therapies are of poor quality and can be judged more efficiently by an independent agent (say, on the basis of randomized, double-blind, placebo controlled trials), then the quality of treatment may rise because the worst products are excluded from the market. The exclusion property of screening regulation then carries two properties: (1) the "worst" tail of the product distribution is excluded or dampened, thereby generating higher average product quality and reduced expectations of the worst quality (which may matter especially when patients are risk averse); (2) a system by which the inclusion and exclusion decisions are based upon randomized-controlled trials will reduce consumer uncertainty about the quality of the drugs that remain—more data is available about them, diffused through medical journals, labels, news, and other features—which can generate an increased likelihood of optimal consumption. Second, by a reversal of the "market for lemons" dynamic by which bad quality products drive out good ones (the famous argument of George Akerlof), the removal of "lemons" from the marketplace may clear space for high-value innovations. The total amount of innovation may actually rise, but the quality-weighted amount of innovation under this scenario will certainly rise.

Debate over these fundamentally divergent views of regulation remains robust. Future empirical work on DESI's long-term market and public health impact has the potential to offer new insight for these long-standing debates and to do so from the perspective of FDA's ability to assert regulatory power on the basis of efficacy. Scholars and policy makers continue to assess ways in which FDA policies and procedures impact drug safety (IOM 2006), and our emerging analysis is uniquely positioned to draw attention to the neglected impact that regulating on the basis of efficacy can have on safety as well. Empirical evidence from FDA that addresses these competing views of regulatory impact, moreover, has the potential to generalize well beyond the FDA case and speak to the impact of regulation through standardization in other policy domains.

For yet another feature, DESI highlights the human impact of product regulation. The ultimate legacies of DESI involved transformations in pharmacology, the practice of experiment, administrative law, and health and safety regulation. Yet we conclude with the observation here that those legacies were tangibly human as well—chemicals coursing through bodies, processes of metabolism, acts and sensations of healing, and side effects. All these processes had occurred for decades and were now, with the removals of hundreds of compounds from the American marketplace, arrested and replaced by others. There are social science and economic implications of these transformations, to be sure, that we will examine in the coming years. It is plausible that the screening mechanism of DESI worked in such a way that approval regulation institutions were employed to rid a market of "lemons," with plausibly beneficial effects on health in the years following the policy intervention. Yet to miss the truly human, emotive, and felt transformations would be also to miss a larger picture in which DESI remade individual and collective experiences of suffering and therapy.

The institutional story of DESI is clear. The joint screening and standardization exercise of DESI combined multiple forms of regulatory power as exercised by the American state: gatekeeping power (exercised retroactively) and conceptual power (patterns of standard and protocol that congealed in the stark rupture of existing patterns of claims making, therapy, and marketing). The legacies of DESI remain even more multidimensional and more durable than has been detected in academic inquiry.

NOTES

A version of this chapter was presented at the STANDEX workshop (*Standard Exchanges Programme; workshop international sur la standardisation en histoire de la santé et de la médicine*), Strasbourg, France, December 7, 2012. Our thanks to Christian Bonah, Nils Kessel, Tricia Koenig, Carsten Timmerman, and others for helpful remarks and criticisms. Kyle Giddon provided helpful research assistance. Please direct all comments, criticisms, and corrections to dcarpenter@gov.harvard.edu.

1. In the simplest sense of microeconomic theory, standardization combined with licensure shapes the population of those agents (whose behavior) and products (whose marketing) determine the "supply function."

2. For the 1938 Federal Food, Drug, and Cosmetic Act, see 21 USC § 301 et seq., as amended by Drug Amendments of 1962, Pub. L. No. 87-781.

3. Rettig, Earley, and Merrill (1992: 49–52, 56) offer a helpful summary of the development of DESI, upon which we draw in this section.

4. Rettig et al. (1992: 51) note that the ineffective designation could be limited to "ineffective as a fixed combination." On the rigor and use of the two-trial standard in the DESI review process, see Carpenter (2010: 348, n. 61). For a rather ideologically tilted (but occasionally informative) critique of the quality of evidence used in the review process, see Shorter (2008).

5. On keeping NAS-NRC reports confidential until FDA decisions had been reached, see Bryan and Stern (1970). On closed meetings, see Rettig et al. (1992: 56).

6. For discussion of closure in FDA committees during this time period, see Moffitt (2014).

7. Goddard noted that the names of the panel members other than the chairmen were even withheld from FDA until after the panel review process was complete.

8. For further discussion on this point and related analysis of the lawsuit brought by the American Public Health Association and the National Council of Senior Citizens, see Carpenter (2010: 353).

9. Edwards noted, however, that designations of "ineffective" were common in the earliest reviews—40 percent of the first 900 rulings. He noted that FDA sought to address ineffective designations first. See p. 8 of speech.

10. In fact, the possibility of fixed combinations of other NDAs created a fifth category, "ineffective as a fixed combination." For purposes here, we fold these into Category D, as this reflects the typological understanding in 1966 when Trexler began to explain the arrangement and procedures to various audiences. Yet the effectiveness of fixed combinations was a principal source

of contention, not least because it raised the possibility that FDA would be judging the relative effectiveness of a fixed combination of previous NDAs to the NDAs themselves.

11. The published orders in the *Federal Register* were not part of conventional notice-and-comment rulemaking either. FDA did not solicit public comment on its list of withdrawal decisions.

12. See Carpenter (2010:329–30). On arguments about the denial of due process, Carpenter refers to Lasagna to Arthur Kallet, September 2, 1971; LSG. See also, Lasagna to Goodman, October 25, 1971; LSG. Lasagna also called for an "evidentiary hearing" and rehearsed the due-process claim in an April 1971 *Wall Street Journal* editorial, "FDA 'Efficacy' Rule: Does It Work?" *WSJ*, April 8, 1971, 12.

13. On Panalba's removal from the market, court cases more generally, and reaction to DESI reports and FDA decisions, see Carpenter (2010:351–57).

14. In *Capitalism, Socialism and Democracy*, 2008, Schumpeter presents creative destruction as a core feature of capitalist economies in which innovation habitually supplants existing structures and processes, yielding (according to theory) more productive economies.

15. Marvin Seife, the first head of FDA's Generic Drug Division, linked the birth of the abbreviated new drug application (ANDA) and the Generic Drug Division to the implementation of DESI in 1970. *FGDAP*, part 2, p. 4.

16. Material in this section adapted with permission from chapters 4 and 7 of Greene (2014).

17. For more on the divisions between these advisory panels on generic equivalence, see also Carpenter and Tobbell (2011).

18. It was unethical, they argued, to expose research subjects to placebo-controlled trials for a drug already known to work; it was economically wasteful to require a full NDA—which could reach thirty or forty volumes of paper for a single submission. For more on the role of FDA in market making and unmaking, see Carpenter (2010).

19. FDA was initially unwilling to play such a role, as noted by Albert Lavender to Robert Shulman, November 23, 1970, vol. 3, AF10 156, FDAAF.

20. "ANDA Inspection," January 17, 19, 22, 23, and 26, 1973, vol. 4 AF10 156, FDAAF.

21. On these concepts, see Carpenter (2010), passim and especially Introduction and chapter 1.

22. For a sense of the stakes involved in one of the conditional designations —"Effective, But"—see "Agenda," Guidelines Conference for Drugs Ranked as "Effective, But," April 5, 1971; "The Drug Efficacy Study of the National Research Council's Division of Medical Sciences, 1966–1969," Series 1, NAS Archives. In a sign of the importance of the meeting, FDA Commissioner Charles C. Edwards and Bureau of Drugs Director Henry E. Simmons

attended, as did PMA representatives (Joseph Stetler and John G. Adams, Director of Scientific Activities), and Joseph Pisani from the Proprietary Association. Former FDA Bureau of Medicine Director Ralph G. Smith reviewed "Categories of Problems Encountered in Evaluation of Claims Ranked as 'Effective, But,'" and also presents "Typical Judgmental Processes."

23. It is possible that a kind of market space was opened by withdrawal, but in the libertarian version of economic theory, this substitution effect is usually more than outweighed by the dampening of incentives for market entry.

REFERENCES

Archives of the Drug Efficacy Study, National Academy of Sciences, Washington, D.C.

Records of the U.S. Food and Drug Administration, Record Group 88, National Archives. "Agenda." Guidelines Conference for Drugs Ranked as "Effective, But." April 5, 1971.

"ANDA Inspection." January 17, 19, 22, 23, and 26, 1973, vol. 4, AF10 156, FDAAF.

Bryan, Paul A. and Lawrence H. Stern. 1970. "The Drug Efficacy Study, 1962–1970." FDA Papers.

Carpenter, Daniel. 2010. *Reputation and Power: Organizational Image and Pharmaceutical Regulation at the FDA*. Princeton, N.J.: Princeton University Press.

Carpenter, Daniel, Jeremy A. Greene, Susan Moffitt, and Jonathan Warsh. "Therapeutic Implications of Efficacy-Based Drug Withdrawals: The Drug Efficacy Study Implementation." Unpublished manuscript.

Carpenter, Daniel and Gisela Sin. 2007. "Policy Tragedy and the Emergence of Regulation: The Food, Drug and Cosmetic Act of 1938." *Studies in American Political Development* 21(2):149–80.

Carpenter, Daniel and Dominique Tobbell. 2011. "Bioequivalence: The Regulatory Career of a Pharmaceutical Concept." *Bulletin of the History of Medicine* 85(1):93–131.

Ciba Corp. v. Weinberger, 412 U.S. 640 (1973).

Interview by Dr. John Swann and Ronald T. Ottes with Dr. J. Richard Crout, Director, Bureau of Drugs, in Rockville, Md. (May 27, 1998), available at http://www.fda.gov/downloads/AboutFDA/WhatWeDo/History/Oral Histories/SelectedOralHistoryTranscripts/UCM263934.pdf.

Edwards, Charles C., MD, Commissioner of Food and Drugs. "Positive and Rational Drug Therapeutics." Presented at the Symposium on Drugs—Hospital Pharmacists. September 26, 1970.

Epstein, Richard. 2006. *Overdose: How Excessive Government Regulation Stifles Pharmaceutical Innovation*. New Haven: Yale University Press.

Food and Drug Administration Press Release, May 1, 1966. 1992. In *Food and Drug Administration Advisory Committees*, ed. Richard A. Rettig, Laurence E. Earley, and Richard A. Merrill, 50. Washington, D.C.: National Academy Press.

Gieringer, Dale. 1985. "The Safety and Efficacy of New Drug Approval." *CATO Journal* (Spring/Summer):177–201.

Goddard, James L., to Keith Cannan, March 31, 1966. 1992. "Efficacy Review of Pre-1962 Drugs." In *Food and Drug Administration Advisory Committees*, ed. Richard A. Rettig, Laurence E. Earley, and Richard A. Merrill, 50. Washington, D.C.: National Academy Press.

Interview by Dr. James Harvey Young with Dr. James L. Goddard, Commissioner, U.S. Food and Drug Administration, in Atlanta, Ga. (June 19, 1969), available at http://www.fda.gov/downloads/AboutFDA/WhatWeDo/History/OralHistories/SelectedOralHistoryTranscripts/UCM264188.pdf.

Greene, Jeremy A. 2011. "What's in a Name? Generics and the Persistence of the Pharmaceutical Brand in American Medicine." *Journal of the History of Medicine & Allied Sciences* 66(4):425–67.

Greene, Jeremy A. 2014. *Generic: The Unbranding of Modern Medicine*. Baltimore: Johns Hopkins University Press.

Greene, Jeremy A., and Aaron S. Kesselheim. 2011. "Why Do the Same Drugs Look Different? Pills, Trade Dress, and Public Health." *New England Journal of Medicine* 365(1):83–89.

Institute of Medicine (IOM). 2006. *The Future of Drug Safety*. Washington, D.C.: National Academy of Sciences.

Jackson, Charles O. 1970. *Food and Drug Legislation in the New Deal*. Princeton, N.J.: Princeton University Press.

Lavender, Albert to Robert Shulman, November 23, 1970, vol. 3, AF10 156, FDAAF.

Marks, Harry M. 1996. *The Progress of Experiment*. New York: Cambridge University Press.

Moffitt, Susan L. 2010. "Promoting Agency Reputation Through Public Advice: Advisory Committee Use in the FDA." *Journal of Politics* 72(3):880–93.

Moffitt, Susan L. 2014. *Making Policy Public: Participatory Bureaucracy in American Democracy*. New York: Cambridge University Press.

Peltzman, Sam. 1973. "An Evaluation of Consumer Protection Legislation: The 1962 Drug Amendments." *Journal of Political Economy* 81(5):1049–91.

Peltzman, Sam. 1974. *Regulation of Pharmaceutical Innovation: The 1962 Amendments*. Washington, D.C.: American Enterprise Institute.

Rettig, Richard A., Laurence E. Earley, and Richard A. Merrill (ed.). 1992. *Food and Drug Administration Advisory Committees.* Washington, D.C.: National Academy Press.

"Report of Intensified Inspection, 4 November 1968." March 20, 1969, vol. 2, AF10 156, FDAAF.

Schumpeter, Joseph. 2008. *Capitalism, Socialism, and Democracy.* New York: Harper Perennial Modern Classics.

Shorter, Edward. 2008. "The Liberal State and the Rogue Agency: FDA's Regulation of Drugs for Mood Disorders, 1950s–1970s." *Journal of International Psychiatry* 31(2):126–35.

Shulman, Robert to Henry Simmons (FDA). October 30, 1970, vol. 3, AF10 156, FDAAF.

Trexler, Duke C. 1966. "The Drug Efficacy Study of the National Academy of Sciences–National Research Council," *Drug Information Bulletin* (October-December):12.

Trexler, Duke C. to DES Panelists. January 24, 1969. Quoted in *Drug Efficacy Study: A Report to the Commissioner of Food and Drugs.* 1969. Washington, D.C: National Academy of Sciences.

USV Pharmaceuticals v. Weinberger, 412 U.S. 455 (1973).

Wall Street Journal editorial. 1971. "FDA 'Efficacy' Rule: Does It Work?" *Wall Street Journal* (April 8), 12.

Warner, Edward. "Violative Inspection—for Information Purposes," May 14–16, 1968, vol. 2, AF10 156, FDAAF.

Weinberger v. Bentex Pharmaceuticals, Inc., 412 U.S. 645 (1973).

Weinberger v. Hynson, Westcott & Dunning, 412 U.S. 609 (1973).

CHAPTER SEVENTEEN

Drug Safety Communication

The Evolving Environment

GEOFFREY LEVITT

WHAT IS the purpose of communicating information about the safety of a drug? From a public health perspective, that purpose can be defined as supporting informed choices about the use of medicines that lead to optimal patient outcomes—i.e., that provide the greatest possible patient benefit with the least possible patient risk. Indeed, the U.S. governmental body charged with analyzing and providing advice on the effective communication of drug safety information—the Food and Drug Administration (FDA) Risk Communication Advisory Committee (2009)—has described its own mission very much along those lines.

So how then do we ensure that drug safety communication actually accomplishes this purpose? Two main ingredients are necessary: (1) there must be valid and adequately substantiated safety information to communicate, and (2) that information must be communicated accurately and meaningfully.

While those two ingredients may seem straightforward enough, in practice each of them has given rise to vast bodies of theory and practice that are still very much in a state of evolution. The process of generating adequately substantiated safety information about a drug in a timely and actionable form—the realm of pharmacovigilance—is the subject of intense medical, statistical, and epidemiological analysis, with a superstructure of comprehensive government regulation and a substructure of human and electronic information processing resources that must cope with thousands of pieces of data flowing in every day. And the process of determining how best to communicate that information—whether in the context of the product's approved labeling or in less formal messaging to patients and health care providers—is a matter of endless revision and active debate among regulators, academics, legislators, and the public.

Alongside all this complexity, however, is a key element without which the drug safety communication system cannot be properly understood, namely, the element of governance. Who has the right, and who has the responsibility, to engage in communications about the safety of a drug? In colloquial terms, who "owns" drug safety communication? Regulators? Industry? Third-party experts? Health care providers? Patients, consumers, or caregivers?

The thesis of this chapter is that we are in a time of unprecedented upheaval in the governance of drug safety communication that has resulted in a diffusion of control away from the parties who traditionally held primacy within this system—in particular, regulators and industry. This upheaval has been driven by three overlapping trends: (1) increasing doubts on the part of a number of vocal observers as to the ability and willingness of regulators and industry to communicate promptly and accurately about drug safety; (2) a related movement toward greater disclosure of certain kinds of health-related information, in particular the results of clinical research, which goes hand in hand with the ever-increasing ease of access to such information through electronic channels; and (3) a broader trend toward increased transparency of scientific and technological information overall (sometimes referred to as the "open-source" or "open-science" approach). Regulators and drug companies, while still central players in the system, now must contend with a constant flow of safety analyses and communications from a wide variety of independent third parties, often putting them in a reactive or even defensive position. This changing environment will require the formulation of new "rules of the road" in order to ensure the quality and integrity of both drug safety information and the manner in which it is communicated to the various stakeholders who rely on it.

I. GENERATING VALID, SUBSTANTIATED DRUG SAFETY INFORMATION

Traditionally, the postmarketing drug safety system has relied on two main sources of information: spontaneous adverse event reporting, and data from organized research studies (21 CFR § 314.80).[1] The limitations of spontaneous reporting are well known. Among other challenges, it is inherently haphazard, arbitrary, and prone to a variety of reporting and interpretation biases. Yet as a generator of possible safety signals and an indicator of broad patterns in the patient population at large following approval of a drug, spontaneous reporting has an undeniable role. Over the years, various additions and modifications to spontaneous reporting processes have been implemented in the United States, Europe, and other major jurisdictions. The current system in the United States is based on MedWatch and dates back roughly twenty years (Kessler 1993). The European Union has only recently overhauled its safety-reporting system to provide more transparency, better communication, and greater engagement of patients and health care professionals, including direct consumer reporting of suspected adverse drug experiences to regulators (European Medicines Agency (EMA) 2014).

Yet even with these reinforcements, spontaneous adverse event reporting remains at best an imperfect source of safety information. Indeed, its primary value often lies not in the information it furnishes directly but in its ability to provide signals that can then be addressed through more systematic research—the paradigmatic process of signal detection followed by hypothesis generation and testing. The testing phase of this process can take a number of different forms depending on the circumstances, ranging from case reviews through epidemiological studies to full-blown randomized safety trials that can last many years and cost hundreds of millions of dollars.

Systematic testing of this kind clearly can provide valuable information on drug safety issues. In recognition of this, regulators have not shied away from requiring drug sponsors to undertake such studies when they perceive safety issues needing to be addressed, whether through postmarketing commitments or the outright imposition of study requirements under legal authorities such as section 505(o) of the U.S. Food, Drug, and Cosmetic Act (FDCA) (21 USC § 355(o)).

Yet such studies have their limitations as well. Exponential increases in clinical study costs mean that only a fraction of the safety issues that may be present with a drug can feasibly be addressed in this fashion. Even when they are, the study in question may take a long time to yield results, during which patients are still being exposed to the drug. Even with the best possible intentions, design and implementation challenges may ultimately prevent a study from providing fully useful data.

For these reasons, there has been longstanding interest in devising a postmarketing safety surveillance system that could supplement the traditional model by offering timely, accurate, and actionable safety information at a reasonable cost. In the United States, this interest led to the creation of the Sentinel System in 2007 as part of legislation reauthorizing FDA to collect drug application user fees (FDA Amendments Act of 2007 (FDAAA), Pub. L. No. 110-85, § 905(a), 121 Stat. 823, 944). Still in its implementation phases, the Sentinel System aims to establish a network of health care data sources spanning government agencies, academic medical centers, and managed-care organizations that will provide targeted real-time safety information and analysis (Behrman et al. 2011; FDA 2014a). One of the innovative features of the Sentinel System is that the information relied upon to support safety analyses is not submitted, gathered, and processed at a central location but remains with its original owners, who become active participants in the network (FDA 2010). This "distributed" model plainly has significant potential time and cost advantages over the traditional centralized approach where the regulator or the sponsor has the burden of gathering, collating, and analyzing masses of safety data. At the same time, for better or worse, it reduces the sponsor's ability to control both the analysis and the reporting of emerging safety issues that may affect one of its products.

II. COMMUNICATING DRUG SAFETY INFORMATION

The effective communication of drug safety information, once such information is generated by the data gathering and analysis processes summarized above, faces a number of challenges. Such information is inherently complex and may be in a form or at a stage that is still subject to further verification, requiring careful calibration and qualification to accurately convey the associated level of uncertainty. In addition, it often must be communicated to multiple audiences with varying levels of sophistication (lay people, caregivers, health care professionals) and with disparate roles in relation to the drug itself (patients, prescribers, payers). For instance, in March 2013, FDA issued a public drug safety communication about new findings indicating an increased risk of pancreatitis and precancerous cellular changes in diabetic patients treated with a commonly used class of antidiabetics known as incretin mimetics. The unpublished findings, which were provided by a group of academic researchers, were based on examination of a small number of pancreatic tissue specimens taken from patients after they died from unspecified causes. The FDA communication noted that the agency had not concluded that these drugs may cause or contribute to the development of pancreatic cancer and emphasized that there should be no change in prescribing or use of the drugs pending further review (FDA 2013). Plainly a message of this nature requires an extremely careful approach in order to appropriately balance the intent to provide meaningful new safety information on a current basis with the need to avoid unnecessarily deterring patients or health care professionals from using a potentially lifesaving treatment.

It is therefore no wonder that an entire discipline has grown up around the process of formulating and transmitting drug safety information. While a detailed look at that discipline is beyond the scope of this chapter, for current purposes we can identify a few major categories of drug safety communication. The most fundamental, of course, is the approved drug label itself, which contains detailed official information about a drug's side effects, precautions, warnings, and contraindications (21 CFR § 201.57). The contents of the label are primarily directed to health care professionals and are accordingly written in technical language, which, at least in principle, requires medical training to fully understand and apply. The primary intent of the drug label is to provide the prescribing physician with sufficient information about the risks and the benefits of the drug to make informed prescribing decisions tailored to the individual patient. In some cases the label may also contain a section specifically directed to the patient to help that patient better understand the use, risks, and benefits of the drug that has been prescribed—the Patient Package Insert (21 CFR § 310.501, 21 CFR § 310.515; FDA 2012).

Many other types of drug safety communication can be considered a supplement, update, or gloss on the approved label. For instance, the approved MedGuide is intended to offer the patient further details about drugs from a user's, as distinct from a prescriber's, perspective (21 CFR Part 208). Similarly, the sheaf of information a patient receives at the drugstore when a prescription is filled is also intended to "translate" available information about drugs into lay terms for the convenience of the patient. Incidentally, this information packet—usually prepared at the behest of the pharmacy and not necessarily possessing official approved status—also poses some risk of inconsistency and/or confusion absent validated quality standards.

At another level entirely are professionally oriented messages conveying newly acquired information about a drug's risks. That new information may emerge from any of the sources discussed above, such as postmarketing surveillance, new analyses, or new studies. If disseminated by the sponsor (whether on its own initiative or as directed by a regulator), such messages typically take the form of a "Dear Healthcare Professional" letter (FDA 2012). Over the past few years, FDA has put in place mechanisms to allow the agency to disseminate drug safety information directly to health care professionals and the public without the involvement of the sponsor. These so-called "Drug Safety Communications" may be used to communicate emerging information about the risks of a drug relatively early in the analytical process, often before the risks in question have been fully analyzed or understood. Because these communications are public, they may also be picked up and rebroadcast by the lay media, Internet services, or other third parties, with or without additional commentary or interpretation.

Finally, there are the much broader and more amorphous kinds of messages about a drug's risks that may be contained in reports put forward by entirely independent actors based upon their own reviews or analyses conducted without any direct input from either regulators or drug sponsors. This category includes everything from summaries in third-party drug databases to entries in drug compendia to comprehensive meta-analyses published in academic medical journals. Again, any of these, but particularly the latter, may be picked up, amplified, and/or modified by other communication channels so they reach a much wider audience than their initial dissemination.

In principle, however, all such messages—if their content proves valid—are ultimately supposed to be incorporated into the authoritative source of information about a drug's risks and benefits, namely the drug label. This incorporation occurs through an ongoing process of dialogue, negotiation, and sometimes debate between sponsor and regulator. While under current law in the United States FDA has considerable direct power over the drug label—including the authority to require a sponsor to include new safety information, subject to specific triggers and criteria—no less an authority than the U.S. Supreme Court has determined that, as a legal matter, ultimate responsibility for the content of the label, as it relates to drug safety, rests with the sponsor, at least for branded drugs (*Wyeth v. Levine* 2009:570–71), as is described in greater depth in chapter 20 in this volume. This view of the world in effect bestows primary control of the communication of authoritative drug safety information upon the drug sponsor. In practice, however, the picture is considerably more complex.

III. THE STRUGGLE FOR CONTROL OF DRUG SAFETY COMMUNICATION

At the center of the traditional model of drug safety communication was a dyad: drug sponsor and health agency. In this model, the drug sponsor was primarily responsible for collecting and analyzing safety information and proposing related labeling changes or other safety updates, while the health agency played the role of neutral arbiter and gatekeeper to ensure that the presentation of safety information in these channels was accurate and timely.[2] Of course, third parties (compendia, researchers, formulary managers) also engaged in drug safety communication, but their roles were largely subordinate to those of the two main players.

Today, while the health agency and the drug sponsor have more or less maintained their traditional roles—with a few interesting adjustments to be discussed following—the world around them has changed dramatically. A variety of third parties have emerged as powerful, at times even dominant, forces in the drug safety communication system. Academic and government medical researchers conduct increasingly ambitious analyses, reanalyses, and meta-analyses of drug safety data, or even clinical studies, and proactively communicate the results to the public and to prescribers, whether through publications, counter-detailing activities,[3] or outright public health advocacy efforts (Loke et al. 2011; Nissen and Wolski 2010; Nissen and Wolski 2007). Independent drug information centers devote substantial resources to the production and dissemination of drug safety information completely outside the influence of sponsors and regulators.[4] Major drug payers, such as government agencies and managed-care organizations, in effect build their own picture of a drug's safety and risk–benefit balance on the basis of a wide array of such sources, among which the information offered by the drug sponsor itself is only one voice among several (Academy of Managed Care Pharmacy 2012).

This shift in the balance of power is in large part the product of a trend of increasingly vocal skepticism among critics of the traditional model, who may express doubts about the ability and willingness of both sides of the traditional dyad to act as effective stewards of drug safety information. These critics may portray the drug sponsor as commercially motivated to downplay, conceal, and delay the communication of accurate and up-to-date safety information (Doshi 2013). At the same time, these critics may accuse the health agency of effectively colluding with the drug sponsor in this behavior, partly out of the institutional imperative to defend its drug approval decisions and partly out of simple lack of resources (Goldacre 2013).

Though these kinds of allegations had surfaced before, the withdrawal of Vioxx (rofecoxib) in 2004 was interpreted in some quarters as reinforcing concerns about the traditional model's ability to handle emerging drug safety issues. The catalyst for the withdrawal was the announcement of study results by both FDA and the sponsor confirming that the users of the drug experienced an increase in cardiovascular risk, including significant numbers of fatal heart attacks (Martinez et al. 2004; FDA 2004). Shortly after the withdrawal, the medical journal *Lancet* published a meta-analysis of rofecoxib studies on the basis of which the authors asserted that the drug should have been withdrawn years earlier (Jüni et al. 2004). Critics charged that the drug sponsor had hidden essential safety information and that FDA, instead of forcing the sponsor to face the truth, had dithered until the weight of the safety evidence became too heavy to ignore—thus needlessly exposing thousands of patients to unacceptable risk (Martinez et al. 2004; Senate Committee on Finance 2004; Berenson et al. 2004).

The Vioxx experience galvanized two main currents of action. The first was a push to reform FDA's own structures for overseeing drug safety and to enhance the agency's authority to force sponsors to address emerging safety issues more promptly. This initiative was encapsulated in a comprehensive 2006 Institute of Medicine (IOM) report on drug safety that called for a number of systemic reforms, including providing FDA with direct legal authority to order safety-related labeling changes and to require sponsors to conduct studies to address emerging safety signals (IOM 2007). The timing was auspicious, as FDA's drug approval user-fee authority was due for legislative renewal the following year, providing a convenient legislative vehicle that allowed the IOM report's major recommendations to be enacted into law relatively quickly (FDAAA, Title IX). In the wake of these legal changes, the balance of power within the traditional dyad of health agency and sponsor shifted distinctly toward the health agency.

This shift was bolstered by operational changes undertaken by FDA itself. Most prominently, the agency decided to move away from its historical caution about publicly communicating emerging drug safety information. If the traditional approach had been to wait for confirmation or near confirmation of a safety concern before communicating it, the new paradigm reversed that presumption. Henceforth, FDA would proactively communicate information about emerging safety issues at a markedly earlier stage—even at the risk of prematurely announcing a potential safety issue that might later prove unfounded. New structures, most notably the Drug Safety Oversight Board, were set up within the agency to provide a forum for diverse internal perspectives on safety issues and to promote effective oversight of drug safety communication (FDCA § 505-1(j); 21 USC § 355-1(j)).

The second major focus of action sparked by the Vioxx experience and other safety issues with marketed drugs—as well as a number of law enforcement proceedings in which major drug companies were accused of improperly concealing negative clinical trial results (Bowe and Dyer 2004; Attorney General of NY 2004)—stemmed from the view that it was necessary not merely to further empower the regulator side of the regulator–sponsor relationship but to actually open up the entire universe of clinical data to public scrutiny. In this view, only by allowing more or less unfettered access to those data would the entire health care community be able to properly judge for itself the risks, as well as the benefits, of the medicines being offered to patients.

Once again, as this perspective emerged and intensified in the United States in the mid-2000s, the 2007 reauthorization of the FDA user-fee law provided a convenient vehicle for a rapid legislative response. An entire title of that reauthorization bill was devoted to requiring drug sponsors to post clinical trial results on a publicly available governmental website, clinicaltrials.gov (FDAAA, Title VIII). Several U.S. states, unwilling to wait for the federal requirements to be implemented, put forward their own (and not always entirely consistent) versions of clinical data disclosure requirements (Me. Rev. Stat. Ann. tit. 22. § 2700-A).[5]

Almost overnight, it seemed, *transparency* became the new watchword (Zarin and Tse 2008; IOM 2013; Rabesandratana 2013). The U.S. trade association of innovative pharmaceutical companies, the Pharmaceutical Research and Manufacturers of America (PhRMA), adopted a set of "Principles on Conduct of Clinical Trials and Communication of Clinical Trial Results" that committed member companies to transparency through the timely submission and registration of all clinical trials involving patients on the database specified in the federal disclosure legislation, regardless of outcome (PhRMA 2009). The association noted in a joint position statement that it "recognize[s] that there are important public health benefits associated with making clinical trial information more widely available to healthcare practitioners, patients, and others" (PhRMA et al. 2009). Posting of clinical trial results quickly became the norm.

But relying on drug sponsors to post results of their own trials still did not go far enough for some transparency advocates. More recently, calls have emerged not just for study results but for actual patient-level data and full clinical study reports to be released to third parties so that they may perform their own analysis of the data. The EMA has announced a policy of providing patient-level data submitted in drug marketing applications to third parties upon request and has indicated that it will affirmatively publish full clinical study reports submitted by drug sponsors.[6] PhRMA and the European Federation of Pharmaceutical Industries and Associations (EFPIA) (2013) outlined their commitment to provide clinical trial information to the public, patients, and private researchers when they released joint "Principles for Responsible Clinical Trial Data Sharing." This document supplements PhRMA's "Principles on Conduct of Clinical Trials and Communication of Clinical Trial Results" (PhRMA 2013). In addition, a number of major pharmaceutical companies, including GlaxoSmithKline, Roche, Pfizer, and Johnson & Johnson, announced their own specific data-sharing platforms (Kmietowicz 2014). From the perspective of drug safety communication, however, the most pressing issue posed by the public or semipublic release of these kinds of information is the specter of unrestricted reanalyses of clinical data by third parties that may lead to disparate results and potentially conflicting messages to the health care community about the risks and benefits of a drug product (EMA 2012): "[I]f inappropriate data analyses were picked up by the media early in a product's lifecycle, it could confuse doctors and prevent patients from receiving effective medicines" (statement of Susan Forda, Eli Lilly& Co.). To address this concern, some observers have called for health agencies to issue quality guidelines or standards to govern third-party reanalysis of clinical data.

Once again, it seems, the issue of trust lies at the heart of the matter—pitting those who believe public health is best served by leaving the formulation and communication of drug safety information primarily in the hands of those responsible for drug development and approval, i.e., sponsors and regulators, against those critics who believe that those parties cannot be relied upon to consistently provide accurate and timely drug safety information. Viewed through this lens, the drug safety communication debate also reflects broader social currents around the control and dissemination of information. Advocates of an open-source or open-science model call for a more transparent approach to all kinds of scientific data and information (Ross and Krumholz 2013). In their view, claims to proprietary ownership of scientific data must generally give way to the rights of the broader community to share in the understanding, analysis, and dissemination of those data. Assertions of superior expertise on the part of drug sponsors or regulators, in this perspective, are trumped by the alleged bias and self-interest of those parties, which supposedly render them incapable of acting as full fiduciaries of the public trust with respect to scientific data of interest to the public. Implicit in this outlook is a view of scientific data as a public or common good, in contrast to the view that those who devote significant resources and expertise to the production of those data—i.e., drug sponsors—or those who are legally entrusted with reviewing and overseeing the use of those data—i.e., regulators—have a legitimate role as a primary (though not necessarily the sole) source for analysis and dissemination of the data.

IV. CONCLUDING THOUGHTS

We are rapidly entering a world in which more and more drug safety information, from more and more sources, will be available to more and more audiences more and more quickly—as some have put it, a world in which we will be "awash in data." As the sources traditionally relied upon for an authoritative interpretation of that information come under increasing challenge, how are practitioners, patients, and other concerned audiences to determine which drug safety messages are validated and which are unconfirmed or even questionable—to separate fact from fiction? The public debate thus far has been dominated by criticism of drug sponsors and regulators. In fact, however, other potential sources of information in the system are not necessarily free of their own self-interest and potential bias. Government and private payers and their agents may have an inherent interest in lowering short-term out-of pocket drug costs—potentially at the expense of other types of health care costs that might have been saved or avoided by using the drugs in question. Plaintiffs' lawyers and their allies may have an inherent interest in undermining confidence in drug sponsors' safety messages or in advancing safety theories that support their litigation interests. In contrast to drug sponsors, these potential sources of bias may not be as apparent.

The goal of drug safety communication—supporting informed choices about the use of medicines that lead to optimal patient outcomes—is not an easy one to attain. As we have seen, choices about how best to promote effective drug safety communication are tied to much broader perspectives about the proper roles of government, industry, prescribers, patients, and independent researchers in creating, communicating, understanding, and acting upon drug safety information. Driven by these perspectives, a new model of drug safety communication is taking shape—one that elevates transparency over certainty and that opens the door for multiple and at times conflicting messages about matters that may be critically important to prescribing decisions and patient care. If this model is to achieve its basic purposes, there is an urgent need for a comprehensive and balanced consensus around system governance, data quality standards, and effective communication processes to support the fundamental public health goals that are at stake. Achieving such a consensus will require not just hard work but also a forthright acknowledgment of the differing perspectives and institutional imperatives that all involved parties bring to the process.

NOTES

The author gratefully acknowledges the assistance of Mark Odynocki and Rachael Martins of the Pfizer Legal Division in the research for this chapter. *The views expressed in this chapter are those of the author only and do not reflect the positions of Pfizer Inc or any of its affiliates.*

1. A third traditional source of safety information, literature reports, is more in the nature of an adjunct (though at times an important one) to these two main categories.

2. The health agency of course also played the "meta" role of regulating the overall process of pharmacovigilance, but that role is less relevant to this discussion.

3. Counter-detailing refers to the practice of representatives of academic institutions or government agencies affirmatively calling on physicians or payers to provide ostensibly independent information about a drug so as to "counter" traditional sales or "detailing" activities of drug company sales representatives.

4. For example, the Alosa Foundation's Independent Drug Information Service (IDIS) works with physicians and drug researchers at Harvard Medical School to collect and synthesize drug information that is presented to physicians by IDIS-trained health care professionals. http://www.alosafoundation.org/independent-drug-information-service/.

5. Clinical trial disclosure requirement repealed by Maine's legislature.

6. In a December 2013 press release, *EMA to Push Ahead in 2014 Towards Publication and Access to Clinical Trial Data*, the agency noted that its policy and implementation plan for access to clinical data would be discussed at a March 2014 management board meeting.

REFERENCES

Academy of Managed Care Pharmacy. 2012. *AMCP Format for Formulary Submissions: A Format for Submission of Clinical and Economic Evidence of Pharmaceuticals in Support of Formulary Consideration, Version 3.1.*

Attorney General of New York. 2004. *In re GlaxoSmithKline, Assurance of Discontinuance.*

Behrman R. E., J. S. Benner, J. S. Brown, M. McClellan, J. Woodcock, and R. Platt. 2011. "Developing the Sentinel System—a National Resource for Evidence Development." *New England Journal of Medicine* 364:498–99.

Berenson, A., G. Harris, B. Meier, and A. Pollack. 2004. "Despite Warnings, Drug Giant Took Long Path to Vioxx Recall." *New York Times* (November 14).

Bowe, C. and G. Dyer. 2004. "GSK Settles Paxil Lawsuit with Spitzer." *Financial Times*, August 26.

Doshi, P. 2013. "Putting GlaxoSmithKline to the Test Over Paroxetine." *British Medical Journal* 347:f6754.

European Medicines Agency (EMA). 2014. "Pharmacovigilance: Regulatory and Procedural Guidance." http://www.ema.europa.eu/ema/index.jsp?curl=pages/regulation/document_listing/document_listing_000199.jsp.

———. 2013. *EMA to Push Ahead in 2014 Towards Publication and Access to Clinical Trial Data.* http://www.ema.europa.eu/ema/index.jsp?curl=pages/news_and_events/news/2013/12/news_detail_001991.jsp&mid=WC0b01ac058004d5c1.

———. 2012. *Agency Moves Towards Proactive Publication of Clinical-Trial Data: Workshop Report.* http://www.ema.europa.eu/docs/en_GB/document_library/Report/2012/12/WC500135841.pdf.

Goldacre, B. 2013. *Bad Pharma: How Drug Companies Mislead Doctors and Harm Patients.* London: Faber and Faber.

Institute of Medicine (IOM), Board on Health Sciences Policy, Board on Health Care Services. 2013. *Sharing Clinical Research Data: Workshop Summary, Rapporteurs.* Steve Olson and Autumn S. Downey. Washington, D.C.: The National Academies Press.

Institute of Medicine (IOM), Committee on the Assessment of the U.S. Drug Safety System, Board on Population Health and Public Health Practice. 2007. *The Future of Drug Safety: Promoting and Protecting the Health of the Public*, ed. Alina Baciu, Kathleen Stratton, and Sheila P. Burke. Washington, D.C.: The National Academies Press.

Jüni, P., L. Nartey, S. Reichenbach, R. Sterch, P. A. Dieppe, and M. Egger. 2004. "Risk of Cardiovascular Events and Rofecoxib: Cumulative Meta-analysis." *Lancet* 364:2021–29.

Kessler, D. A. 1993. "Introducing MEDWatch: A New Approach to Reporting Medication and Device Adverse Effects and Product Problems." *Journal of the American Medical Association* 269:2765–68.

Kmietowicz, Z. 2014. "Johnson & Johnson Appoints Yale Project Team to Run Data Sharing Scheme." *British Medical Journal* 348:g1361.

Loke, Y. K., C. S. Kwok, and S. Singh. 2011. "Comparative Cardiovascular Effects of Thiazolidinediones: Systematic Review and Meta-Analysis of Observational Studies." *British Medical Journal* 342:d1309.

Martinez, B., A. W. Matthews, J. S. Lublin, and R. Winslow. 2004. "Merck Pulls Vioxx from Market After Link to Heart Problems." *Wall Street Journal* (October 1).

Nissen. S. E. and K. Wolski. 2010. "Rosiglitazone Revisited: An Updated Meta-Analysis of Risk for Myocardial Infarction and Cardiovascular Mortality." *Archive of Internal Medicine* 170:1191–1201.

———. "Effect of Rosiglitazone on the Risk of Myocardial Infarction and Death from Cardiovascular Causes." *New England Journal of Medicine* 356:2457–71.

Pharmaceutical Research and Manufacturers of America (PhRMA). 2013. *EFPIA and PhRMA Release Joint Principles for Responsible Clinical Trial Data Sharing to Benefit Patients.*

PhRMA and European Federation of Pharmaceutical Industries and Associations (EFPIA). 2013. *Principles for Responsible Clinical Trial Data Sharing.*

PhRMA, EFPIA, Japan Pharmaceuticals Manufacturers Association and the International Federation of Pharmaceutical Manufacturers and Associations. 2009. *Joint Position on the Disclosure of Clinical Trial Information via Clinical Trial Registries and Databases.*

Rabesandratana, T. 2013. "Drug Watchdog Ponders How to Open Clinical Trial Data Vault." *Science* 339:1369–70.

Ross, J. S. and H. M. Krumholz. 2013. "Ushering in a New Era of Open Science Through Data Sharing: The Wall Must Come Down." *Journal of the American Medical Association.* 309:1355–56.

Senate Committee on Finance. 2004. *FDA, Merck, and Vioxx: Putting Patient Safety First? Hearing Before the Senate Committee on Finance, United States Senate, 108th Congress, 2d sess.* (November 18).

U.S. Food and Drug Administration (FDA). 2014. "Sentinel Initiative." http://www.fda.gov/Safety/FDAsSentinelInitiative/default.htm.

———. 2013. *Drug Safety Communication: FDA Investigating Reports of Possible Increased Risk of Pancreatitis and Pre-Cancerous Findings of the Pancreas From Incretin Mimetic Drugs For Type 2 Diabetes.* (March 14).

———. 2012. *Draft Guidance, Drug Safety Information—FDA's Communication to the Public.*

———. 2010. *The Sentinel Initiative: Access to Electronic Healthcare Data for More Than 25 Million Lives.*

———. 2004. *FDA Issues Public Health Advisory on Vioxx as Its Manufacturer Voluntarily Withdraws the Product.*

U.S. Food and Drug Administration, Risk Communication Advisory Committee. 2009. *Charter of the Risk Communication Advisory Committee to the FDA.*

Wyeth v. Levine, 555 U.S. 555 (2009).

Zarin, D. A. and T. Tse. 2008. "Moving Towards Transparency of Clinical Trials." *Science* 319:1340–42.

CHAPTER EIGHTEEN

Innovation Policy Failures in the Manufacturing of Drugs

W. NICHOLSON PRICE II

M&M CANDIES are made with an efficiency and precision far surpassing the capabilities of many drug manufacturers (Helferich 2013; Hussain 2013). This is surprising—both are regulated by the U.S. Food and Drug Administration (FDA), and the drug industry's operations have major implications for human health and for the health care system in general. But manufacturing in the drug industry generally is inefficient and expensive and shows little of the innovation that has characterized other manufacturing sectors over the past few decades (Price 2014a).

The lack of innovation in manufacturing might not be worrisome if it did not create major problems; unfortunately, it does. Drug recalls are rising, with more than 2,000 in 2011, many due to manufacturing quality problems and contamination (Shanley 2012). Drug shortages are also increasing, with hundreds of ongoing shortages, including common drugs and frontline cancer therapies; the majority of them are caused by manufacturing problems (Gatesman and Smith 2011; Kweder and Dill 2013). And more generally, expenditures on drugs make up over 15 percent of health care costs (Thomas 2013), and billions of dollars are wasted annually on excessive drug manufacturing costs. Reducing those costs would generate tens or hundreds of billions of dollars in annual consumer surplus, depending on how the savings were used.

The absence of innovation—and the attendant presence of inefficiency and waste—is especially surprising in the context of an industry that has long been viewed as a relatively successful target of innovation policy. Both regulatory and intellectual property factors are aimed at getting the industry to develop newer and better drugs, and despite some notable failures, the drug industry is generally seen as a success story for patents and innovation policy as a whole.

That success story, however, does not apply to pharmaceutical manufacturing. While strong intellectual property and regulatory incentives drive the costly development of new drugs, those incentives largely fail in the context of drug manufacturing. Instead, the combination of costly regulatory barriers to innovative change and ineffective intellectual-property incentives to develop new technology mean that pharmaceutical manufacturing is largely stagnant, with tremendous attendant costs.

Improving the situation requires policy change to lower regulatory hurdles or increase incentives for innovation—ideally, both. This chapter concludes by briefly discussing a few possible paths for positive change, including direct regulatory changes and ways that the regulatory apparatus can be used to change the incentives in the intellectual-property system without making wholesale changes to that system itself.

I. PHARMACEUTICAL MANUFACTURING TODAY

Manufacturing is either the first or second highest expense for drug firms. Nonetheless, it is surprisingly inefficient, lagging significantly behind the modernized manufacturing techniques of other industries; the industry was recently characterized as being "in the dark ages with respect to . . . efficiency" (Eglovitch 2012). If the industry could modernize its manufacturing practices, it could not only reduce the health care problems that arise from drug shortages and recalls but could also save tens of billions of dollars annually by modernizing manufacturing, with even larger social-welfare benefits.

Drug manufacturing is perceived as inexpensive because the most salient costs—marginal costs of blockbuster small-molecule drugs—are in fact typically very low (Outterson 2005). However, other drugs frequently have much higher marginal costs (Berndt et al. 1995), and the industry has high fixed costs, including capital costs and quality costs. Overall manufacturing costs, measured as "cost of goods sold" (COGS), averaged 26 percent of brand-name small-molecule drug manufacturer sales between 1994 and 2006 and 52 percent for generic manufacturers (Basu et al. 2008). Though the story for biologics is more complex, they typically face higher absolute manufacturing costs on both fixed and per-unit bases (Price 2014a). Throughout the drug industry, manufacturing is quite expensive. Surprisingly, given the potential for competitive advantage—and contributing to the initial expense—manufacturing is largely noninnovative and relies on outdated techniques and processes.

While other industries have developed and adopted modern manufacturing processes over the last few decades, the pharmaceutical industry has lagged far behind (Friedli et al. 2006). Modern manufacturing includes not only new technology but also processes like constant monitoring of production parameters, systematic waste reduction, continuous process improvement, and continuous quality management. These techniques have spread through many industries but not generally drug manufacturing (Friedli et al. 2010; Yu 2008).

As a result, pharmaceutical manufacturing suffers from poor operational performance, manifest in several ways (Herlant 2010). Process rigidity is frequent and problematic; instead of embracing flexibility and continuous improvement, drug companies view the initial regulatory commitments on manufacturing methods as approved processes, which should only be changed in exceptional circumstances (Friedli et al. 2010). Although regulatory barriers are certainly significant, as discussed later, they are not the only source of process rigidity, which also results from inadequate internal or external incentives for change. A linked problem is the frequent use of outdated equipment and production lines, including factories several decades old with limited upgrades (Woodcock and Wosinska 2013) that have been described as "in terrible shape" (Thomas 2012). Other problems with process management result in surprisingly loose control over processes, with much higher error rates than other industries to be caught in intermediate- or final-stage testing (Friedli et al. 2006). This emphasis on quality through testing creates major inefficiencies and slows the production of drugs tremendously.

Fixing these inefficiencies would make a major contribution to welfare. Studies have estimated that the industry could save $15 to $90 billion annually worldwide by reducing manufacturing inefficiency (Suresh and Basu 2008). But the potential consumer welfare gains from these savings are much larger. If manufacturing expenses decreased by 20 percent and that decrease resulted in lower drug prices, as would be expected in a competitive market equilibrium, the expected annual consumer surplus gain in the United States alone would be $47.4 billion (Vernon, Hughen, and Trujillo 2007). If instead the savings resulted in increased research and development (Vernon 2005), the health gains could be worth as much as $574 billion annually.

In addition to lower costs, manufacturing innovation can improve drug quality and can decrease the number of drug recalls and shortages. Poorly controlled processes and outdated equipment contribute directly to quality problems, contamination, and drug recalls (Shanley 2012). The growing problem of drug shortages—estimated to cost $416 million a year in economic costs (Cherici et al. 2011; Kaakeh et al. 2011) and unknown human health costs (Born 2012)—is also highly linked to manufacturing problems. Forty-six percent of drug shortages in 2011 were caused by quality issues, "including bacterial or mold contamination, tablet disintegration, and the presence of foreign particles such as glass or metal in vials" (Kweder and Dill 2013). Delays or capacity issues caused another 19 percent. Manufacturing innovation, to the extent that it results in closer monitoring, greater control, and higher quality of drugs, could help decrease both recalls and shortages.

II. INNOVATION POLICY FAILURES

The significant problems in drug manufacturing and the major benefits from fixing them lead to the obvious question: Why is drug manufacturing in such poor shape? Two contrasts are illustrative. First, the drug industry as a whole certainly has the capacity to innovate; new drugs are constantly developed, despite the extremely high regulatory hurdles and costs of the drug development-and-approval process. Second, in other industries, manufacturing has been notably innovative in the last few decades, yielding major advances in efficiency and quality. Drug manufacturing is different because it suffers from two interlinked challenges: high regulatory barriers to innovation and low intellectual-property incentives for innovation, both of which will be discussed in detail following. In contrast, each of the comparison situations faces only one challenge: drug development has high regulatory hurdles but strong patent protection for new drugs, while other industries' manufacturing faces lower regulatory barriers but little intellectual property protection.[1] The combination of these two factors has resulted in the stagnation of manufacturing in the pharmaceutical industry.

A. Regulatory Barriers

Innovation in the pharmaceutical industry takes place against a backdrop of pervasive regulation designed to ensure that drugs are safe and effective. In the context of drug manufacturing, all manufacturing methods and changes to those methods must be approved by FDA (21 USC § 255(b)(1)(D)). Throughout this approval process, as has been discussed elsewhere in this volume, FDA imposes significant limits on innovation in three primary ways. First, institutional resistance to approving novel technologies restrains innovation during the New Drug Application (NDA) process, so that firms avoid innovative technologies in NDAs for fear of delays in receiving marketing approval. Second, de facto technical standards arise from current Good Manufacturing Processes (cGMP) regulations and associated guidances, to which firms frequently adhere overly closely in innovation-decreasing ways. Third, FDA imposes procedural hurdles to postapproval manufacturing changes, in the form of required filings of supplemental NDAs (sNDAs). Fourth and finally, FDA also imposes substantive hurdles to postapproval change, where manufacturing methods from early drug development are effectively locked in by regulation. These regulatory constraints are generally imposed without taking into account their effect on manufacturing innovation or efficiency.

The first and perhaps most pervasive hurdle to innovation comes before approval, where agency practice and market dynamics make firms reluctant to include new technologies in NDAs. FDA has historically been leery of approving new technologies, sometimes requiring years to accept them. For instance, FDA took several years to accept the shift from thin-layer chromatography to high-performance liquid chromatography, despite its superior performance and subsequent wide adoption in the industry (Hussain 2013). Firms have learned from this experience and understand that even though FDA may eventually be persuaded to accept a new technology, inclusion of such a technology within the context comes with the risk of significant delay.

The second hurdle to innovation comes from technological standards. Technological standards, when specific technologies are required, are recognized as problematic for innovation because they impose a cost, sometimes very high, on shifting technologies because the standard needs to change. The unusual aspect of technological standards in pharmaceutical manufacturing is that they are not actually mandated by FDA. FDA cGMP regulations are rigorous and wide-ranging but typically are goal-oriented performance standards rather than technological standards (21 CFR § 211). Instead, the industry has effectively created de facto technology mandates for itself by adhering tightly to technical examples found in cGMP guidance documents. In the largest example, the industry-wide practice of using three batches to validate a process apparently arose from FDA's 1987 "Guideline on General Principles of Process Validation," which stated that processes should be repeated "a sufficient number of times to assure reliable and meaningful results." The guidance also mentioned a single example that required three repetitions; from this and a few other contemporary examples, the industry adopted a three-batch rule rather than the broad principle described in the guidance (FDA 2013a). Similarly, FDA noted that an approved-equipment addendum to a guidance was "misinterpreted as equipment required by FDA" (FDA 2013b). FDA has tried to limit this type of industry adherence, recently replacing the 1987 guideline (FDA 2011) and removing tables listing specific approved equipment, but the risk-averse industry continues setting its own de facto technical standards, which limit innovation.

The third hurdle is the set of procedural barriers to receive FDA approval of changes to manufacturing methods after the drug has been approved for the market. Drug sponsors must notify FDA of any changes to an approved application, which include manufacturing changes (21 CFR § 314.70(b)–(d)). Major changes, which include any change that "may affect the impurity profile and/or the physical, chemical, or biological properties of the drug substance," require FDA preapproval (21 CFR § 314.70(b)). The procedure of getting changes approved is costly in terms of money to prepare and file the documents, time to prepare the submission and await a reply, and the risk both of nonapproval and of other questions about the drug being reopened by the agency. These procedural costs, even if small, can effectively block continuous improvement, which relies on small changes, and help create a change-averse culture, which also blocks larger changes (Friedli et al. 2006).

The fourth and final hurdle is a set of drug-specific substantive barriers to innovative change that arrive through a process of regulatory lock-in. Put briefly (Price 2014a; discussing in further detail), drugs are approved on the basis of clinical trials; accordingly, the specifications of the approved drug are based on the characteristics of the drug at the time it was used for clinical trials. Once specifications are set, the drug must continue to meet those specifications. However, manufacturers typically wait until late in development to optimize manufacturing; by that time, it may be too late to change many aspects of the drug, and associated manufacturing methods are locked in by the specifications (Friedli et al. 2006). This makes sense for drug parameters that are poorly understood but is just as true for parameters that are well understood and for which larger deviations from specifications would not be clinically relevant (Yu 2008).

B. Inadequate Intellectual-Property Incentives

Regulation significantly reduces innovation in pharmaceutical manufacturing. Regulatory hurdles alone, however, cannot fully explain manufacturing stagnation because firms are able to overcome the high hurdles for getting drugs initially approved. The lack of innovation in pharmaceutical manufacturing also arises from a lack of sufficient innovation incentives. Patents, which are dominant in drug discovery and development, are less valuable for drug manufacturing and consequently less important. Trade secrecy plays a greater role for manufacturing methods but has major drawbacks as an innovation incentive. Finally, regulatory exclusivity is important for drug development but does not exist for manufacturing methods.

1. Patents

Patents reward invention by letting inventors recoup high initial costs by creating a temporary monopoly and allowing correspondingly higher prices and are strikingly important in drug discovery and development, to the extent that most firms avoid developing drugs unless they can ensure strong patent protection (Roin 2008). Although patents are available throughout the development process, they are particularly prominent in protecting early investments (Kitch 1977). Postapproval, patents also protect drug innovations by preventing generic-drug entry until they expire, and that protection is often extended by "evergreening" tactics (Hemphill and Sampat 2012).

For drug manufacturing, by contrast, patents are worth much less than in other industries because they require costly disclosure but are difficult to enforce. Patents require that the patented invention be disclosed to the public. However, manufacturing processes are hard to observe and difficult to reverse engineer from the final product; this means that without disclosure, competitors must generally develop manufacturing process innovations independently. Thus, the disclosure of manufacturing process patents is particularly costly.

This costly disclosure is negatively complemented by the difficulty of enforcing manufacturing process patents. Process patents are easier to invent around than patents on drug products because a manufacturing process can sometimes be tweaked slightly to avoid a process patent in a way that drug products cannot. More fundamentally, knowing when a process patent is infringed is especially difficult because "no one outside the potential infringer knows how the product was made" (Lewis and Cody 2002). Identifying the use of a patented process is likely to be especially difficult for general patents, like ways to more closely monitor ongoing production. In addition, a significant subset of manufacturing patents, those focused on monitoring and gathering information, have become effectively impossible to enforce since the Federal Circuit held in the 2012 case *Momenta v. Amphastar* (2012:1348) that they fall within the "safe harbor" provision of the Hatch-Waxman Act (35 USC § 271(e)(1)) because they generate information used for regulatory submissions (Price 2014a). Overall, patents are frequently too costly and too hard to enforce for firms to rely on them to protect manufacturing innovations. As a consequence, firms turn instead to trade secrecy.

2. Trade Secrecy

Trade secrecy matters much more for innovation in drug manufacturing than patents do. Trade secrecy, grounded in state law, provides protection from misappropriation of information that is reasonably kept secret and derives value from its secrecy (Uniform Trade Secrets Act § 1(4)). Like patents, trade secrecy creates incentives for innovation by keeping others from copying the innovation and therefore allowing supracompetitive pricing. Trade secrecy can act at a range of strengths, from creating a competitive advantage when one firm can use a more efficient technique than another (*Norbrook Labs. v. H.C. Manuf'g* 2003:463) to excluding competitors from the market entirely if creating the product is impossible without the secret process, which is especially likely for biologics (*Wyeth v Natural Biologics* 2003:*1). But trade secrets play a bigger role than patents in protecting manufacturing processes for at least three reasons. First, while enforcing manufacturing process patents is difficult, as long as trade secrets can be kept secret, their effectiveness does not depend on monitoring other firms' activities. Second, trade secrets, by definition, do not require disclosure of information to competitors, who might derive significant benefit from the information. Finally, trade secrets, unlike patents or statutory exclusivity, do not have a predetermined lifespan but can continue indefinitely.

Trade secrecy creates some incentives for innovation but also has major problems as innovation policy. First, the maintenance of secrecy means that trade-secret protection blocks cumulative innovation because neither other firms nor society in general can develop new knowledge based on the secret. Second, further innovation even within the firm is likely lessened because maintaining trade secrecy requires security measures that limit the dissemination of information to other potential innovators in the same firm. Third and finally, trade secrecy likely creates insufficient incentives for broadly applicable forms of innovation—which may be too hard to protect—or for forms of innovation that require wide licensing or network effects to function profitably. Unfortunately, the type of innovation most needed in drug manufacturing—innovations reflecting greater understanding and process knowledge—are particularly poorly suited to protection as trade secrets.

3. Regulatory Exclusivity

FDA also directly governs innovation incentives by offering a variety of forms of regulatory exclusivity to approved drugs. For five years after a new drug is approved, no generic version can be approved; biologics receive 12 years of exclusivity, and rewards are also available for conducting pediatric trials or developing drugs for rare conditions (so-called "orphan drugs") (Eisenberg 2006). However, regulatory exclusivity is completely unavailable for manufacturing methods, and therefore does not serve as an incentive for innovation.

Overall, although the pharmaceutical industry is a major focus of innovation policy, the policy levers and rules, patents, and regulatory exclusivity, which work well for drug discovery and development work poorly for pharmaceutical manufacturing. Even the one form of incentive that has the most effect, trade secrecy, provides limited and flawed incentives.

III. PROPOSALS FOR CHANGE

The state of innovation in pharmaceutical manufacturing is poor, with major consequences, and that lack of innovation arises from a failure of innovation policy. Regulatory barriers to innovation are high, and incentives for firms to overcome those barriers are low. Potential solutions to this complex problem could act on both of those levers, lowering barriers—while maintaining safety—or raising incentives for innovation, or ideally both. This section briefly discusses possibilities for reform along those lines. One other set of reforms, not discussed here, focuses on increasing the information about drug quality available to the market, aiming to harness competitive forces (Woodcock and Wosinska 2013; Price 2014b).

A. Regulatory Changes

Regulatory changes to lower barriers to innovation could come in several forms (Price 2014a). FDA could offer increased regulatory flexibility as a reward for demonstrated manufacturing excellence and deeper understanding, by, for instance, allowing more significant changes to be made with only notice rather than preapproval, modeled on the Occupational Safety and Health Administration's Voluntary Protection Programs.

Perhaps most promisingly, FDA could create an independent pathway to validate new technologies outside the context of the NDA process. A major reason new technologies do not enter drug manufacturing is the risk of delay in getting NDAs approved. If firms could demonstrate the efficacy and advantages of a new technology outside the context of an NDA, they might be significantly more willing to devote resources to developing that technology. Perhaps more importantly, other innovators—firms with special manufacturing excellence, such as fine chemical manufacturers or food manufacturers—could develop drug-applicable technologies, get them independently approved by FDA, and then market them to drug manufacturers.

Other possibilities include devolving regulatory authority to the states or privatizing it entirely, mandating innovation through FDA's extant but slow-moving Quality-by-Design program, or shifting requirements for deeper manufacturing understanding earlier in the drug development process to reduce the problematic effects of substantive lock-in. FDA is taking some steps to mandate quality more directly through the establishment of a unified Office of Pharmaceutical Quality. Congress acted in 2013 by passing the Drug Quality and Security Act, though this focuses primarily on compounding pharmacies and increasing the traceability of the drug supply chain and not on driving quality or innovation specifically. Each approach has potential benefits, and together they might drive sufficient innovation. It seems more likely, however, given the continuing necessity of strong regulatory oversight, that additional incentives would be required.

B. Changing Intellectual Property Through Regulation

The intellectual-property system could provide significant incentives for manufacturing process innovation but does not currently do so for the reasons described above. Rather than changing the entire system—which would have major effects on other industries—regulatory structures could change the way intellectual-property incentives function to drive drug manufacturing innovation. Alternately, FDA could operate a system of regulatory exclusivity for manufacturing innovation in parallel to the intellectual-property system.

The first scenario seeks to change the major misalignment of innovation incentives in pharmaceutical manufacturing caused by the dominance of trade secrecy over difficult-to-enforce patents. FDA could profoundly alter this dynamic by requiring that manufacturing method regulatory submissions be public disclosures. Firms could much more easily police their manufacturing patents if they could observe their competitors' processes, and such patents would consequently be more valuable and would provide greater incentives for innovation. Other potential benefits include innovation to increase efficiency and quality, as well as increased cumulative innovation and greater overall transparency, which would ease oversight.

The change from secrecy to disclosure and transparency would be challenging to implement retrospectively, raising constitutional Takings Clause concerns and practical concerns based on prior use rights found in the America Invents Act. However, implementing a disclosure regime prospectively would be more straightforward, and, in fact, the Biologics Price Competition and Innovation Act of 2009 (BPCIA) already contains a very limited version of this idea for biologics: under the law, biosimilar applicants are required to disclose their entire application package—including manufacturing details—to the reference biologic sponsor so that the sponsor can determine which, if any, of its patents are infringed (42 USC § 262(l)). A broader regime of that nature—and one that covered not only biosimilars but all drug manufacturers—could have major systemic benefits.

The second scenario involves FDA taking a more active role in rewarding manufacturing innovation by imposing regulatory exclusivity for such innovation. FDA already does this for drug-discovery-related investments like winning drug approval, developing an orphan drug, or conducting pediatric studies. A similar approach could apply to innovative manufacturing methods, and since FDA approval is already required to implement a manufacturing method, FDA could expand this gatekeeping role to reward innovation.

Implementation concerns are significant. Principally, FDA currently lacks the institutional experience to manage the line-drawing, overlapping-innovation, and novelty concerns that are familiar in patent law but relatively absent in FDA's current regulatory exclusivity decisions. However, FDA does already manage innovation incentives to some extent (Eisenberg 2006) and complements that experience with deep scientific expertise. In a way largely unavailable to the patent system, FDA could potentially tune rewards much more closely to innovation value, focusing innovation on the most important areas of drug manufacturing.

IV. CONCLUSION

Academic and policy studies of pharmaceutical innovation policy have, until now, missed a crucial piece of the industry puzzle: the costs and complexities of pharmaceutical manufacturing. This gap has had major practical consequences: the current combination of regulatory barriers to manufacturing innovation and poorly aligned intellectual property incentives results in immense economic and human costs.

The problem, however, is not insoluble. Discovery and development of new drugs is a paradigm area where regulation is designed to encourage innovation, and manufacturing those drugs is another target for such design, whether focused directly on regulatory hurdles or on changing intellectual-property incentives. Furthermore, such regulatory levers are not limited to the pharmaceutical industry and may be especially appropriate for substantively related industries with tight regulation, like medical devices or biomedical diagnostics. Whatever the particular solutions, actively shaping the interaction of regulatory forms and intellectual property has far-reaching implications for innovation policy.

NOTES

This chapter is based in large part on a previously published, longer treatment of these issues (Price 2014a).

1. The optimal balance between hurdles and incentives is not obvious. The classical justification for intellectual property assumes a low-regulation baseline and justifies intellectual property as necessary to incentivize the creation of public knowledge goods, the value of which is difficult for firms to capture—a low-regulation, high-IP situation. Drug development is high regulation, high IP, and nondrug manufacturing is typically low regulation, low IP; both demonstrate significant innovation. Drug manufacturing, a high-regulation, low-IP situation, shows very little innovation. Whether the classical situation

is preferable to that of nondrug manufacturing or drug development is questionable, and the existence of patent thickets may suggest that the intermediate situations might be preferable.

REFERENCES

Basu, P., G. Joglekar, S. Rai, P. Suresh, and J. Vernon. 2008. "Analysis of Manufacturing Costs in Pharmaceutical Companies." *Journal of Pharmaceutical Innovation* 3(1):30–40.

Berndt, E. R., L. Bui, D. R. Reiley, and G. L. Urban. 1995. "Information, Marketing, and Pricing in the US Antiulcer Drug Market." *American Economic Review*, 85(1): 100–5.

Born, K. 2012. "Time and Money: An Analysis of the Legislative Efforts to Address the Prescription Drug Shortage Crisis in America." *Journal of Legal Medicine* 33(2):235–51.

Cherici, C., J. Frazier, M. Feldman, and others. 2011. "Navigating Drug Shortages in American Healthcare: A Premier Healthcare Alliance Analysis." *Premier Inc.* (March).

Eglovitch, Joanne. 2012. "Regulatory Relief Explored for QbD Use in Post-Approval Changes." *Gold Sheet*, August. http://www.elsevierbi.com/publications/the-gold-sheet/46/8/regulatory-relief-explored-for-qbd-use-in-postapproval-changes.

Eisenberg, R. S. 2006. "The Role of the FDA in Innovation Policy." *Michigan Telecommunications and Technology Law Review* 13:345.

Friedli, Thomas, Prabir K. Basu, Thomas Gronauer, and Juergen Werani. 2010. *The Pathway to Operational Excellence in the Pharmaceutical Industry*. Aulendorf: ECV Editio Cantor Verlag GmbH.

Friedli, Thomas, Michael Kickuth, Frank Stieneker, Peter Thaler, and Juergen Werani. 2006. *Operational Excellence in the Pharmaceutical Industry*. Thomas Friedli, ed. Aulendorf: ECV Editio Cantor Verlag GmbH.

Gatesman, Mandy L. and Thomas J. Smith. 2011. "The Shortage of Essential Chemotherapy Drugs in the United States." *New England Journal of Medicine* 365(18):1653–55.

Helferich, John (Former Senior Vice President of Research and Development at Mars/Masterfoods) in discussion with the author, May 2013.

Hemphill, C. Scott and Bhaven N. Sampat. 2012. "Evergreening, Patent Challenges, and Effective Market Life in Pharmaceuticals." *Journal of Health Economics* 31(2):327–39.

Herlant, Mark. 2010. "Restoring the Balance: A Strategic Role for Operations." In *The Pathway to Operational Excellence in the Pharmaceutical Industry*, 64–76. Aulendorf: ECV Editio Cantor Verlag GmbH.

Hussain, Ajaz (Former Deputy Dir. of the Office of Pharm. Sci., U.S. Food & Drug Admin.) in discussion with the author, April 2013.

Kaakeh, Rola, B. V. Sweet, C. Reilly, C. Bush, S. DeLoach, B. Higgins, A. M. Clark, and J. Stevenson. 2011. "Impact of Drug Shortages on US Health Systems." *American Journal of Health-System Pharmacy* 68(19):1811–19.

Kitch, E. W. 1977. "The Nature and Function of the Patent System." *Journal of Law and Economics* 20(2):265–90.

Kweder, Sandra L. and Susie Dill. 2013. "Drug Shortages: The Cycle of Quantity and Quality." *Clinical Pharmacology & Therapeutics* 93(3):245–51.

Lewis, Jeffrey I. D. and Art C. Cody. 2002. "Unscrambling the Egg: Pre-Suit Infringement Investigations of Process and Method Patents." *Journal of the Patent and Trademark Office Society* 84:5–37.

Momenta Pharmaceuticals v. Amphastar Pharmaceuticals, Inc., 686 F.3d 1348 (Fed. Cir. 2012).

Norbrook Laboratories Ltd. v. H.C. Manufacturing Co., 297 F. Supp. 2d 463 (N.D.N.Y. 2003).

Outterson, Kevin. 2005. "Pharmaceutical Arbitrage: Balancing Access and Innovation in International Prescription Drug Markets." *Yale Journal of Health Policy, Law, and Ethics* 5:193–292.

Price, W. Nicholson. 2014a. "Making Do in Making Drugs: Innovation Policy and Pharmaceutical Manufacturing." *Boston College Law Review* 55:491–562.

———. 2014b. "Generic Entry Jujitsu: Innovation and Quality in Drug Manufacturing." *IP Theory* 4(1):1–8.

Roin, Benjamin N. 2008. "Unpatentable Drugs and the Standards of Patentability." *Texas Law Review* 87:503–70.

Shanley, Agnes. 2012. "cGMP Judgment Day." *PharmaManufacturing*. http://www.pharmamanufacturing.com/articles/2012/159.html.

Suresh, P. and P. K. Basu. 2008. "Improving Pharmaceutical Product Development and Manufacturing: Impact on Cost of Drug Development and Cost of Goods Sold of Pharmaceuticals." *Journal of Pharmaceutical Innovation* 3(3):175–87.

Thomas, Katie. 2012. "Drug Makers Stalled in a Cycle of Quality Lapses and Shortages." *New York Times*, October 17, sec. Business Day. http://www.nytimes.com/2012/10/18/business/drug-makers-stalled-in-a-cycle-of-quality-lapses-and-shortages.html.

———. 2013. "Use of Generics Produces an Unusual Drop in Drug Spending." *New York Times*, March 18, sec. Business Day. http://www.nytimes.com/2013/03/19/business/use-of-generics-produces-an-unusual-drop-in-drug-spending.html.

U.S. Food and Drug Administration (FDA). 2011. "Guideline for Industry: Process Validation: General Principles and Practices."

———. 2013a. "Guidance for Industry: SUPAC: Manufacturing Equipment Addendum."

———. 2013b. "Questions and Answers on Current Good Manufacturing Practices, Good Guidance Practices, Level 2 Guidance—Production and Process Controls."

Vernon, J. A., W. K. Hughen, and A. J. Trujillo. 2007. "Pharmaceutical Manufacturing Efficiency, Drug Prices, and Public Health: Examining the Causal Links." *Drug Information Journal* 41(2):229–39.

Vernon, John A. 2005. "Examining the Link Between Price Regulation and Pharmaceutical R&D Investment." *Health Economics* 14(1):1–16.

Woodcock, Janet and Marta Wosinska. 2013. "Economic and Technological Drivers of Generic Sterile Injectable Drug Shortages." *Clinical Pharmacology & Therapeutics* 93(2):170–76.

Wyeth v. Natural Biologics, Inc., No. Civ. 98-2469, 2003 WL 22282371 (D. Minn. Oct. 2, 2003), *aff'd* 395 F.3d 897 (8th Cir. 2005).

Yu, Lawrence. 2008. "Pharmaceutical Quality by Design: Product and Process Development, Understanding, and Control." *Pharmaceutical Research* 25(4):781–91.

PART SIX

Regulatory Exclusivities and the Regulation of Generic Drugs and Biosimilars

Introduction

BENJAMIN N. ROIN

FDA'S PRIMARY mission is to protect consumer safety in the markets it regulates. When Congress first created the U.S. Food and Drug Administration (FDA) in 1906, its mission was to keep adulterated food and drug products off the market. Legislators expanded FDA's mission in 1938 to include protecting consumers from unsafe drugs and gave the agency authority to block new drugs from the market until manufacturers establish their safety. In 1962, Congress again broadened FDA's mission to protecting consumers from ineffective drugs and gave the agency additional premarket regulatory authority to demand evidence of efficacy before allowing new drugs onto the market.

This premarket regulatory authority gives FDA economic powers that extend far beyond its primary mission of protecting consumers from unsafe and ineffective drugs. By setting the evidentiary standards for establishing a new drug's safety and efficacy, FDA controls which drugs enter the market and their development costs. Since FDA also sets the safety and efficacy standards for any later-arriving drugs that would compete against those earlier drugs, it controls the degree and timing of price competition in the market for new drugs. Consequently, FDA exerts tremendous influence over the profits from new drug development and thus over the incentives for investing in pharmaceutical R&D.

Congress formalized FDA's dual role of protecting consumer safety and setting innovation incentives when it passed the Drug Price Competition and Patent Term Restoration Act of 1984 (Hatch-Waxman Act). Legislators wrote the Hatch-Waxman Act to give FDA two additional functions beyond regulating the safety and efficacy of new drugs: first, to encourage the production of safe (and low-cost) generics of brand-name drugs; and second, to provide adequate incentives for innovative companies to invest in pharmaceutical R&D through lengthy monopoly periods over the new drugs they develop. Both objectives involve the regulation of generic drugs.

The Hatch-Waxman Act accomplishes its first objective—encouraging the production of generic drugs—by creating a regulatory shortcut for generic manufacturers to introduce their products onto the market. Normally FDA will not allow drug manufacturers to sell their products without first establishing their safety and efficacy through extensive clinical-trial testing—a process that usually takes the better part of a decade and can cost hundreds of millions of dollars. Under Hatch-Waxman, FDA will approve a generic drug for sale to the public based on the clinical-trial data submitted in support of the brand-name drug. Hatch-Waxman therefore saves generic manufacturers from the time and expense of clinical trials, but only if they can show that their drug contains the same active ingredient as the brand-name drug, that it is "bioequivalent" to that brand-name drug, and that it will have the same label. In short, if generic manufacturers can copy the brand-name drug and its label, FDA will allow them to avoid the clinical-development stage of pharmaceutical R&D.

The Hatch-Waxman Act has been a remarkable success in accomplishing this first objective. Generic manufacturers rely heavily on the abbreviated regulatory pathway, so much so that it essentially defines their products—a generic drug is one that enters the market based on the clinical-trial data of the brand-name drug it imitates. This abbreviated regulatory pathway also gives generic manufacturers a distinct economic advantage over pharmaceutical companies in the marketplace. While pharmaceutical companies reportedly spend over $1 billion on R&D to successfully develop a single novel drug compound, generic manufacturers can imitate those products for only a few million dollars on average. Consequently, the typical retail price for a generic drug ranges from 15 to 25 percent of the brand-name drug price. By some estimates, pharmaceutical companies generally lose about 80 percent of their sales on average within four to six weeks following generic entry. Not surprisingly, generic drugs have taken on an ever-larger share of the prescription-drug market in the United States. The market share for generic drugs has gradually increased from 19 percent of prescriptions filled in 1984, when Congress passed Hatch-Waxman, to 80 percent in 2012—currently saving the U.S. health care system an estimated $1 billion every two days.

The Hatch-Waxman Act accomplishes its second objective—providing pharmaceutical companies with lengthy (but temporary) monopoly periods over new drugs to incentivize their development—through a complex set of legal protections for new drugs. Pharmaceutical companies receive an automatic five-year monopoly term over any novel drug compound (i.e., a New Molecular Entity, NME) following its FDA approval, during which time FDA bars generics from entering the market based on the brand-name drug's clinical-trial data. But these regulatory exclusivity periods cover only a small portion of the product life cycle for most new drugs. Pharmaceutical companies can usually receive a longer monopoly term from the patent system and thus generally rely on patents to delay generic entry for long enough to recoup their R&D investments. Unfortunately, patents are an awkward incentive mechanism for pharmaceutical R&D. Firms file their patents early in R&D, and the twenty-year patent term starts running on their patent applications' filing date. Consequently, firms typically lose much of their drugs' patent life during development. The Hatch-Waxman Act lessens—but does not eliminate—this distortion by providing pharmaceutical companies with patent-term extensions that partially compensate for their patent life lost during R&D. Firms can extend their patent term by the sum of half of the time the drug spent in clinical trials and the full time the drug spent under FDA review, not to exceed a five-year extension or an effective patent life of more than fourteen years. Pharmaceutical companies may try to extend their monopoly protection further by acquiring secondary patents filed later in R&D—a practice known as evergreening. These later-filed patents are usually much weaker than the drug's initial patents, in that they are narrower and/or more likely to be invalid. To minimize evergreening, the Hatch-Waxman Act creates a legal framework for generic manufacturers to challenge drug patents, and rewards successful generic challengers with six months as the exclusive generic supplier. The goal of this complex legal framework described above is to provide pharmaceutical companies with adequate protection to encourage the development of new drugs but not excessive monopoly protection that unnecessarily delays generic entry.

The Hatch-Waxman Act's success in accomplishing this second objective has been more limited. The seemingly endless patent litigation between generic and pharmaceutical companies gives the impression—sometimes accurate—that one or both parties are abusing the system. Overall, the average effective patent life for new drugs has held steady at eleven to twelve years since Hatch-Waxman passed, which suggests that most of these perceived abuses are either exaggerated or part of the normal protection for new drugs. The Hatch-Waxman Act's more serious problems involve gaps in the available monopoly protection for new drugs. Many of the most promising new medical treatments involve new uses for existing drugs, which firms rarely develop because the Hatch-Waxman Act provides little or no incentive to invest in drugs once generics are on the market. Moreover, many potentially valuable new drugs are left undeveloped because the idea for the drug was previously disclosed in a prior patent or journal publication, thereby rendering the drug unpatentable. Others go undeveloped because they would require a lengthy R&D period that would consume too much of the drug's effective patent life. Of course all new drugs, including drugs that are unpatentable or take too long to develop, receive a minimum period of monopoly protection through regulatory exclusivity periods. However, the current regulatory exclusivity periods appear to be too short to motivate the development of most such drugs, since firms remain reluctant to invest in them.

All of this is familiar territory for the academic literature on the regulation of generic drugs. The following chapters move into the unfamiliar, discussing a host of questions that Hatch-Waxman never addressed. Henry Grabowski and Erika Lietzan's chapter, "FDA Regulation of Biosimilars," discusses the economics of regulating biosimilars. Biosimilars are imitations of biologic drugs, which were excluded from the Hatch-Waxman Act because their structural complexity makes them difficult to imitate. Congress established a new regulatory regime for generic biologics (or biosimilars) in 2010 as part of the overall health care reforms. Professors Grabowski and Lietzan argue that, at least in the short run, the economics of biosimilars are likely to differ substantially from the economics of traditional generic drugs because of the significantly greater barriers to entry for biosimilars. Arti Rai's chapter, "The 'Follow-On' Challenge: Statutory Exclusivities and Patent Dances," analyzes a particular component of the new biosimilar regulatory framework—the procedures for patent disclosure and litigation between brand-name and biosimilar manufacturers—and expresses concern that these procedures may create opportunities for collusion. Kate Greenwood's chapter, "From 'Recycled Molecule' to Orphan Drug: Lessons from Makena," discusses possible strategies FDA might use to reign in abusive pricing practices by orphan drug manufacturers, including allowing early generic competition. Finally, Marie Boyd's chapter, "FDA, Negotiated Rulemaking, and Generics: A Proposal," argues that FDA should use negotiated rulemaking to address the issues raised by the preemption of state law failure to warn claims against generic drug manufacturers and to amend its drug labeling regulations.

CHAPTER NINETEEN

From "Recycled Molecule" to Orphan Drug

Lessons from Makena

KATE GREENWOOD

I. INTRODUCTION

In the nearly thirty years since the Orphan Drug Act was passed, its package of user fee exemptions, grants and tax credits for clinical trial costs, and exclusive marketing rights has been controversial. The spotlight on orphan drugs has intensified in recent years as the number of orphan drugs that cost hundreds of thousands of dollars a year and bring in hundreds of millions of dollars in revenue continues to grow. Disputes have arisen over the Orphan Drug Act's application to so-called "recycled molecules"—older drugs that may have already been available to patients in compounded or generic form before they were designated "orphan drugs" and approved for marketing as such. Perhaps the most high profile of these is Makena, a branded version of the synthetic hormone 17-alpha hydroxyprogesterone caproate (17P), which the U.S. Food and Drug Administration (FDA) approved in February 2011 to treat pregnant women at high risk of giving birth prematurely. Makena was doubly controversial because, in addition to involving a recycled molecule that was already available to patients from compounding pharmacies, the drug's sponsor relied on government-funded research to secure FDA's approval.

After FDA approved Makena, its manufacturer, K-V Pharmaceutical, set the price of the drug at $1,440 per injection, for a cost per pregnancy of approximately $30,000; the cost of the compounded version was $15 per injection, for a cost per pregnancy of approximately $300 (Patel and Rumore 2012:406). Under pressure from advocacy groups, health care providers, Congress, and, reportedly, the White House, FDA took an unprecedented step, announcing that it would exercise its "enforcement discretion" and allow pharmacies to continue compounding 17P (FDA 2011). In so doing, the agency eased access to the drug but thwarted K-V's expectation of exclusivity, sparking since-settled litigation. FDA also reinvigorated a perennial debate about how to strike the balance between, on the one hand, incentivizing research and development of new orphan drugs and, on the other hand, ensuring that the drugs, once developed, are accessible to patients.

The balance struck by the Orphan Drug Act could be adjusted in favor of access in a number of ways, including shortening the exclusivity period, allowing for limited competition during the exclusivity period, and implementing a cap on drug prices. This chapter reviews these policy levers, focusing on the degree to which they address concerns about access, the adoption and implementation challenges they pose, and their potential for collateral effects. I also raise the alternative, suggested by the Makena case, of giving FDA the authority to decide whether to award market exclusivity on a case-by-case basis, taking into account factors such as the need to provide a reasonable reward for a company's investment, the degree to which the application for marketing approval relies on government research, the public health benefits of the drug at issue, and whether the drug is already available in compounded or generic form.

I conclude by recommending against amending the Orphan Drug Act to modify the exclusivity period or to allow FDA to take a case-by-case approach to awards of exclusivity. The first recommendation would not account for differences in value of different orphan drugs, while the second would come at a significant cost to the predictability of the regulatory process. Both would likely reduce the level of investment in orphan drug research and development. Potentially preferable are incremental reforms such as an amendment establishing a formal mechanism through which patients or others could challenge as inadequate a company's patient assistance program or other efforts to ensure that individual patients who cannot afford a drug are nonetheless able to access it. Such a mechanism could have a number of advantages over the status quo, especially in cases in which the patient population lacks the political appeal and salience of expectant mothers. If the Orphan Drug Act is amended to allow access-based challenges, an agency or entity other than FDA, with its historic focus on safety and efficacy, should be charged with adjudicating the claims.

II. THE CURRENT APPROACH TO INCENTIVIZING AND SUPPORTING ORPHAN DRUG RESEARCH AND DEVELOPMENT IN THE UNITED STATES

The National Institutes of Health (NIH) has identified close to 7,000 rare diseases, defined as those that affect fewer than 200,000 people in the United States (NIH 2013). Altogether, such diseases affect 25 to 30 million—nearly one in ten—people within the United States (NIH 2013). For a number of reasons, including that a drug's potential profitability hinges in part on the size of the population affected by the disease the drug treats, these diseases were not historically the focus of the pharmaceutical industry's research and development pipeline.

When the Orphan Drug Act (Pub. L. No. 97-414, 96 Stat. 2049 (1983) (codified principally at 21 USC §§ 360aa–360ee)) was signed into law in 1983, only thirty-eight orphan drugs had been developed (Rare Diseases Orphan Product Development Act of 2002, Pub. L. No. 107-281, § 2, 116 Stat. 1992 (2002)). In the decade before the Act's passage, FDA had approved just ten orphan drugs (Murphy et. al. 2012:482). In the decade afterward, the agency approved over 220 new orphan drugs and there were another 800 being researched (Pub. L. No. 107-281, § 2). By the middle of 2014, FDA had issued 3,127 orphan drug designations and 458 orphan drug approvals (FDA 2014). There is broad agreement that the act has succeeded in incentivizing and supporting research into, and commercialization of, orphan drugs.

The Orphan Drug Act takes a multipronged approach to encouraging the private sector to develop treatments for rare diseases.

The Act provides for grants and a tax credit for sponsors of orphan drugs, and it exempts them from the user fees that would otherwise apply for investigational new drug and new drug applications (Murphy et al. 2012:481–82). Most important from the perspective of the pharmaceutical industry, the Act also provides for a seven-year period of market exclusivity, which takes effect when and if FDA approves a drug for sale to treat a rare disease (ibid., 482). Unlike the five-year-long grant of data exclusivity that is provided for in the Drug Price Competition and Patent Term Restoration Act of 1984 (commonly known as the Hatch-Waxman Act), the Orphan Drug Act's grant of exclusivity does not hinge on the drug being a new chemical entity. Congress was not only concerned about the failure to discover new treatments to treat orphan diseases; it was also concerned about the failure to bring such treatments to market once discovered (Kux 2011). The act was intentionally designed to incentivize manufacturers to make and sell "old" drugs for orphan uses (ibid.).

The Orphan Drug Act's grant of exclusivity is subject to a limited number of narrow exceptions. First, if a sponsor can show that the drug it has developed, while the same as a designated orphan drug, is nonetheless clinically superior, the second sponsor's drug can also be so designated (21 CFR § 316.20(a)). Second, the act permits FDA to approve additional applications "if . . . the Secretary finds, after providing the holder notice and opportunity for the submission of views, that in such period the holder of the approved application or of the license cannot assure the availability of sufficient quantities of the drug to meet the needs of persons with the disease or condition for which the drug was designated" (21 USC § 360cc(b)(2)). Finally, the statute provides that FDA can approve additional applications for permission to manufacture an orphan drug if the holder of the original license consents (21 USC § 360cc(b)(1)).

III. COST AND ACCESS CONCERNS ARISING OUT OF THE CURRENT APPROACH

In 2001, the Office of Inspector General (OIG) of the Department of Health and Human Services issued a report in which it concluded that the implementation of the Orphan Drug Act had not raised significant access concerns (OIG 2001:9). Drug shortages rarely occurred, and although some orphan drugs were very costly, patients were nonetheless able to access them. The OIG noted that the very high prices of certain orphan drugs, and the very high profits reported by certain orphan drug manufacturers, had spurred Congress to consider amendments to the Act, but no such amendments were signed into law (ibid., 10).

In recent years, concerns about the price of orphan drugs have again come to the fore. In 2011, the "increasingly hot orphan drug market" was worth more than $50 billion and it was estimated to "turn . . . out blockbusters at the same rate as the broader industry" (Thomas 2013). In 2008, according to an analysis conducted by André Côté and Bernard Keating (2012:1186), there were forty-three drugs with at least one orphan designation that earned over $1 billion in revenue each year.

The price of an orphan drug is determined in the same way that the price of a non-orphan drug is. Among the factors companies may consider are (1) their own investment in researching and developing the drug, (2) the cost of manufacturing the drug, which can turn in part on the complexity of the molecule, (3) the cost of marketing and distributing the drug, (4) the medical benefit of the drug, (5) whether there are competing drugs and, if so, their price, (6) the prevalence of the disease or diseases the drug can be used to treat, and (7) relevant government regulation. Côté and Keating argue that these factors do not explain the high price of orphan drugs. They claim that the research and development costs of orphan drugs are significantly less than for standard drugs, in part because orphan drugs are often approved on the basis of fewer, smaller clinical trials. Marketing costs are reduced as well because "[p]atients with rare diseases are, for the most part, referred to and followed by teams of specialists, doctors, and pharmacists in tertiary hospitals[,]" and "[s]pecialists are exposed to the marketing of orphan drugs on a regular basis through their clinical activities, teaching activities, and research activities, and through their participation in international meetings" (ibid., 1189).

Côté and Keating (2012:1189) contend that in fact orphan drug "pricing is based on what patients and/or third-party payers are willing to pay." They explain that "[b]ecause orphan molecules are targeted at a captive market and have no therapeutic equivalents, third-party payer organizations have little room for maneuver and often resign themselves to accepting the manufacturer's suggested price, all the more so because they are subjected both to the influence of the media and to pressure from patient associations" (ibid.).

Eline Picavet and colleagues (2011:275) have taken Côté and Keating's argument a step further. Using data from Belgium from 2010, they analyzed "whether orphan drug designation status has an influence on the price setting of drugs for rare disease indications" (ibid.). To determine this, they compared the prices of designated orphan drugs to the prices of drugs that were used to treat rare diseases but were not so designated. They found that "[t]he median price per [dose] was higher for designated orphan drugs . . . than for non-designated drugs" (ibid., 277).

In many cases, individual patients are able to avoid paying an orphan drug's high price. In an article about the orphan drug Gattex, a treatment for short bowel syndrome marketed by NPS Pharmaceuticals, the *New York Times* reported that "[d]espite the high cost of the drug, NPS executives say few, if any, patients will have to pay much for it" (Thomas 2013). If a patient is privately insured, the company will cover his or her out-of-pocket expenses in excess of $10 a month. Foundations supported by the company will cover out-of-pocket expenses incurred by patients on Medicare and Medicaid. Finally, the company will provide patients who do not have insurance of any kind with Gattex free of charge. Payers are, of course, not provided with similar protections, and the pressure on them is increasing.

IV. THE CURRENT APPROACH DISRUPTED: THE CONVOLUTED REGULATORY HISTORY OF 17-ALPHA HYDROXYPROGESTERONE CAPROATE

Price and access concerns are perhaps most intense when orphan drug exclusivity is granted to "recycled molecules" such as 17-alpha hydroxyprogesterone caproate (17P), which was first approved in 1956 (Kim 2013:548). Marketed under the brand name Delalutin, 17P was used to treat various conditions at various times. In 2000, FDA withdrew its approval of Delalutin at the request of the drug's manufacturer, Bristol-Myers Squibb, which had not sold it for several years (ibid., 548–49).

In 2003, the *New England Journal of Medicine* published an article reporting on the results of a clinical trial of 17P funded by the National Institute of Child Health and Human Development (Meis et al. 2003). The clinical trial showed that 17P was effective in preventing preterm labor and delivery in pregnant women with a documented history of a spontaneous premature birth. A follow-up study published in 2007 demonstrated that children exposed to 17P in the second and third semesters, which is when the drug is given to prevent prematurity, had no adverse health outcomes (Northen et al. 2007). The drug, which is administered via intramuscular injection, became the standard of care for pregnant women with a history of preterm birth, but because it was not commercially available in the wake of Delalutin's withdrawal, prescriptions for it had to be filled at compounding pharmacies (Patel and Rumore 2012:405).

Not surprisingly, interest arose in commercializing 17P. On May 4, 2006, a new drug application was filed seeking approval to market it for the prevention of recurrent premature birth (Adeza Biomedical Corporation 2006). The drug received orphan drug designation on January 25, 2007, and was finally approved for marketing, under the name Makena, on February 3, 2011 (FDA 2014).

Also on February 3, 2011, FDA denied an October 24, 2007 Citizen Petition from the Sidelines National Support Network seeking revocation of Makena's orphan-drug designation. Sidelines argued that orphan drug designation was inappropriate because, among other reasons, (1) should the drug "be approved it would have achieved that status and the attainment of a seven-year monopoly . . . via utilization, not of its own original research, but of data licensed from a *government-funded* clinical trial" and (2) "if marketing exclusivity is assigned to [the] product, all chance of competitive pricing will be lost for the exclusivity period" (Sidelines 2007).

In its response, FDA emphasized that although it administers the orphan drug designation and approval process, its discretion is very limited (Kux 2011). Under current law, if a drug is being developed to treat a "rare disease or condition," FDA must grant a request from the sponsor that the drug be designated an orphan drug; if an orphan drug is subsequently approved for marketing, orphan-drug exclusivity occurs automatically. FDA cannot revoke a drug's designation unless "(1) [t]he request for designation contained an untrue statement of material fact; or (2) [t]he request for designation omitted material information required by this part; or (3) FDA subsequently finds that the drug in fact had not been eligible for orphan-drug designation at the time of submission of the request therefor" (21 CFR § 316.29(a)).

FDA responded to Sidelines's argument that 17P's orphan drug designation should be revoked because its sponsor relied on government-funded research by explaining that "new research on its own is not what the [Orphan Drug Act] seeks to achieve" (Kux 2011). Relying on government-funded research is consistent with the Act's aim of encouraging commercialization of orphan drugs. To Sidelines's argument that a grant of market exclusivity to 17P would result in a period of unjustified cost increase, FDA responded that this too comported with the intent of Congress (ibid.).

Its professed lack of discretion notwithstanding, FDA found a way to take action. On March 30, 2011, the agency issued a press release in which it explained in the first paragraph that "KV Pharmaceuticals, the drug's owner, received considerable assistance from the federal government in connection with the development of Makena by relying on research funded by the National Institutes of Health to demonstrate the drug's effectiveness" (FDA 2011). The agency went on to announce that it would not enforce K-V's statutory right to seven years of marketing exclusivity. The Centers for Medicare & Medicaid Services (CMS) issued a statement the same day, announcing that compounded 17P could be reimbursed by Medicaid (CMS 2011).

On April 1, 2011, K-V issued a press release of its own, announcing that it had decided to, among other things, reduce the list price of Makena from $1,440 to $690 and expand the company's patient assistance program (K-V Pharmaceutical 2011). K-V noted that "85 percent of patients will pay $20 or less per injection for FDA-approved Makena, and patients whose financial need is greatest would receive FDA-approved Makena at no out-of-pocket cost" (ibid.). It was widely viewed as too little too late. On August 4, 2012, the manufacturer declared bankruptcy, blaming in its press release "a lack of enforcement of the orphan drug exclusivity granted" to Makena[1] (K-V Pharmaceutical 2012). On July 5, 2012, K-V sued FDA, challenging the agency's decision to decline to take enforcement action against pharmacies that compound 17P (Karst 2014). Two years later, in July 2014, K-V agreed to dismiss the case, without making public what, if anything, FDA promised in return (ibid.).

V. ADJUSTING THE ORPHAN DRUG ACT'S BALANCE IN FAVOR OF ACCESS

A. Shortening the Exclusivity Period

One way that Congress could adjust the Orphan Drug Act's balance between encouraging innovation, on the one hand, and ensuring access, on the other, would be to shorten the exclusivity period, whether across the board or on the basis of predetermined criteria. A reduction in the exclusivity period could be paired with an alternative incentive that does not affect access, such as the priority review vouchers that are already available for rare pediatric diseases and for certain tropical diseases (Mueller-Langer 2013:192–93).

An across-the-board reduction in the exclusivity period would have the advantage of ease of administration for the government and predictability for industry. There is the possibility, though, that manufacturers faced with a shorter amount of time in which to recoup their investments and earn profits would set prices proportionately higher. An across-the-board reduction would also fail to account for factors such as whether the drug was already available in compounded or generic form.

In 1992, Congress considered an amendment to the Orphan Drug Act that would have allowed for two years of guaranteed exclusivity, after which competing drugs could be approved if, at any point in the next seven years, sales of the originator drug topped $200 million (Pulsinelli 1999:332). The European Union has adopted an approach akin to this. There, the orphan-drug exclusivity period can be shortened from ten years to six if the orphan-drug designation criteria are no longer met (Michaux 2010:661). A regime in which Congress, or FDA with Congress's authorization and guidance, provided for a shortened exclusivity period when a drug reached a predetermined level of sales, profit, or return on investment, or met other criteria, would be more calibrated than an across-the-board reduction, but it could be difficult to administer and would have the downside of undermining predictability. Perhaps a more fundamental concern is that, after Makena, payers may balk at any period of monopoly pricing for drugs that are already available to patients in compounded or generic form.

B. Allowing for Limited Competition During the Exclusivity Period

Over the years, a number of amendments to the Orphan Drug Act have been proposed that would authorize two or more companies to sell an orphan drug during the seven-year exclusivity period (Pulsinelli 1999:324). Instead of a monopoly, the market would take the form of an oligopoly. Prices would still be supracompetitive during the exclusivity period, but they would not be as high as they are under the current system. In 1990, Congress passed a set of amendments that, among other things, would have allowed companies that simultaneously developed the same orphan drug to share exclusivity (ibid., 325). The amendments were vetoed by then-President George H. W. Bush, who was concerned about their potential to undermine the act's incentive structure (ibid., 337).

Gary Pulsinelli has raised the possibility that, to a manufacturer, shared exclusivity might be "practically indistinguishable from straight competition," which could in turn "lead to investment at less than the optimal level" (ibid., 328). On the other hand, there is the possibility that, to consumers, shared exclusivity might look a lot like exclusivity. Pulsinelli gives the example of human growth hormone, which at one time was manufactured by two different companies that "essentially" shared exclusivity (ibid., 326). Human growth hormone was nonetheless "one of the highest priced and most profitable orphans, for both companies" (ibid.).

One way to address concerns that the incentive for sponsors to engage in orphan drug research and development be preserved would be to lengthen the exclusivity period in those cases in which there is limited competition. Another variant would be to allow for limited competition postapproval but only after a drug reached a certain predetermined level of sales, profit, or return on investment. Both of these variations would have a cost in terms of access, though.

C. Implementing a Cap on Drug Prices

A cap on the price that an orphan drug's manufacturer can charge during the period of exclusivity is the most straightforward approach to adjusting the balance between innovation and access. Price caps would also have the advantage of moderating prices from day one. Setting, applying, and enforcing them could pose political and practical difficulties for Congress and FDA, however. Before 1995, the NIH included a term in the contracts it entered into with partners in the life-sciences industry that required that there be "a reasonable relationship between the pricing of a licensed product, the public investment in that product, and the health and safety needs of the public" ("PHS/NIH Abandons . . ." 1995). The term was widely believed to be a deal breaker, preventing public–private partnerships from forming at all (ibid.).

In an article in *Health Affairs*, Ana Valverde (2012:2530) and colleagues proposed what they called the "grant-and-access pathway," under which grants would be available to support early phase orphan-drug development. Companies that opted in to this pathway and were awarded grants would not be able to claim the Orphan Drug Act's tax credits (ibid.). Of more significance, they would have to agree to price caps, which would be set "based on how long drug development took and how much it cost, expected market size, and the target internal rate of return" (ibid.). The authors' analysis revealed that "[i]n the absence of a robust grant program, the ultra-orphan drug developer in our example would need to increase drug prices to $329,000 per patient for an annual course of treatment to yield a 20 percent rate of return" (ibid., 2531). By contrast, under the grant-and-access pathway, 75 percent of the development costs, estimated at $4.5 million in their model, would be covered. This would mean that the ultra-orphan drug developer would yield a 20 percent return on its investment at "a treatment price of $84,750 per patient—roughly one-fourth of the cost estimated under the current pathway" (ibid., 2531–32). Because the grant-and-access pathway proposed by Valverde and colleagues is optional, it might not pose some of the political difficulties price caps usually present. Another advantage of the grant-and-access pathway is that it would leave undiluted the incentive provided by the Orphan Drug Act's grant of seven years of market exclusivity. A significant downside of the pathway is that it would only increase access in those cases in which orphan drug developers agreed to its terms.

D. Authorizing FDA to Address Access Concerns on a Case-by-Case Basis

A fourth possibility would be to preserve as the default the Orphan Drug Act's current balance between incentivizing innovation and ensuring access, while permitting FDA to make adjustments on a case-by-case basis. FDA could be given the authority to decide in each case whether and for how long to award market exclusivity, taking into account factors such as the need to provide a reasonable reward for a company's investment, the degree to which the application for marketing approval relies on government research, the public health benefits of the drug at issue, and whether the drug is already available in compounded or generic form. Alternatively, or in addition, FDA could be given the authority to decide in each case whether an orphan drug, given its price, reimbursement profile, and other factors, is sufficiently accessible to patients.

Currently, FDA does not consider the questions of price and access that the Makena case posed, not typically at least, or through a mechanism other than the exercise of its enforcement discretion. The agency does administer the statutory exclusivity period that gives orphan drug developers monopoly-pricing power, but the statute affords it little discretion. As noted earlier, the regulations implementing the Orphan Drug Act provide that "a sponsor of a drug that is otherwise the same drug as an already approved orphan drug may seek and obtain orphan-drug designation for the subsequent drug for the same rare disease or condition if it can present a plausible hypothesis that its drug may be clinically superior to the first drug" (21 CFR § 316.20(a)). When the regulations were adopted, several commenters "argued that FDA must recognize the effect of price on access to patient care and urged that cost considerations must be used in determining whether a subsequent drug makes a major contribution to patient care" (Orphan Drug Regulations, 57 Fed. Reg. 62076, 62078 (Dec. 29, 1992)). FDA declined to so hold, explaining that "[a]lthough FDA understands that costs can indeed have a major impact on access to a drug, FDA has no authority over drug pricing or any authority to consider it in drug approval" (ibid., 62079).

FDA was also asked to factor in cost when determining whether "sufficient quantities" of the drug are not available "to meet the needs of persons with the disease or condition for which the drug was designated" (21 USC § 360cc(b)(1)). The agency responded that it "does not have the authority under existing law to equate high cost with lack of sufficient quantities, even though cost may affect access to a drug" (Orphan Drug Regulations, 57 Fed. Reg. 62076, 62084 (Dec. 29, 1992)).

A case-by-case approach to awards of exclusivity would allow for calibration, but it would come at a cost to the predictability of the regulatory process. This could in turn reduce the level of investment in orphan drug research and development. A formal mechanism for challenging the high price of orphan drugs would have the advantage of preserving the basic structure of the current system, but it, too, could reduce companies' incentive to bring orphan drugs to market. It also seems highly unlikely to attract political support. If patients and providers were limited to challenging the adequacy of companies' patient assistance programs, the effect on incentives would be minimized. Allowing for such challenges could be particularly valuable in cases in which the nature of the drug or the patient population were such that it would be difficult to mount an advocacy campaign of the sort that was successful in the Makena case.

If the Orphan Drug Act is amended to allow access-based challenges, an agency or entity other than FDA, with its historic focus on safety and efficacy, should be charged with adjudicating the claims. In the European Union, the decision to grant market authorization is made by one agency and the decision to grant reimbursement is made by another. Eline Picavet and colleagues (2012) note that "[m]arket authorization does not automatically mean central funding of, and access to, an orphan drug."

VI. CONCLUSION

The Makena case drew attention to the high price that patients, providers, the government, and other payers pay for the relatively small gains that accrue when a recycled molecule becomes an FDA-approved orphan drug. Eliminating or shortening the period of exclusivity would result in orphan drugs that might otherwise have been adopted by a private-sector sponsor going unstudied or remaining unapproved by FDA. Adjusting the period for some but not all drugs leads to the difficult question of where to draw the line and would undermine predictability, which is key to encouraging investment. A formal mechanism to challenge the adequacy of a company's patient assistance program is potentially a more promising reform.

There can be gains to public health associated with FDA approval even of "recycled molecules" like Makena that were previously available to patients in compounded form. For one, FDA-approved drugs are available through standard commercial channels, which eases access. In addition, drugs manufactured according to good manufacturing practices are more likely to be pure and to have the desired potency than drugs that are compounded (Patel and Rumore 2012:410). The risk of contamination is higher for sterile injectables like Makena than it is for other drug forms. While these are real benefits, they are not invaluable. Even if the status quo is maintained, going forward companies will surely keep the Makena case in mind when pricing recycled molecules.

NOTE

1. In May 2013, the *Wall Street Journal* reported that sales of Makena were surging (Gleason 2013). In September 2013, the *St. Louis Post-Dispatch* reported that K-V Pharmaceutical had emerged from bankruptcy (Brown 2013).

REFERENCES

Adeza Biomedical Corporation. 2006. "Adeza Submits New Drug Application for Gestiva to Prevent Preterm Births" (May 4). http://www.drugs.com/nda/gestiva_060504.html (accessed July 12, 2014).

Brown, L. "KV Pharmaceutical Emerges from Bankruptcy." *St. Louis Post-Dispatch*, Sept. 16, 2013.

Centers for Medicare & Medicaid Services. 2011. "Makena" (March 30). http://www.medicaid.gov/Federal-Policy-Guidance/downloads/Makena-CMCS-Info-Bulletin-03-30-2011.pdf (accessed Mar. 6, 2015).

Côté, A. and B. Keating. 2012. "What Is Wrong with Orphan Drug Policies?" *Value in Health* 15:1185–91.

Department of Health and Human Services (DHHS), Office of Inspector General (OIG). 2001. "The Orphan Drug Act: Implementation and Impact." OEI-09–00–00380.

Gleason, S. 2013. "How a Pricey Pregnancy Medicine Made a Comeback." *Wall Street Journal* (May 31).

Karst, K. R. 2014. "KV Lawsuit Involving MAKENA and Compounded 17p Concludes. . . . In Sopranos Style." *FDA Law Blog* (July 7).

Kim, S. 2013. "The Orphan Drug Act: How the FDA Unlawfully Usurped Market Exclusivity." *Northwestern Journal of Technology & Intellectual Property* 11:541–57.

Kux, L. Letter from Leslie Kux, Acting Assistant Commissioner for Policy, U.S. Food and Drug Administration, to Candace Hurley, Executive Director/Founder, Sidelines National Support Network, Laguna Beach, CA (February 3, 2011).

K-V Pharmaceutical Company. 2012. "K-V Pharmaceutical Company Files Voluntary Petitions for Reorganization to Restructure Financial Obligations" (August 4). http://www.prnewswire.com/news-releases/k-v-pharmaceutical-company-files-voluntary-petitions-for-reorganization-to-restructure-financial-obligations-164981076.html (accessed Mar. 6, 2015).

———. 2011. "Ther-Rx Corporation Takes Action to Further Ensure High-Risk Women Are Able to Access FDA-Approved Makena™" (April 1). http://www.prnewswire.com/news-releases/ther-rx-corporation-takes-action-to-further-ensure-high-risk-women-are-able-to-access-fda-approved-makena-119056354.html (accessed July 12, 2014).

Meis, P. J. et al. 2003. "Prevention of Recurrent Preterm Delivery by 17 Alpha-Hydroxyprogesterone Caproate." *New England Journal of Medicine* 348:2379–85.

Michaux, G. 2010. "EU Orphan Regulation—Ten Years of Application." *Food and Drug Law Journal* 65:639–69.

Mueller-Langer, F. 2013. "Neglected Infectious Diseases: Are Push and Pull Incentive Mechanisms Suitable for Promoting Drug Development Research?" *Health, Economics, Policy and Law* 8:185–208.

Murphy, S. M. et al. 2012. "Unintended Effects of Orphan Product Designation for Rare Neurological Diseases." *Annals of Neurology* 72:481–90.

National Institutes of Health (NIH). "Frequently Asked Questions." http://rarediseases.info.nih.gov/about-ordr/pages/31/frequently-asked-questions (accessed April 23, 2013).

Northen, A. T. et al. 2007. "Follow Up of Children Exposed in Utero to 17a-Hydroxyprogesterone Caproate Compared with Placebo." *Obstetrics and Gynecology* 110:865–72.

Patel, P. and M. M. Rumore. 2012. "Hydroxyprogesterone Caproate Injection (Makena) One Year Later." *Pharmacy and Therapeutics* 37:405–11.

"PHS/NIH Abandons 'Reasonable Pricing' Licensing/CRADA Policy." *Antiviral Agents Bulletin* (April 1995).

———. 2012. "Market Uptake of Orphan Drugs—A European Analysis." *Journal of Clinical Pharmacy and Therapeutics* 37:664–67.

Picavet, E. et al. 2011. "Drugs for Rare Diseases: Influence of Orphan Designation Status on Price." *Applied Health Economics & Health Policy* 9:275–79.

Pulsinelli, G. A. 1999. "The Orphan Drug Act: What's Right with It." *Santa Clara Computer & High Technology Law Journal* 15:299–345.

Sidelines National Support Network. 2007. Citizen Petition (October 24).
Thomas, K. 2013. "Making 'Every Patient Counts' a Business Imperative." *New York Times* (January 30).
U.S. Food and Drug Administration (FDA). 2014. "Orphan Drug Designations and Approvals." http://www.accessdata.fda.gov/scripts/opdlisting/oopd/index.cfm (accessed July 12, 2014).
———. 2011. "FDA Statement on Makena." (March 30). http://www.fda.gov/NewsEvents/Newsroom/PressAnnouncements/2011/ucm249025.htm (accessed March 27, 2014).
Valverde, A. M., S. D. Reed, and K. A. Schulman. 2012. "Proposed 'Grant-and-Access' Program with Price Caps Could Stimulate Development of Drugs for Very Rare Diseases." *Health Affairs* 31:2528–35.

CHAPTER TWENTY

FDA, Negotiated Rulemaking, and Generics: A Proposal

MARIE BOYD

I. INTRODUCTION

The potential legal remedies available to a person injured by a generic drug differ from those available to a person injured by the corresponding brand-name drug. In *Wyeth v. Levine*, the U.S. Supreme Court held that federal law does not preempt state failure-to-warn claims against the manufacturer of a brand-name drug (*Wyeth v. Levine*, 2009:581). In *PLIVA, Inc. v. Mensing*, it held that federal law does preempt such claims against the manufacturer of a generic drug (*PLIVA, Inc. v. Mensing* 2011:2581). This result is due to differences in how brand-name and generic drugs are regulated under federal law (ibid., 2582).

By holding that state failure-to-warn claims against generic drug manufacturers are preempted, the Court eliminated the protections that state tort law can provide consumers of generic drugs through the law's compensation and information disclosure functions (Rabin 2007:301; Kessler and Vladeck 2008:483). It also exposed a gap in the federal regulation of generic drug labeling in which no manufacturer is responsible for updating the labeling of generic drugs if the corresponding brand-name drug is no longer marketed (*PLIVA, Inc. v. Mensing* 2011:2592–93; Sotomayor, J., dissenting; Lee 2012:239–41).[1] The preemption of failure-to-warn claims against generic drug manufacturers could have widespread effect as generic drugs account for approximately 80 percent of the prescriptions dispensed in the United States (Dicken 2012:2, 9; IMS Inst. for Healthcare Informatics 2012:26), and approximately 23 to 32 percent of drugs are available solely as generics (Brief for Marc T. Law et al. 2011:18). There are approximately 106,000 deaths per year from "nonerror, adverse effects of medications," and the actual magnitude of adverse drug effects is likely greater because that estimate does "not include adverse effects that are associated with disability or discomfort" (Starfield 2000:484).

In apparent recognition of the gravity of these issues, the U.S. Food and Drug Administration (FDA) published a notice of proposed rulemaking (NPRM), which if finalized would permit generic drug manufacturers to update their product labeling in certain circumstances (FDA's proposed rule) (Supplemental Applications Proposing Labeling Changes for Approved Drugs and Biological Products, 78 Fed. Reg. 67,985 (Nov. 13, 2013)). This chapter argues that rather than proceed with the informal, notice-and-comment rulemaking procedure, FDA should use negotiated rulemaking to work with stakeholders to address the issues raised and exposed by *Mensing*, build consensus, and amend the regulations. Negotiated rulemaking may foster the development of a more effective, enforceable, and legitimate rule as compared to informal rulemaking.

II. PREEMPTION AND THE REGULATION OF DRUGS

Both *Wyeth* and *Mensing* involved patient injuries that followed the administration of a prescription drug and allegations that the drug's manufacturer failed to adequately warn the plaintiff of the risk of the injuries suffered. In *Wyeth*, a brand-name drug was administered, whereas in *Mensing*, a generic drug was administered, and the preemption results differed because the regulations for generic drugs are "meaningfully different" from those for brand-name drugs (*PLIVA, Inc. v. Mensing* 2011:2582).

A. *Wyeth v. Levine* and the Regulation of Brand-Name Drugs

In *Wyeth*, the Court held that federal law did not preempt a plaintiff's state-law claim that brand-name drug labeling did not contain an adequate warning (*Wyeth v. Levine* 2009:564–65, 581). According to the Court, Wyeth could have unilaterally strengthened its warning under FDA's "changes-being-effected" (CBE) regulation, which permits manufacturers to make certain changes (including "[t]o add or strengthen . . . a warning") to the FDA-approved labeling of a drug when FDA receives a CBE supplement from the manufacturer (*Wyeth v. Levine* 2009:571, 573; 21 CFR § 314.70(c)(6)(iii) (2014); FDA 2004:3–4, 24–26). Thus, it was not impossible for Wyeth to comply with both the federal requirements—the Federal Food, Drug, and Cosmetic Act (FDCA) and FDA's regulations, which Wyeth argued required it to keep the drug labeling the same as that in its approved new drug application (NDA)—and the state requirements, which Wyeth argued required it to change the drug's labeling (*Wyeth v. Levine* 2009:571, 573; Brief for Petitioner 2009:33–34; Reply Brief for Petitioner 2009:1–4). The Court also held that the failure-to-warn claims did not obstruct the "purposes and objectives" of the federal regulatory scheme (*Wyeth v. Levine* 2009:581). The Court stated that "it has remained a central premise of federal drug regulation that the manufacturer bears responsibility for the content of its label at all times" (ibid., 570–71, 581).

B. *PLIVA, Inc. v. Mensing* and the Regulation of Generic Drugs

Approximately two years after *Wyeth*, the Court in *Mensing* considered whether similar state failure-to-warn claims against the manufacturers of generic drugs were preempted (*PLIVA, Inc. v. Mensing* 2011:2572–73). The Court held that they were because it was "impossible" for the manufacturers to independently comply with both their federal law duty that the labeling of their generic drug products be the same as the corresponding brand-name drug labeling and their state-law duty to change the labeling to strengthen their warnings (ibid., 2578).

The Drug Price Competition and Patent Term Restoration Act of 1984 (the "Hatch-Waxman Act") created an abbreviated approval process for generic drugs—the abbreviated new drug application (ANDA) process, in which generic drugs are approved on the basis of information showing that they are bioequivalent to a reference listed drug (RLD), generally a brand-name drug (21 USC § 355(j)(2)(A)(iv), (j)(4)(F), (j)(7) (2012); FDA Orange Book). FDA's regulations define RLD as "the listed drug identified by FDA as the drug product upon which an applicant relies in seeking approval of its abbreviated application" (21 CFR § 314.3 (2014)). The statute and regulations provide that a generic drug manufacturer must show, among other things, that the proposed generic drug labeling "is the same as" the RLD's approved labeling (21 USC § 355(j)(2)(A), (j)(4) (2012); 21 CFR §§ 314.94(a)(8)(iii), 314.127(a)(7) (2014)).[2] The regulations also add that FDA may act to withdraw an approved ANDA if the product's labeling "is no longer consistent with that" of the RLD (21 CFR § 314.150(b)(10) (2014)).

The *Mensing* Court deferred to FDA's interpretation of the regulations as requiring that the generic manufacturers' labeling "always be the same" as that of the brand-name drug and as precluding generic manufacturers from unilaterally strengthening their generic drug's warnings using the CBE process or sending "Dear Doctor" Letters (*PLIVA, Inc. v. Mensing* 2011:2575–76; Brief for the United States as Amicus Curiae Supporting Respondents 2011:14–19). FDA interpreted its regulations as permitting generic manufacturers to use the CBE process to change the labeling of a generic drug only when the change is to match the corresponding brand-name drug's labeling or to follow FDA's instructions (*PLIVA, Inc. v. Mensing* 2011:2575–76; Brief for the United States as Amicus Curiae Supporting Respondents 2011:15, 16 nn7–8).

III. FDA'S PROPOSAL TO REFORM THE REGULATION OF GENERIC DRUGS

In November 2013, FDA issued an NPRM that would amend FDA's current labeling regulations to create a modified CBE process that would permit generic manufacturers to independently make certain labeling changes at the time a CBE supplement is received by FDA

(Supplemental Applications Proposing Labeling Changes for Approved Drugs and Biological Products, 78 Fed. Reg. 67,985, 67,989 (Nov. 13, 2013)). The proposal contained procedures that would make submitted changes publicly available during FDA's review and alert the manufacturer of the corresponding brand-name drug, if any, of the supplement (ibid., 67,990). The manufacturers of other versions of the generic drug and the corresponding brand-name drug would be permitted to submit CBE supplements and comment on a submitted supplement (ibid., 67,991–92). If FDA approved a labeling change, it would do so "for the generic drug and the corresponding brand drug at the same time, so that . . . [these] products [would] have the same FDA-approved labeling" (Howard 2014:2). Within thirty days of the approval, manufacturers of other versions of the generic drug would be required to submit CBE supplements with conforming labeling changes (Supplemental Applications Proposing Labeling Changes for Approved Drugs and Biological Products, 78 Fed. Reg. 67,985, 67,993, 67,999 (Nov. 13, 2013)). When this chapter was written, FDA's proposed rule had not been finalized.[3]

While FDA's proposal was a step toward addressing the issues raised and highlighted by *Mensing*, rather than proceed with the conventional informal rulemaking process, FDA should utilize negotiated rulemaking.

IV. NEGOTIATED RULEMAKING

In 1982, the Administrative Conference of the United States recommended procedures for negotiating proposed regulations and urged Congress to pass legislation authorizing agencies to conduct regulatory negotiation (Recommendations of the Administrative Conference, 47 Fed. Reg. 30,701, 30,709 (July 15, 1982)). It based its recommendation on a report by Philip Harter (Harter 1982:1). Harter also authored an article on negotiated rulemaking as an alternative to the notice-and-comment rulemaking set forth by section 553 of the Administrative Procedure Act (APA), which he argued had become mired in a "malaise" (ibid., 5–6, 28).

Harter identified several factors that may help guide the determination of whether negotiations are appropriate (ibid., 42–52): The parties must believe that participation is in their best interests and no party should have the power to impose its will on the others; the number of parties should be limited; the issues to be resolved concrete and ready for resolution; a decision inevitable or even imminent; the dispute capable of being "transformed into a 'win/win' situation" for the parties; the parties able to agree on fundamental principles; the number of issues sufficient to permit trade-offs; the "[r]esearch [n]ot [d]eterminative of [the o]utcome"; and the implementation of the negotiated agreement likely (ibid., 45–52). Harter emphasized the importance of identifying the interests that should be represented in the negotiations, identifying appropriate representatives of such interests, and obtaining their participation (ibid., 52–57).

The Negotiated Rulemaking Act of 1990 (NRA) created a framework for using a "negotiated rulemaking committee . . . to consider and discuss issues for the purpose of reaching a consensus in the development of a proposed rule" and "encourage[s] agencies to use [negotiated rulemaking] when it enhances the informal rulemaking process" (5 USC § 562(6), (7) (2012); Negotiated Rulemaking Act 1990). Before an agency can establish a negotiated rulemaking committee, it must consider whether (1) the rule is needed; (2) "there are a limited number of identifiable interests that will be significantly affected by the rule"; (3) it is reasonably likely "that a committee can be convened with a balanced representation of persons who . . . can adequately represent the [identified] interests . . . and . . . are willing to negotiate in good faith to reach a consensus on the proposed rule"; (4) it is reasonably likely that a committee will reach such a consensus "within a fixed period of time"; (5) the "procedure will not unreasonably delay the notice of proposed rulemaking and the issuance of the final rule"; (6) "the agency has adequate resources . . . [that it] is willing to commit . . . to the committee"; and (7) "the agency, to the maximum extent possible consistent with [its] legal obligations . . . , will use the consensus of the committee with respect to the proposed rule as the basis for the rule proposed by the agency for notice and comment" (5 USC § 563 (2012)).

If the agency determines that negotiated rulemaking "is in the public interest" and decides to establish a negotiated rulemaking committee, it must publish a notice in the *Federal Register* announcing its intention to do so and provide a period for the submission of comments and committee membership applications (5 USC § 564 (2012)). The agency may establish a negotiated rulemaking committee if it determines that such a committee "can adequately represent the interests that will be significantly affected by a proposed rule and that it is feasible and appropriate in the particular rulemaking" (5 USC § 565 (2012)). The committee, including the agency representatives, must negotiate to attempt to reach a consensus on a proposed rule (5 USC § 566(a) (2012)): If the committee reaches a consensus on a proposed rule, it must provide the agency with a report and the proposal; if it does not, it may provide a report on any areas of consensus (5 USC § 566(f) (2012)). A rule based on the committee's consensus that is proposed by an agency is subject to the APA's informal rulemaking requirements (5 USC § 561 (2012); Pritzker and Dalton 1995:1, 2).

Although used infrequently (Coglianese 1997:1276, 1277 table 2; Lubbers 2008:1007–17), negotiated rulemaking has been the subject of ongoing debate. There is substantial academic literature supporting negotiated rulemaking (Funk 1997:1353); however, negotiated rulemaking is not without critics (Coglianese 1997:1316–17; Funk 1997:1356; Funk 1987:66–78, 92–96; Rose-Ackerman 1994:1211). Commentators have divided over questions of the legitimacy, benefits, and effectiveness of negotiated rulemaking. Supporters have argued that it may further legitimacy and accountability, "improve relationships among repeat players," and create better and more widely accepted rules (Freeman 2000:548–49, 656–57, 666; Freeman 1997:30–33; Freeman and Langbein 2000; Harter 2000:52–54; Harter 1982:22, 31, 69, 84, 94; Susskind and McMahon 1985:133). Critics have countered that it lacks legitimacy, undermines the public interest with private bargaining, and has not been successful in decreasing rulemaking time and judicial challenges to rules (Funk 1987:57; Rose-Ackerman 1994:1208–12; Coglianese 1997:1335–36).

To date, FDA has not convened a negotiated rulemaking committee or been required by Congress to do so.[4] Nevertheless, FDA expressed openness to considering the use of negotiated rulemaking in one of its food labeling regulations, which requires certain petitions to include a statement on the feasibility of using negotiated rulemaking (21 CFR § 101.12(h) (2014)). Also, it has been reported that FDA has considered using negotiated rulemaking in other contexts (FDA Waiver of User Fees 1994; OTC Label Reform 1995; Device Software Policy Revisions 1996).

V. THE CASE FOR NEGOTIATED RULEMAKING

The issues raised and highlighted by the *Mensing* decision appear well suited to negotiation, and the use of negotiated rulemaking may further the public interest. Although FDA has already promulgated an NPRM, it is not too late for FDA to employ negotiated rulemaking, as nothing in the NRA prohibits an agency from using negotiated rulemaking after an NPRM so long as the requirements of the NRA are met (5 USC §§ 561–70a (2012)). The Administrative Conference specifically recognized that "negotiating the terms of a final rule could be a useful procedure even after publication of a proposed rule" (1 CFR § 305.85-5(3) (1986)), and several agencies have created negotiated rulemaking committees after publication of an interim or proposed rule (Vehicles Built in Two or More Stages, 64 Fed. Reg. 27,499 (May 20, 1999); Paleontology; Negotiated Rulemaking, 54 Fed. Reg. 48,647 (November 24, 1989); Varroa Mite Regulations, 54 Fed. Reg. 15,217 (Apr. 17, 1989)).

A. The Need for a Rule

There is a need for new drug labeling regulations. By finding that state failure-to-warn claims against the manufacturers of generic drugs are preempted under FDA's current regulatory regime, the *Mensing* Court removed the protections that state tort law can provide to consumers. As discussed earlier, that decision also highlighted a gap in the regulation of generic drug labeling where no manufacturer may be responsible for updating a drug's labeling. This is concerning given that "[m]any serious [adverse drug reactions] are discovered only after a drug has been on the market for years" (Lasser et al. 2002:2218) and FDA "faces significant resource constraints that limit its ability to protect the public from dangerous drugs" (*Mutual Pharmaceutical Co. v. Bartlett* 2013:2485; Sotomayor, J. dissenting). Furthermore, the different potential remedies for injured consumers of generic versus brand-name drugs are inconsistent with FDA's policy of promoting the "sameness" of these products (*PLIVA, Inc. v. Mensing*, 2011:2593; Sotomayor, J., dissenting).

The Court in *Mensing* noted that FDA retains the authority to change its regulations if it so desires (ibid., 2582), and in apparent recognition of the need for regulatory change, FDA has proposed new regulations using the informal rulemaking process (Supplemental Applications Proposing Labeling Changes for Approved Drugs and Biological Products, 78 Fed. Reg. 67,985 (Nov. 13, 2013)).

B. The Issues Are Concrete, Ready for Decision, and Sufficient to Permit Trade-Offs

The issues for consideration are concrete and ready for decision;[5] issues related to the regulation of drug labeling were considered in several opinions (*Mutual Pharmaceutical Co., Inc. v. Bartlett* 2013:2476–77; *PLIVA, Inc. v. Mensing* 2011:2574–77; *Wyeth v. Levine* 2009:568–73; *Mutual Pharmaceutical Co., Inc. v. Bartlett* 2013:2480, Breyer, J., dissenting; *Mutual Pharmaceutical Co., Inc. v. Bartlett* 2013:2483, Sotomayor, J., dissenting; *PLIVA, Inc. v. Mensing* 2011:2582, Sotomayor, J., dissenting), briefs (Brief for Marc T. Law et al. 2011; Brief for the United States as Amicus Curiae Supporting Respondents 2011; Brief for the United States as Amicus Curiae 2011), proposed legislation (Patient Safety and Generic Labeling Improvement Act, S. 2295 2012; Patient Safety and Generic Labeling Improvement Act, H.R. 4384 2012), FDA's proposed rule (Supplemental Applications Proposing Labeling Changes for Approved Drugs and Biological Products, 78 Fed. Reg. 67,985, 67,989 (Nov. 13, 2013)), a Citizen Petition (Public Citizen 2011:9–10) and FDA's response to that petition (FDA Response to Public Citizen 2013:2), and a growing body of academic literature including proposals for legislative or regulatory change (Lee 2012:252–54; Duncan 2010:209–10; Kazhdan 2012:919; Stoddart 2012:1993–96; Weeks 2012:1259, 1289). The existing proposals suggest that in addition to the issues noted earlier, any proposal should consider and address: (1) who should be able to make labeling changes and under what circumstances; (2) how to encourage appropriate and timely warnings; (3) whether and how to reconcile differences between the labeling of different versions of a drug after a labeling change; and (4) whether there is a need for increased information sharing, reporting, or producing in order for manufacturers to fulfill any new regulatory responsibilities

(Boyd 2014; discussing proposals). While FDA's regulation of drug labeling is likely to be central to any rulemaking, the existence of multiple potential issues and differences in the participants' values may permit trade-offs among the parties to maximize their interests (Harter 1982:50; Susskind and McMahon 1985:152).[6] The parties' interests may include the shared value of consumer access to safe and effective drugs,[7] and this value may serve as the foundation for regulatory negotiation (Harter 1982:49).

C. Interests Likely to Be Impacted and Representation

There appears to be a limited number of identifiable interests that would be significantly affected by a rule to address the issues implicated by the *Mensing* decision. Pursuant to the NRA, an "interest" is "multiple parties which have a similar point of view or which are likely to be affected in a similar manner" with respect to an issue (5 USC § 562(5) (2012)). So, for example, all generic drug companies that must comply with new regulations may be affected by a change in generic drug labeling regulation. Many of these companies may be affected in a similar manner and represent one interest, which should be represented on the negotiated rulemaking committee.

Brand-name drug manufacturers, consumers, and health care providers may also be significantly impacted by a new rule and should be represented. For example, regulatory change may impact brand-name manufacturers by requiring them to update their labeling following generic drug labeling updates or to provide information to facilitate labeling updates. It may impact consumers by affecting the safety and efficacy of generic drugs and the potential remedies available to consumers injured by such drugs. It may also impact health care providers who are licensed to administer prescription drugs and who utilize drug labeling to make prescription decisions (21 USC §§ 352(f)(1), 353(b)(1)–(2) (2012); 21 CFR § 201.5 (2014); Brief of the AMA et al. 2011:29, 30). Other potential interests that may be impacted by a new rule include states, biologic manufacturers, and pharmacists. The negotiated rulemaking committee must include at least one FDA representative (5 USC § 565(b) (2012)). This discussion suggests that the "number of identifiable interests that will be significantly affected by the rule" (and thus, the number of committee members needed to represent such interests) is limited and appears likely to be less than the twenty-five-member limit generally provided by the NRA (5 USC §§ 563(a)(2), 565(b) (2012)).

It seems reasonably likely that FDA could convene a negotiated rulemaking committee with a balanced representation of persons who (1) can "adequately represent" the interests identified as "significantly affected by the rule"; and (2) "are willing to negotiate in good faith to reach a consensus on the proposed rule" (5 USC §§ 563, 565 (2012)). For example, trade associations may be able to represent the interests of brand-name and generic drug manufacturers, and consumer and professional organizations may be able to represent the consumer and health care providers' interests. FDA may be able to draw on its experience in convening advisory committees to facilitate this process (FDA Draft Guidance 2008:3; 21 CFR §§ 14.1–14.174 (2014); FDA Advisory Committees 2011; Sherman 2004:99–102).

D. Potential Gains

There are several reasons why the significantly affected interests may be "willing to negotiate in good faith to reach a consensus on the proposed rule" (5 USC § 563(a)(3)(B) (2012)) and believe that negotiated rulemaking would be for their benefit.

First, FDA has proposed a rule that would revise the procedures for changes to the labeling of an approved drug, suggesting that a new rule is inevitable, if not imminent (Supplemental Applications Proposing Labeling Changes for Approved Drugs and Biological Products, 78 Fed. Reg. 67,985 (Nov. 13, 2013)). This may create a sense of urgency on the part of the participants in the proposed negotiated rulemaking and speed up negotiations (Harter 1982:47–48). These participants may view the proposed negotiated rulemaking as a beneficial opportunity for meaningful participation in and some control over the creation of a new regulatory system for drugs, which may further encourage negotiation (Freeman and Langbein 2000:62–69). Participants would be deprived of this opportunity if negotiations failed and FDA continued with informal rulemaking (Freeman and Langbein 2000:67, 81, 84, 124). Second, the fact that the drug industry is a "highly regulated industry, in which all the players—including the agency, the drug companies, and even the representatives of consumers—are repeat players" (Rakoff 2000:169–70) may encourage the participants to negotiate in good faith, as they are likely to have future interactions. Third, although the current regulatory system's impact on the various interests is highly complex (and empirical evidence would be needed to make any definitive statements about its impact), certain aspects of the current system may harm each of the interests, which may further encourage negotiation. For example, the current system may harm the market for generic drugs, encourage innovator liability suits against brand-name manufacturers (Rostron 2011:1123, 1135), present an "ethical dilemma" for doctors (Brief of the AMA et al. 2011:29–30), and have negative impacts on consumers.

E. Countervailing Power

A balance of power is one of the criteria that Harter identified as predictive of successful negotiations (Harter 1982:45). Power appears to be divided among the interests in the proposed negotiation. For example, the importance to the public health of the drugs that brand-name and generic drug manufacturers produce and the extent and characteristics of those markets (Snyder 2012:4–5; Snyder 2012a:4–5); FDA's broad "authority to promulgate regulations for the efficient enforcement" of the FDCA and its ability to proceed with informal rulemaking if no consensus is reached (21 USC § 371(a) (2012)); health care providers' roles as prescribers and learned intermediaries (21 USC § 353 (2012); Noah 2009:890); and consumers' ability to request drugs and make purchasing choices (Campbell et al. 2013:237, 238) suggest that each of these interests has significant power. The use of negotiated rulemaking may be appropriate even if the parties' power were unequal because the process may empower and constrain each of the parties as no party may "want to appear responsible for" a failure to reach consensus (Susskind and McMahon 1985:154–55).[8]

F. Potential Benefits

Using negotiated rulemaking to create new drug regulations may be in the public interest (5 USC § 563 (2012)). First, while FDA has not used negotiated rulemaking and the discussions of this process have been based on the experiences of other agencies (Coglianese 1997:1273), empirical data and commentators suggest that negotiated rulemaking may be "faster than traditional rulemaking" (Harter 2000:49) and may save time (Freeman and Langbein 2000:75). But even if negotiated rulemaking is not faster, it may hold other benefits (Harter 1982:28–31).

By engaging persons who can adequately represent the interests that will be significantly affected by new drug labeling rules, negotiated rulemaking may produce "better rules" (ibid., 115) and result in a process that functions better than the existing process (Harter 1997:1402–4; Freeman and Langbein 2000:66–67; Langbein and Kerwin 2000:605–8). For example, the current regulatory procedure for labeling updates (under which a manufacturer must update its generic drug labeling to match that of the corresponding brand-name drug following an update to the brand-name labeling) may not be functioning optimally; this may result in differences between the labeling of the brand-name and generic versions of a drug (Duke et al. 2013:299–300). While the impact of these differences on patient safety is not known, there may need to be "changes in the labeling cascade . . . to ensure ongoing synchronization of drug safety warnings" (ibid., 300). The existing labeling regime was created by FDA regulations promulgated through notice-and-comment rulemaking (57 Fed. Reg. 17,950 (Apr. 28, 1992); 50 Fed. Reg. 7,452, 7,466–70, 7,498–99 (Feb. 22, 1985)) and supplemented by the agency's interpretations in preambles, briefs, and guidance. Using negotiated rulemaking to create new drug labeling and postmarket safety rules may increase the legitimacy of FDA's final regulations (Freeman and Langbein 2000:124–27; Harter 1997:1403–4, 1407; Harter 1983:480, 489; Langbein and Kerwin 2000:625) and may lead to better relationships among the participants, who are likely to be repeat players in the world of drug regulation (Freeman 2000:656–57; Rakoff 2000:169–70). The perceived legitimacy of the final rule and the interactions among participants in the rulemaking may be significant because, while promulgation of a new final rule is an important first step in reform, once a new rule on drug labeling and postmarket safety goes into effect, the success of any new regulatory regime will depend on the participation of FDA and the stakeholders.

G. Response to Anticipated Criticisms

Despite the potential benefits of negotiated rulemaking, there may be critiques of the proposal that FDA use negotiated rulemaking to address the issues flowing from the *Mensing* decision. First, critics may argue that FDA does not need to use negotiated rulemaking because it already provides for public participation through its use of advisory committees and public meetings (21 CFR §§ 10.65, 14.1–174 (2014); DHHS and FDA 2013; Kobick 2010:439–40). Specifically, critics may argue that negotiated rulemaking is unnecessary because FDA has scheduled a public meeting on its proposed rule and alternatives (Supplemental Applications Proposing Labeling Changes for Approved Drugs and Biological Products; Public Meeting; Request for Comments; Reopening of the Comment Period, 80 Fed. Reg. 8577 (Feb. 18, 2015)).

These critiques, however, do not account for the important ways in which negotiated rulemaking would differ from FDA's general use of advisory committees and public meetings. Unlike those other processes, negotiated rulemaking focuses on negotiation to generate consensus among stakeholders for use as the basis for a proposed rule. Indeed, many of the potential benefits of negotiated rulemaking discussed in Part V.F may flow from the negotiation and consensus building that characterize the process—benefits that other advisory committee processes or public meetings may not produce. For example, a negotiated rulemaking committee's purpose would be to produce a consensus among stakeholders concerning a proposed rule and not simply to "provide advice and recommendations to the [FDA] Commissioner" (21 CFR § 14.5(a) (2014)) or "to listen to comments" (Supplemental Applications Proposing Labeling Changes for Approved Drugs and Biological Products; Public Meeting; Request for Comments; Reopening of the Comment Period, 80 Fed. Reg. 8577 (Feb. 18, 2015)). Furthermore, the agency's commitment to use the consensus of the committee as the basis for a proposed rule, "to the maximum extent possible consistent with the legal obligations of the agency" (5 USC § 563(a) (2012)), is an "essential ingredient of the success" of the process (Harter 1982:100).

Second, critics may argue that negotiated rulemaking could cost both FDA and participants more than conventional informal rulemaking (Kobick 2010:438–39; Lubbers 2008:997–98). Negotiated rulemaking, however, may save costs because the negotiated rulemaking committee members may bring to the table important information that FDA would otherwise have to speculate about or invest resources in locating or developing (Freeman 2000:641). Negotiated rulemaking may save FDA and stakeholders costs at the end (i.e., through fewer comments and court challenges) (Lubbers 2008:997; Harter 2000:56). It may also save costs in the implementation of, compliance with, and enforcement of a new rule by creating a more effective rule (Harter 1997:1403–4; Merrill 1999:1180 n.137).[9]

A third anticipated criticism is that negotiated rulemaking may create rules that are no less subject to litigation than conventional rules (Coglianese 1997:1286–1309; Harter 2000:55, quoting Langbein and Kerwin 2000:625–26; Kobick 2010:441–42). But even if rules produced using negotiated rulemaking have a similar rate of judicial review as those produced by conventional rulemaking, using negotiated rulemaking to create new regulations may still be valuable in light of the potential benefits that negotiated rulemaking may offer compared to informal rulemaking as discussed in Part IV.F (Harter 2000:52–56).

VI. CONCLUSION

While negotiated rulemaking is not appropriate for all rulemaking, it may be appropriate and offer benefits in the current environment. To date, FDA has not used the NRA's negotiated rulemaking process, but, to quote Harter, "[a]t the very least, regulatory negotiation is worth a try" (Harter 1982:113).

NOTES

Adapted from Marie Boyd, 2014, "Unequal Protection Under the Law: Why FDA Should Use Negotiated Rulemaking to Reform the Regulation of Generic Drugs," *Cardozo Law Review* 35: 1525–85. Thanks to the organizers and participants of the 2013 Annual Conference of the Petrie-Flom Center for Health Law Policy, Biotechnology, and Bioethics at Harvard Law School for their feedback and questions, to Chan Mo Ahn and Jessica Kelly for their research assistance, and to Vanessa Byars and Carol Young for their support. All views expressed herein are my own.

1. The manufacturer of a brand-name drug must ensure that the labeling is appropriately updated as long as the drug is marketed. When the brand-name drug labeling is updated, manufacturers are required to update the labeling of their corresponding generic drugs accordingly. But if the brand-name drug leaves the market, leaving only the generic versions, there is no manufacturer responsible for updating the labeling in light of newly acquired information because generic manufacturers cannot independently change their generic drug labeling under the current regulatory framework.

2. The FDCA permits some differences between the labeling of the RLD and generic drugs, 21 USC § 355(j)(2)(A)(v), (j)(4)(G) (2012), and FDA's regulations provide a nonexclusive list of permissible differences, 21 CFR § 314.94(a)(8)(iv) (2014).

3. On February 18, 2015, FDA extended the comment period for the proposed rule (Supplemental Applications Proposing Labeling Changes for Approved Drugs and Biological Products; Public Meeting; Request for Comments; Reopening of the Comment Period, 80 Fed. Reg. 8577 (Feb. 18, 2015)).

4. A search of the *Federal Register*, the FDCA, and the *United States Statutes at Large* revealed no notices of FDA's intent to establish a negotiated rulemaking committee and no instances in which Congress had required FDA to conduct negotiated rulemaking. See also Kobick 2010:425.

5. Harter identifies "Mature Issues" as one of the criteria for determining when negotiation is likely to be fruitful (Harter 1982:47).

6. Such negotiation may be consistent with the public interest; as Harter has noted, "there is some indication that rules that emerge from [negotiated rulemakings] are more stringent than those the agency would have been able to issue on its own" (Harter 1997:1403–4).

7. While this chapter does not seek to identify particular representatives for the proposed negotiated rulemaking, the mission statements of FDA, brand-name and generic pharmaceutical industry trade associations, organizations of health care professionals, and consumer advocacy groups suggest that this may be a shared value. See, e.g., "About FDA: What We Do," Food and Drug Administration, last updated August 5, 2014, http://www.fda.gov/AboutFDA/WhatWeDo/default.htm; "About PhRMA," Pharmaceutical Research and Manufacturers of America (PhRMA), accessed March 28, 2014, http://www.phrma.org/about; "About: The Association," Generic Pharmaceutical Association (GPhA), accessed March 28, 2014, http://www.gphaonline.org/about/the-gpha-association; "AMA Mission & Guiding Principles," American Medical Association, accessed March 28, 2014, http://www.ama-assn.org/ama/pub/about-ama/our-mission.page?; "Health and Safety," Public Citizen, accessed March 28, 2014, http://www.citizen.org/Page.aspx?pid=524.

8. As noted earlier, if no consensus is reached, FDA may proceed with informal rulemaking.

9. In addition, FDA likely does not have the resources to effectively monitor and update generic drug labeling (*Wyeth v. Levine* 2009:578, 578 n.11). Accordingly—although not unique to negotiated rulemaking—investing in the creation of a better regulatory system, in which drug manufacturers are responsible for labeling updates and state failure-to-warn claims are not preempted, may be especially important in promoting drug safety.

REFERENCES

American Medical Association. "AMA Mission & Guiding Principles." http://www.ama-assn.org/ama/pub/about-ama/our-mission.page? (accessed March 28, 2014).

Boyd, Marie. 2014. "Unequal Protection Under the Law: Why FDA Should Use Negotiated Rulemaking to Reform the Regulation of Generic Drugs." *Cardozo Law Review* 35:1525–85.

Brief for Marc T. Law et al. as Amici Curiae in Support of Respondents, *PLIVA, Inc. v. Mensing*, 131 S. Ct. 2567 (2011) (Nos. 09-993, 09-1039, 09-1501).

Brief for Petitioner, *Wyeth v. Levine*, 555 U.S. 555 (2009) (No. 06-1249).

Brief for the United States as Amicus Curiae, *PLIVA, Inc. v. Mensing*, 131 S. Ct. 2567 (2011) (Nos. 09-993, 09-1039).

Brief for the United States as Amicus Curiae Supporting Respondents, *PLIVA, Inc. v. Mensing*, 131 S. Ct. 2567 (2011) (Nos. 09-993, 09-1039, 09-1501).

Brief of the American Medical Association et al. as Amici Curiae in Support of Respondents, *PLIVA v. Mensing*, 131 S. Ct. 2567 (2011) (Nos. 09-993, 09-1039, 09-1501).

Campbell, Eric G. et al. 2013. "Physician Acquiescence to Patient Demands for Brand-Name Drugs: Results of a National Survey of Physicians." *Journal of the American Medical Association* 173:237–39.

Coglianese, Cary. 1997. "Assessing Consensus: The Promise and Performance of Negotiated Rulemaking." *Duke Law Journal* 46:1255–1350.

Department of Health and Human Services (DHHS) and U.S. Food and Drug Administration (FDA). Fiscal Year 2014 Justification for Estimates for Appropriations Committees. http://www.fda.gov/downloads/AboutFDA/ReportsManualsForms/Reports/BudgetReports/UCM347422.pdf (accessed March 28, 2014).

"Device Software Policy Revisions via Negotiated Rulemaking Under Consideration by FDA." *Gray Sheet* (December 23, 1996).

Dicken, John E., Director of Health Care, U.S. Gov't Accountability Office (GAO) to Senator Orrin G. Hatch, January 31, 2012, http://www.gao.gov/assets/590/588064.pdf.

Duke, Jon et al. 2013. "Consistency in the Safety Labeling of Bioequivalent Medications." *Pharmacoepidemiology and Drug Safety* 22:294–301.

Duncan, Sarah C. 2010. Note. "Allocating Liability for Deficient Warnings on Generic Drugs: A Prescription for Change." *Vanderbilt Journal of Entertainment and Technology Law* 13:185–216.

"FDA Waiver of User Fees." *Pink Sheet* (November 7, 1994).

Food and Drug Administration. "About FDA: What We Do." Last updated August 5, 2014. http://www.fda.gov/AboutFDA/WhatWeDo/default.htm.

———. 2011. "Advisory Committees, Committees & Meeting Materials." Last updated December 16, 2011. http://www.fda.gov/AdvisoryCommittees/CommitteesMeetingMaterials/default.htm.

———. 2008. *Draft Guidance: Guidance for the Public and FDA Staff on Convening Advisory Committee Meetings.*

———. 2004. *Guidance for Industry: Changes to an Approved NDA or ANDA.*

———. "Orange Book: Approved Drug Products with Therapeutic Equivalence Evaluations." http://www.accessdata.fda.gov/scripts/cder/ob/docs/queryai.cfm (accessed March 24, 2014).

———. Response to Public Citizen, Citizen Petition. Docket No. FDA-2011-P-0675 (November 8, 2013).

Freeman, Jody. 1997. "Collaborative Governance in the Administrative State." *University of California Los Angeles Law Review* 45:1–98.

———. 2000. "The Private Role in Public Governance." *New York University Law Review* 75:543–675.

Freeman, Jody and Laura I. Langbein. 2000. "Regulatory Negotiation and the Legitimacy Benefit." *New York University Environmental Law Journal* 9:60–151.

Funk, William. 1997. "Bargaining Toward the New Millennium: Regulatory Negotiation and the Subversion of the Public Interest." *Duke Law Journal* 46:1351–88.

———. 1987. "When Smoke Gets in Your Eyes: Regulatory Negotiation and the Public Interest—EPA's Woodstove Standards." *Environmental Law* 18:55–98.

Generic Pharmaceutical Association (GPhA). "About: The Association." http://www.gphaonline.org/about/the-gpha-association (accessed March 28, 2014).

Harter, Philip J. 2000. "Assessing the Assessors: The Actual Performance of Negotiated Rulemaking." *New York University Environmental Law Journal* 9:32–59.

———. 1997. "Fear of Commitment: An Affliction of Adolescents." *Duke Law Journal* 46:1389–1428.

———. 1982. "Negotiating Regulations: A Cure for Malaise." *Georgetown Law Journal* 71:1–118.

———. 1983. "The Political Legitimacy and Judicial Review of Consensual Rules." *American University Law Review* 32:471–96.

Howard, Sally, Deputy Commissioner FDA to Joseph R. Pitts, House of Representatives, February 26, 2014.

IMS Institute for Healthcare Informatics. *The Use of Medicines in the United States: Review of 2011* (Parsippany, NJ: IMS Institute for Healthcare Informatics, April 2012), http://www.imshealth.com/ims/Global/Content/Insights/IMS%20Institute%20for%20Healthcare%20Informatics/IHII_Medicines_in_U.S_Report_2011.pdf.

Kazhdan, Daniel. 2012. "*Wyeth* and *PLIVA*: The Law of Inadequate Drug Labeling." *Berkeley Technology Law Journal* 27:893–926.

Kessler, David A. and David C. Vladeck. 2008. "A Critical Examination of the FDA's Efforts to Preempt Failure-to-Warn Claims." *Georgetown Law Journal* 96:461–96.

Kobick, Julia. 2010. "Negotiated Rulemaking: The Next Step in Regulatory Innovation at the Food and Drug Administration?" *Food and Drug Law Journal* 65:425–46.

Langbein, Laura I. and Cornelius M. Kerwin. 2000. "Regulatory Negotiation Versus Conventional Rule Making: Claims, Counterclaims, and Empirical Evidence." *Journal of Public Administration Research and Theory* 10:599–632.

Lasser, Karen E. et al. 2002. "Timing of New Black Box Warnings and Withdrawals for Prescription Medications." *Journal of the American Medical Association* 287:2215–20.

Lee, Stacey B. 2012. "*PLIVA v. Mensing*: Generic Consumers' Unfortunate Hand." *Yale Journal of Health Policy, Law, and Ethics* 12:209–63.

Lubbers, Jeffrey S. 2008. "Achieving Policymaking Consensus: The (Unfortunate) Waning of Negotiated Rulemaking." *South Texas Law Review* 49:987–1018.

Merrill, Thomas W. 1999. "The Constitution and the Cathedral: Prohibiting, Purchasing, and Possibly Condemning Tobacco Advertising." *Northwestern University Law Review* 93:1143–1204.

Mutual Pharmaceutical Co., Inc. v. Bartlett, 133 S. Ct. 2466 (2013).

Noah, Lars. 2009. "This Is Your Products Liability Restatement on Drugs." *Brooklyn Law Review* 74:839–926.

"OTC Label Reform, Supplement GMPs Seen as Candidates for Negotiated Rulemaking—HHS." *Tan Sheet* (September 11, 1995).

Pharmaceutical Research and Manufacturers of America (PhRMA). "About PhRMA." http://www.phrma.org/about (accessed March 28, 2014).

PLIVA, Inc. v. Mensing, 131 S. Ct. 2567 (2011).

Pritzker, David M. and Deborah S. Dalton. 1995. *Negotiated Rulemaking Sourcebook*. Washington, D.C.: U.S. Government Printing Office.

Public Citizen. Citizen Petition. Docket No. FDA-2011-P-0675 (August 29, 2011).

———. "Health and Safety." http://www.citizen.org/Page.aspx?pid=524 (accessed March 28, 2014).

Rabin, Robert L. 2007. "Poking Holes in the Fabric of Tort: A Comment." *DePaul Law Review* 56:293–306.

Rakoff, Todd D. 2000. "The Choice Between Formal and Informal Modes of Administrative Regulation." *Administrative Law Review* 52:157–74.

Reply Brief for Petitioner, *Wyeth v. Levine*, 555 U.S. 555 (2009) (No. 06-1249).

Rose-Ackerman, Susan. 1994. "Consensus Versus Incentives: A Skeptical Look at Regulatory Negotiation." *Duke Law Journal* 43:1206–20.

Rostron, Allen. 2011. "Prescription for Fairness: A New Approach to Tort Liability of Brand-Name and Generic Drug Manufacturers." *Duke Law Journal* 60:1123–92.

Sherman, Linda Ann. 2004. "Looking Through a Window of the Food and Drug Administration: FDA's Advisory Committee System." *Preclinica* 2:99–102.

Snyder, Sophia. 2012. *IBISWorld Industry Report 32541a, Brand Name Pharmaceutical Manufacturing in the US*.

———. 2012a. *IBISWorld Industry Report 32541b: Generic Pharmaceutical Manufacturing in the US*.

Starfield, Barbara. 2000. "Is US Health Really the Best in the World?" *Journal of the American Medical Association* 284:483–85.

Stoddart, Allison. 2012. Note. "Missing After *Mensing*: A Remedy for Consumers of Generic Drugs." *Boston College Law Review* 53:1967–2003.

Susskind, Lawrence and Gerard McMahon. 1985. "The Theory and Practice of Negotiated Rulemaking." *Yale Journal on Regulation* 3:133–66.

Weeks, Wesley E. 2012. Comment. "Picking up the Tab for Your Competitors: Innovator Liability After *PLIVA, Inc. v. Mensing*." *George Mason Law Review* 19:1257–92.

Wyeth v. Levine, 555 U.S. 555 (2009).

CHAPTER TWENTY-ONE

The "Follow-On" Challenge

Statutory Exclusivities and Patent Dances

ARTI RAI

I. INTRODUCTION

The Biologics Price Competition and Innovation Act (BPCIA) (Pub. L. No. 111-148, §§ 7001–03, 124 Stat. 119, 804–21 (2010)), enacted by Congress in March 2010 as part of the Patient Protection and Affordable Care Act, sets up a pathway for so-called "follow-on" biologics. This chapter discusses two key features of the BPCIA's exclusivity regime for branded biologics, in part through a contrast with Hatch-Waxman's exclusivity regime for branded small molecules.

One widely discussed disparity is that between the BPCIA's twelve-year statutory exclusivity for originator firms and Hatch-Waxman's five-year term. Less discussed but also notable is the significant contrast between the regimes' different mechanisms for addressing questions regarding patent validity and infringement. The BPCIA enunciates a highly complex set of procedures through which originator and follow-on firms exchange information regarding patents and commercial marketing. These "patent dance" procedures are quite different from the Hatch-Waxman regime's grant to originator patentees of the equivalent of an automatic preliminary injunction.

Both aspects of the BPCIA are justifiably controversial. Even though the correct term of statutory exclusivity may be difficult to ascertain, the large divergence between the BPCIA and Hatch-Waxman terms seems arbitrary. As for patents, the BPCIA's patent dance may represent an improvement over Hatch-Waxman. Even so, if courts read the relevant case law on standing to allow follow-on manufacturers who have filed marketing applications with FDA to bring declaratory judgment actions challenging the validity and infringement of originator patents, eliminating the dance might be advisable.

Part 1 briefly summarizes the BPCIA pathway and lays out its statutory and patent exclusivities. Part 2 evaluates the BPCIA exclusivities.

II. AN OVERVIEW OF BPCIA EXCLUSIVITIES

A. The Basics of the BPCIA Pathway

While the Hatch-Waxman Act of 1984 established a pathway for generic small molecule drugs, a pathway for "follow-on" biologics did not emerge until 2010. In part, the delay resulted from difficulties associated with applying to most biologics the relatively straightforward analytic processes used for proving bioequivalence between branded and generic small molecules. In contrast to small molecules, biologics are typically large, complex molecules produced by living cells. Slight variations in the manufacturing process can change the quality, safety, or efficacy of the final product. Even for an individual company producing a product, inadvertent changes in processes can lead to product changes from batch to batch. Moreover, these changes may not always be detectable through current techniques for analyzing end products. Thus, at least for some biologics, the "product is the process" (Jeske et al. 2013).

Notably, however, unlike the end products themselves, manufacturing processes are often closely guarded trade secrets. Absent access to this originator trade secret information (an issue the BPCIA does not address), showing similarity to the originator biologic may be a complex, relatively expensive undertaking for which the U.S. Food and Drug Administration (FDA) will require preclinical and clinical studies.

These real-world divergences are reflected in legislative differences. Under Hatch-Waxman's single pathway, the generic manufacturer needs to demonstrate the same chemical structure and bioavailability as the originator drug. In contrast, the BPCIA sets up two different pathways. A biologic can be approved as "biosimilar" if it is "highly similar to the reference product, notwithstanding minor differences in clinically inactive components" and "there are no clinically meaningful differences between the biological product and the reference product in terms of the safety, purity, and potency of the product" (42 USC § 262(i)(2)). A biologic is deemed "interchangeable" if it meets the standards of biosimilarity and also shows that it "can be expected to produce the same clinical result as the reference product in any given patient" and "the risk in terms of safety or diminished efficacy of alternating or switching between use of the biological product and the reference product is not greater than the risk of using the reference product without such alternation or switch" (42 USC § 262(k)(4)(B)). While a biosimilar product can be marketed, only an interchangeable product can be subject to the automatic generic substitution that is routine with small molecules—that is, supplied by a pharmacy without approval from the prescribing physician (42 USC § 262(i)(3)).

B. BPCIA Exclusivities, and the Comparison to Hatch-Waxman

As with other patented inventions, the patent term for biologics runs for twenty years from the time of filing. Biologics also benefit from a provision similar to that found in the Hatch-Waxman statute, which provides for patent term extension based on any patent term lost while the product is going through during FDA approval. As many as five years of patent life can be restored, up to a maximum of fourteen years of patent life from the product's FDA approval date.

Additionally, the BPCIA provides a twelve-year statutory exclusivity that begins after FDA approval. In contrast, the Hatch-Waxman statute provides five years of exclusivity. Controversy persists over whether the BPCIA's twelve-year exclusivity is a market exclusivity or a data exclusivity (Gitter 2013). If it is a data exclusivity (as generally argued by originator firms), then a follow-on competitor cannot rely during the relevant period on the originator firm's preclinical and clinical data. If it is a market exclusivity, the follow-on firm may rely upon the data during the twelve-year period but may not market it until the twelve years have expired. In either event, the twelve years provided by the BPCIA substantially exceeds the five years of data exclusivity available to small molecules.

The BPCIA's intricate provisions for exchanging patent information also raise possible barriers to competition. In ongoing litigation discussed further below, one district court has held that before FDA approval, follow-on manufacturers must follow this system in order to secure a judicial determination regarding the validity and scope of originator patents. This court has also held that under the terms of the BPCIA commercial marketing cannot begin until 180 days after FDA approval. If these conclusions are sustained on appeal to the Court of Appeals for the Federal Circuit (which hears all appeals in patent cases), the result will be additional protection against competition even after data exclusivities have expired and FDA approval has been attained.

The patent information exchange system established by the BPCIA, sometimes called the "patent dance," runs as follows. Within twenty days after the FDA publishes a notice that a follow-on application has been accepted for review, the follow-on applicant must disclose the contents of this application to the originator. Within sixty days, the brand-name patent holder must then identify patents it believes it could assert against the follow-on manufacturer. Next, the follow-on applicant has sixty days to state, with respect to each patent, whether it will wait to market its product until after patent expiry or whether it believes the patent is invalid or would not be infringed. If the follow-on applicant makes arguments regarding validity or infringement, the brand name must respond within sixty days (42 USC § 262(l)(b)(2)-(3)).

Following this exchange of views on patents, the parties are required to engage in good-faith negotiations for up to fifteen days on which patents will be the subject of a patent suit. If the parties reach agreement, the brand name must bring a patent infringement action in thirty days. If there is no agreement, the follow-on applicant must notify the brand name of the number (but not the identity) of the patents it wants to litigate. The parties then have five days to exchange a list of patents to be litigated. The number of patents on each list identified by the brand name may not exceed the number identified by the follow-on, except that if the follow-on application identifies no patents, then the brand name may identify one. The brand-name firm must then commence patent infringement litigation within thirty days (42 USC § 262(b)(l)(4)-(6)).

The BPCIA also requires the follow-on applicant to give the brand name "notice of commercial marketing" at least 180 days before such marketing commences. After receiving this notice, the brand name may seek a preliminary injunction based on any patent that had previously been identified but that was not subject to initial litigation (42 USC § 262(b)(8)(A)(B)). The BPCIA precludes declaratory judgment actions on patents identified as relevant by either party in the "patent dance" after the follow-on application has been filed but before the notice of commercial marketing (42 USC § 262(b)(9)(A)).

For purposes of precluding declaratory judgments, and for determining when a drug can be marketed, the issue of how early "notice of commercial marketing" can occur is critical. Indeed, it is already the subject of litigation. In the 2013 case *Sandoz, Inc. v. Amgen, Inc. et al.*, the district court held that because the BPCIA's "notice of commercial marketing" provision specifically refers to a follow-on biological product "licensed" by FDA, FDA must have already approved the application before "notice of commercial marketing" can occur. Through this decision, which has been appealed to the Federal Circuit, the district court established an additional 180-day period *after* FDA approval during which the follow-on cannot market. The district court also interpreted the BPCIA as precluding any declaratory judgment actions before FDA approval, thereby establishing the patent dance, and accompanying infringement action, as the only acceptable procedure for going to court before FDA approval.

III. EVALUATING THE EXCLUSIVITIES

The twelve-year statutory exclusivity period available for biologics is obviously much longer than the five-year statutory period available for small molecules. Supporters of this longer exclusivity have argued that it is justified by the weaker protection that patents purportedly offer. For example, in the run-up to passage of the BPCIA, the Biotechnology Industry Organization (BIO) focused on the allegedly narrower scope afforded to patents on biological macromolecules relative to patents on other products (BIO 2008).

Historically, it did appear that the so-called written description requirement, which regulates patent scope, might apply with greater vigor to the biopharmaceutical industry than to other industries. However, a 2010 en banc decision by the Federal Circuit makes it clear that the written description requirement not only applies to all technologies but also that it is perhaps less stringent than originally estimated (*Ariad v. Eli Lilly* 2010). Moreover, since at least 2004, it has been clear that the written description requirement applies to both small and large molecules, and thus small molecule patents also have narrow scope (*University of Rochester v. Pfizer* 2004). Indeed, scope can be so narrow that noninfringing "me-too" products that work by the same biological mechanism as the first-in-class patent are often able to enter a few years after first-in-class entry.

Other arguments about biologics patents also apply equally to small molecule patents. For example, the nonobviousness requirement may not comport well with the reality of how biologics are commercialized. Specifically, the fact that it may be obvious to show that a biological molecule has efficacy in preclinical models may say nothing about whether a patent is nonetheless to spur the further innovation necessary to show safety and efficacy in humans, as required by the FDA (Benjamin and Rai 2007). However, this is hardly a problem unique to macromolecules; it applies as well to small molecules (Roin 2009).

Similarly, both small molecules and biologics are affected by the Supreme Court's recent jurisprudence on what constitutes subject matter eligible for patenting (Rai 2013). In several cases, the Supreme Court has indicated that molecules claimed by patents must represent more than a mere "product of nature." Although the Court's guidance on the level of additional transformation required has been less than clear, the rulings clearly implicate all molecules derived from nature.

Another potential problem with patents is that they guarantee exclusivity from the date the patent is granted, not from the date the biologic is approved. Because firms usually seek their "primary" patent on a biologic when they begin clinical trials, products that spend many years in the FDA approval process may enjoy less patent life than products with shorter clinical trial times. Moreover, although the BPCIA (like Hatch-Waxman before it) provides for patent term extension based upon patent period lost during regulatory review, this extension is capped at five years. Again, however, this argument applies to small molecules as well as biologics.

In support of its argument for longer statutory exclusivity, the biotechnology industry also cited a number of studies that relied on the proposition that biologics development takes longer than small molecule development and is also more costly. For example, using numbers such as a $1.24 billion cost for bringing a biologic to market, a series of studies from Duke University calculated that a brand name biologic needed at least thirteen years of exclusivity to recoup R&D costs (Grabowski 2009, Grabowski et al. 2011). However, the assumptions behind these studies have been contested (FTC 2009).

Moreover, unlike generic manufacturers of small molecules, follow-on biologics manufacturers will, in most cases, have to undertake preclinical and clinical trials to show their product is biosimilar. Trial requirements to show interchangeability are likely to be even more onerous. Some analysts have estimated trial costs, particularly for complex biologics, may escalate to as much $100 or $150 million. Additionally, as the Grabowski and Lietzan contribution to this volume notes, the cost of building or leasing appropriate manufacturing facilities could be even higher. In comparison, the cost of completing bioequivalence studies for a small molecule may be as low as $1–2 million (Grabowski et al. 2011).

Because of this significant barrier to entry, true generic competition in biologics will, at least for the foreseeable future, be difficult to achieve. Instead, as a prominent Federal Trade Commission (FTC) report that analyzed the issue concluded, competition from biosimilars and even interchangeables may resemble competition between branded biologics (FTC 2009). With two or three biosimilar entrants, prices may decrease by 10–30 percent.

Notably, although comparisons across jurisdictions have familiar limitations, data from the European Union do bolster the FTC's case. The European Medicines Agency started allowing biosimilars in 2006. Since that time, European countries have seen only modest price reductions for biosimilars when compared to the reference product (25–30 percent reduction in price) (Engelberg 2009).

If anything, originator biologics firms are likely to be better situated, and hence need *less* government-granted exclusivity, than originator small molecule firms. The seven-year difference in exclusivity is likely to further bias firms in favor of biologics development, even where the health benefit from the biologic is not as large as that from the small molecule.

Notably, in the European Union, branded biologics and small molecules are both granted ten years of exclusivity (eight years of data exclusivity and an additional two years of market exclusivity). While perhaps difficult to achieve as a political matter, a rebalancing of U.S. exclusivities along the lines of the European model would be normatively desirable.

As noted, the BPCIA departs substantially from Hatch-Waxman not only in its approach to statutory exclusivity but also in the manner in which it addresses resolution of patent disputes. In the case of Hatch-Waxman, the patent holder is allowed an automatic thirty-month stay of FDA marketing approval merely by asserting in litigation a patent it had previously listed on the FDA's Orange Book as purportedly claiming its drug (or a method of using the drug). In other words, the originator patent holder secures the equivalent of a preliminary injunction without showing *any* likelihood of success on the merits with respect to either validity or infringement. Given the well-known limitations faced by the U.S. Patent and Trademark Office (USPTO) in its initial examination of patent applications (Rai 2013), patent validity is by no means clear. Additionally, because FDA disavows evaluating in any manner whether the patents put on the Orange Book are properly listed under the relevant statutory provision (21 USC § 355(c)(2)), and because certain patents can be "invented around," infringement may also be in doubt. In general, when generic firms do challenge patents other than the "primary" molecule patent, they fare well on questions of invalidity and/or non-infringement (Hemphill and Sampat 2013).

Generic challenges are by no means guaranteed, however. To the contrary, the mechanism Hatch-Waxman sets up for challenging patents placed on the Orange Book creates opportunities for the branded and generic firm to collude. The first generic filer to allege invalidity or noninfringement of an Orange Book patent secures a 180-day exclusivity period. Although this exclusivity is intended to act as an incentive for generics to challenge dubious patents, the generic challenger does not actually have to win the suit to secure exclusivity. Rather, even if the first challenger settles the lawsuit and delays entry, the exclusivity not only fails to transfer to a second challenger but is forfeited by the first challenger only seventy-five days after a second challenger succeeds in showing the patent is invalid or not infringed.

The consequence of this peculiar structure has been so-called "reverse payment" settlements. In these settlements, the originator can often use a single payment to the first, most advantaged, challenger to buy protection against all challengers.

Following a 2013 Supreme Court decision, *FTC v. Actavis*, these reverse payment settlements will now receive scrutiny as a matter of antitrust law. However, because the Court rejected not only per se rules but also a quick look rule of reason, precisely how different settlements will fare is quite unclear. Given errors in the judicial process, a small reverse payment that is above litigation costs but still represents a high level of confidence in the patent could be seen as legitimate.

More importantly, antitrust is being deployed to address a problem that would be addressed more directly by fixing Hatch-Waxman itself. For example, Congress could establish an administrative apparatus, perhaps involving both the FDA and the PTO, to adjudicate the validity and scope of all Orange Book patents (Rai 2012). Alternatively, if the 180-day exclusivity accrued only to generics that succeeded in challenging patents or otherwise made it to market without securing a reverse payment, reverse payment settlements would be less attractive (Hemphill and Lemley 2011). In that circumstance, the originator could be less likely to buy patent peace simply by settling with the first challenger. Unfortunately, neither Congress nor the relevant agencies appears inclined toward such these changes.

As contrasted with Hatch-Waxman, the BPCIA may set up fewer obvious opportunities for strategic behavior that would delay or otherwise limit follow-on competition. Most obviously, the BPCIA contains no provisions for automatic injunctive relief based on questionable Orange Book listings or for a 180-day exclusivity that can be parked by a settling defendant.

Moreover, contrary to the district court's decision in *Sandoz v. Amgen*, nothing in the statute precludes a follow-on manufacturer from filing a declaratory judgment action *before* it files an FDA application. Although the follow-on manufacturer that seeks declaratory judgment before filing with the FDA may be unable to satisfy Article III standing requirements, standing issues can arise in any declaratory judgment suit. The district court's view that the follow-on manufacturer cannot provide notice of commercial marketing until it has secured FDA approval, such that it then has to wait another 180 days before it actually begins marketing, also appears misplaced. Nothing in the BPCIA's structure suggests that Congress wanted an additional half-year of exclusivity even after FDA approval had been secured and all other exclusivities had expired.

That said, the BPCIA is hardly a model of clear drafting. At least in part, the BPCIA's problems may be a consequence of the fact that the legislation passed as part of the Affordable Care Act (ACA). The Obama administration apparently wanted significant changes to the version of the BPCIA that had passed the Senate in December 2009. However, the threat of a Senate filibuster that emerged after the January 19, 2010, election of Republican Senator Scott Brown to fill the seat left open by Democratic Senator Edward Kennedy's death in 2009 sharply limited possibilities to amend the ACA, and the December 2009 version was enacted into law (Carver et al. 2010).

The problem of bad drafting is particularly acute with respect to declaratory judgment actions arising *after* an application for FDA approval has been filed but before approval has been granted. One could persuasively argue that because notice of commercial filing can occur before FDA approval, and because the statutory language does not prohibit declaratory judgment actions after notice of commercial filing, declaratory judgment actions before FDA approval are permissible. However, because the BPCIA does make it clear that originator and follow-on firms must engage in the patent dance after an FDA application has been filed, this interpretation sets up a system in which declaratory judgment filings will occur simultaneously with the patent dance. Such parallel proceedings will exacerbate possibilities for strategic behavior in determining which patents are brought to the dance.

In setting up the patent dance, Congress was attempting to promote patent dispute resolution ex ante so that follow-on manufacturers did not have to launch "at risk" of significant infringement liability. Whether the patent dance is necessary for achieving this goal is unclear, however. The relatively liberal approach toward standing in patent-related declaratory judgment actions that the Supreme Court took in its 2007 *Medimmune v. Genentech* decision represents a signal that the Court understands the need for competitors to achieve patent clarity without incurring infringement liability. Given the Supreme Court's approach, it seems entirely possible that follow-on manufacturers who have filed applications for FDA approval will have standing to challenge originator patents.

The BPCIA could also have done more to ensure that patent dances do not become opportunities for collusion. For example, Congress might have required greater supervision of the patent dance by a regulatory authority. Indeed, it might even have required the FDA and PTO to cooperate in making some initial determinations with respect to the validity and scope of asserted patents.

As a practical matter, given the many uncertainties that continue to surround the BPCIA, follow-on manufacturers might seriously consider the new post-grant administrative proceedings now available at the newly created PTO Patent Trial and Appeals Board under the America Invents Act of 2011. Through these new proceedings, which should not be affected by how courts interpret the BPCIA, follow-on manufacturers can secure an expert, trial-type evaluation of patent validity questions. The most relevant proceeding is likely inter partes review, which allows anyone to challenge a patent on grounds of lack of novelty or obviousness and is available throughout the life of the patent (Rai 2013). If this expert valuation cancels relevant claims of a patent, and this cancellation is affirmed by the Court of Appeals for the Federal Circuit, no further litigation with respect to those claims should be necessary.

Additionally, as case law evolves with respect to the Article III standing of follow-on manufacturers to file declaratory judgment actions, Congress should watch this law closely. As noted, a robust standing doctrine would obviate the need for the patent dance.

REFERENCES

Ariad v. Eli Lilly, 598 F.3d 1336 (Fed. Cir. 2010) (en banc).
Benjamin, S. M. and A. K. Rai. 2007. "Who's Afraid of the APA: What the Patent System Can Learn from Administrative Law." *Georgetown Law Journal* 95:269–336.
Biotechnology Industry Organization (BIO). 2008. "A Follow-On Biologics Regime Without Strong Data Exclusivity Will Stifle the Development of New Medicines."
Carver, K. H., J. Elikan, and E. Lietzan. 2010. "An Unofficial Legislative History of the Biologics Price Competition and Innovation Act of 2009." *Food and Drug Law Journal* 65(4):671–818.
Engelberg, A.B. 2009. "Balancing Innovation, Access, and Profits—Market Exclusivity for Biologics." *New England Journal of Medicine* 361: 1917–19.

FTC, Emerging Health Care Issues: Follow-On Biologic Drugs Competition (2009).

FTC v. Actavis, 570 U.S. __ (2013).

Gitter, D. M. 2013. "Biopharmaceuticals Under the Patient Protection and Affordable Care Act." *Texas Intellectual Property Law Journal* 21:213–244.

Grabowski, H. 2008. "Follow-On Biologics: Data Exclusivity and the Balance Between Innovation and Competition." *Nature Reviews Drug Discovery* 7:479–88.

Grabowski, H., G. Long, and R. Mortimer. 2011. "Data Exclusivity for Biologics." *Nature Reviews Drug Discovery* 10:15–16.

Hemphill, C. S. and M. A. Lemley. 2011. "Earning Exclusivity: Generic Drug Incentives and the Hatch-Waxman Act." *Antitrust Law Journal* 77:947–89.

Hemphill, C. S. and B. N. Sampat. 2013. "Drug Patents at the Supreme Court." *Science* 339:1386–87.

Jeske, W., J. M. Walenga, D. Hoppensteadt, and J. Fareed. 2013. "Update on the Safety and Bioequivalence of Biosimilars." *Drug Healthcare Patient Safety* 5:133–41.

Medimmune v. Genentech, 549 U.S. 118 (2007).

Rai, A. K. 2012. "Use Patents, Carve-Outs, and Incentives—A New Battle in the Drug-Patent Wars." *New England Journal of Medicine* 367:491–93.

———. 2013. Biomedical Patents at the Supreme Court: A Path Forward. 66 Stanford Law Review Online 111.

Roin, B. N. 2009. "Unpatentable Drugs and the Standards of Patentability." *Texas Law Review* 87:1–55.

Sandoz Inc. v. Amgen, Inc. et al., No. C-13-2904-MMC (N.D. Cal. 2013).

University of Rochester v. Pfizer, 358 F.3d 916 (2004).

CHAPTER TWENTY-TWO

FDA Regulation of Biosimilars

HENRY GRABOWSKI AND ERIKA LIETZAN

I. INTRODUCTION

A recent development in Europe and the United States is the establishment of an abbreviated pathway to market for biosimilars—biological products that are similar but not identical in purity, safety, and efficacy to a reference biological product. The U.S. law, the Biologics Price Competition and Innovation Act (BPCIA), was enacted in March 2010 as part of the Affordable Care Act (ACA). The objective of the BPCIA, like the Hatch-Waxman Act of 1984 before it for generic chemically synthesized drugs, was to encourage increased price competition and cost savings for biological products, while maintaining incentives for innovation.

Biological drugs are manufactured in living systems, typically using recombinant DNA technology. The resulting molecules are large, complex, and often difficult to characterize analytically. The drug substance may be heterogeneous, and its mechanism(s) of action can be poorly understood. Even seemingly minor changes to the raw materials or manufacturing process can lead to differences—sometimes analytically undetectable—that have enormous clinical consequences. It is generally understood that it will not be scientifically possible for a biosimilar manufacturer to prove that its product is "the same as" a reference biological product. Determining instead that the product is "similar enough" to justify—as a scientific and regulatory matter—an abbreviated application that relies on approval of the earlier product is a challenging task for regulators and industry. FDA is currently implementing the BPCIA through Draft Guidance, public forums, and meetings with potential applicants. FDA's decisions with respect to how biosimilarity may be shown (i.e., the content and nature of abbreviated applications)—and other related regulatory factors discussed below, such as the interchangeability standard—will have a profound impact on the number of biosimilar entrants for any particular biological product and on the evolution of market competition.

In this chapter, we consider the U.S. Food and Drug Administration's (FDA) evolving guidelines and a number of open issues that could have important implications for development of biosimilar competition.

A. Biosimilarity

To reach the market, a biosimilar firm must demonstrate that its product is "biosimilar" to a reference product. This entails showing that (1) the candidate product is "highly similar to the reference product notwithstanding minor differences in clinical inactive components," and (2) there are "no clinically meaningful differences" between the two, in terms of safety, purity, and potency (42 USC § 262(i)). The statute does not define these terms, leaving FDA discretion to tailor application requirements to the particular product class as well as the particular applicant and candidate product.

B. Innovation Incentives and Exclusivity Provisions

In order to maintain incentives for new biologics, the BPCIA provides biological product innovators with exclusivity periods that preclude biosimilar competition for a fixed period of time. Specifically, biosimilar applications may not be submitted until four years after

FDA licensure of the reference biological product. Also, FDA cannot approve a biosimilar application until twelve years after licensure of the reference product. An additional six months of exclusivity are available for the reference product if the innovator satisfies pediatric study requirements. The six months in question are added to both the four-year and the twelve-year periods.

The statute limits which innovative biologic license applications (BLAs) are eligible for their own four-year and twelve-year terms of exclusivity. Specifically, there are no new terms for (1) a supplemental BLA for the reference product; (2) a subsequent BLA filed by the same sponsor, manufacturer, or other related entity for a new indication, route of administration, dosing schedule, dosage form, delivery system, delivery device, or strength; or (3) a subsequent BLA filed by the same sponsor, manufacturer, or other related entity for a structural change that does not result a change in safety, purity, or potency (42 USC § 262(k)(7)). Stakeholders generally assume that these "second generation" applications, while not receiving four-year and twelve-year terms of their own, will be protected under the "umbrella" of the initial application's exclusivity terms.

C. Patent Provisions

The BPCIA and Hatch-Waxman Act differ significantly with respect to premarket patent litigation. Unlike the Hatch-Waxman Act, the BPCIA does not require a public listing of relevant innovator patents. Nor does it provide a thirty-month stay of biosimilar application approval when the innovator brings patent litigation. Nor does it offer 180-day exclusivity to the first firm to file an abbreviated application with a patent challenge. The BPCIA instead requires a series of complex private information exchanges between the biosimilar applicant and reference product sponsor, followed by negotiated identification of patents for litigation, and then typically two separate phases of premarket litigation (42 USC § 262(l); compare 21 USC § 355(j)).

Potential biosimilar entrants have raised concerns about the information disclosure required by the BPCIA premarket patent provisions, specifically, the obligation to provide a copy of their pending marketing applications to the reference product sponsors. This concern may prompt some companies to submit full BLAs, rather than abbreviated applications under the new statute. This is discussed further in section V.

II. FDA REGULATORY STANDARDS

A. FDA Draft Guidance

FDA has issued a number of Draft Guidance documents beginning in 2012 to implement the BPCIA, and several more are expected in 2014. The drafts to date indicate that the agency "intends to consider the *totality of the evidence* provided by a sponsor to support a demonstration of biosimilarity" (FDA Draft Guidances 2013). FDA also intends a stepwise approach, in which, for a given biosimilar application, "(t)he scope and magnitude of clinical studies will depend on the extent of residual uncertainty about the biosimilarity of the two products after conducting structural and functional characterizations and possible animal studies" (ibid.).

Eventually, FDA may permit applicants to use a "fingerprint-like analysis algorithm" to justify significant reductions in—or perhaps even omission of—preclinical and clinical testing. Fingerprinting means comparing a large number of product attributes and combinations of attributes, and in particular quantifying the similarity of the resulting patterns in two different products. Agency officials described this approach in a peer-reviewed journal article that is generally understood to represent the agency's position (Kozlowski et al. 2011). Fingerprinting strategies also played a key role in FDA's approval of the Sandoz abbreviated new drug application (ANDA) for enoxaparin sodium, a complex product mixture of polysaccharides that is derived from natural sources (Grabowski, Guha, and Salgado 2014b).

Advances in the fingerprinting approach could substantially lower the cost of developing and obtaining FDA approval of biosimilars. However, these advances may not be available or acceptable to regulatory authorities as sufficient for approval for a number of years, especially in the case of more complex biologics, such as monoclonal antibodies and interferons, which account for an increasing proportion of expenditures on biopharmaceuticals.

Development times and costs of FDA submissions for U.S. biosimilars could be substantially reduced for at least some biosimilars already on the market in Europe, if the applicants can rely on the trials that supported European approval. In its Draft Guidance, FDA noted it will accept studies using foreign comparator products (i.e., as a practical matter, studies undertaken for approval in other jurisdictions) as supportive of FDA approval under certain circumstances, when justified scientifically and when accompanied by "bridging" data. The scientific justification may need to address issues such as whether the foreign comparator product is made in a facility inspected by a regulatory authority comparable to FDA. The scientific bridge between the products should include comparative physicochemical characterization, bioassays, and functional assays, and it may include comparative nonclinical or even clinical data (FDA Draft Guidance 2013).

Implementation of the Biosimilar User Fee Act of 2012 (BsUFA) will also allow biosimilar applicants to streamline their development plans through early and regular meetings with FDA about the size and scope of their applications. Specifically, FDA has committed to hold various types of biosimilar product development meetings with applicants. During these meetings, FDA will provide input regarding biosimilar development programs, including with respect to the similarity of the proposed biosimilar and reference product, the need for additional studies, and the design of those studies (FDA, Biosimilar User Fee Act, 2013a).

B. Expected Costs of Developing and Marketing Biosimilars

If FDA requires significant evidence from clinical trials to support the approval of biosimilars, biosimilar manufacturers will need to make much bigger premarket investments than do generic drug manufacturers. The investment necessary will depend on the number and size of the necessary clinical trials, the number of indications involved, and other FDA requirements. The average clinical costs to develop a new biologic entity have been estimated to be several hundred million dollars with more than a billion dollars for the complete development process (DiMasi and Grabowski 2007). Although out-of-pocket clinical development costs for biosimilars will be significantly less, they have been estimated to be up to $100 million or more (FTC 2009). By contrast, the cost of completing bioequivalence studies for a generic drug is estimated to be only $1–$5 million (FTC 2009).

The clinical data requirements are expected to vary with the complexity of the biological product and for monoclonal antibodies may well be at the upper end of the investment time and cost spectrum. These products tend to be approved for a variety of indications, sometimes in unrelated therapeutic areas with different pathophysiologies. Their mechanisms of action may be poorly understood and can differ by indication. Although the EMA has now approved one biosimilar monoclonal antibody with indication extrapolation (albeit with a commitment from the applicant to conduct an additional comparative study in the postmarket period), it remains to be seen whether FDA will be comfortable with complete indication extrapolation for these more complex products. Limiting extrapolation would increase both the cost and the time necessary to put together an application (Schneider et al. 2012). If the agency conditioned approval on verifying clinical efficacy in some or all of the extrapolated indications, this could slow market uptake and would in any case increase costs in the postmarket period.

Biosimilar manufacturers will also need to construct expensive plants or obtain long-term lease or purchase agreements with third parties that already have FDA-approved biologics manufacturing facilities if they do not already have suitable manufacturing capacity. Manufacturing processes for biologics involve numerous challenging cell culture and purification steps, and the use of living cells introduces unavoidable variability into the process. The FTC has estimated that plant investment expenditures for biosimilars will range from $250 million to $1 billion, and other more recent estimates are within this range (FTC 2009).

In sum, the total cost of entry for biosimilars in terms of clinical development expenditures and plant capacity investment is likely to be many times higher than the comparable cost for generic drug products. The high costs of entry—particularly, the substantial development and capital requirements—are likely to restrict the number and types of biosimilar entrants, at least initially (Grabowski, Guha, and Salgado 2014b). Furthermore, initial entry is likely to be limited to the biologics with the largest revenues and to companies with the capacity to pursue global investment strategies. If competition is restricted to a small number of competitors, and with most products expected to compete as therapeutic alternatives rather than equivalents (as explained in the next section, for the foreseeable future automatic substitution is not anticipated), then competition is likely to resemble brand-to-brand differentiated product competition rather than brand-to-generic competition. Biosimilar entrants are thus expected to compete in terms of both price and quality and to engage in promotional activities.

As the science improves and in particular as analytical methodologies advance, and as FDA and its peer regulators grow more comfortable with biosimilars of the more complex biologics, biosimilars will reach the market faster, and more entrants can be expected. But it will take several years for the market to evolve in this fashion.

III. THE POSSIBILITY OF INTERCHANGEABILITY DESIGNATIONS

The BPCIA contemplates that biosimilar applicants may show their products to be not only biosimilar but also interchangeable. This contrasts with Hatch-Waxman, under which generic chemical drugs are ordinarily A-rated (therapeutically equivalent) without any additional showing needed. Interchangeability represents a higher bar than biosimilarity. In particular, FDA may deem a biosimilar interchangeable with its reference product if (1) it "can be expected to produce the same clinical result as the reference product in any given patient" and (2) "the risk in terms of safety or diminished efficacy of alternating or switching between use of the biological product and the reference product is not greater than the risk of using the reference product without such alternation or switch" (42 USC § 262 (k)(4)). The first biosimilar shown to be interchangeable is entitled to a one-year exclusivity period during which no other biosimilar product may be deemed interchangeable with the same reference product.

A key regulatory issue will be the analytical and clinical evidence that FDA will require to deem a biosimilar interchangeable with its reference product, thus signaling its conclusion that substitution without physician approval (subject to state laws) is appropriate. For biosimilar products used more than once by patients (the majority of biologics), the biosimilar sponsor will essentially need to demonstrate that switching between the biosimilar and reference product poses no additional risk of reduced safety or efficacy beyond that posed by taking the reference product alone. This may require crossover trials in which subjects switch between the products repeatedly over time. It might be difficult and expensive to recruit enough subjects to obtain statistical significance for such trials. It is also unclear what factors FDA will consider in evaluating the potential risks related to alternating or switching between the biosimilar(s) and the reference product. Finally, the agency has signaled that at least for the time being, biosimilarity and interchangeability are likely to be sequential showings, which may indicate not only the need for switching studies but also the need to evaluate postmarket data (Kozlowski et al. 2011; FDA Draft Guidances, 2013).

It is reasonable to expect that FDA will license, on the basis of relatively small applications, less complex biosimilars that have already been approved and marketed in Europe for some time, like the granulocyte colony stimulating factors (G-CSFs), such as filgrastim, and the erythropoiesis-stimulating agents (ESAs), such as epoetin alfa. These products may be approved on the basis of studies conducted with the foreign reference product, combined with bridging data, or perhaps on the basis of somewhat smaller comparative clinical trials than initially required in Europe a half-dozen years ago. Interchangeability designations will presumably come later, although earlier for these products than for the more complex products. Biosimilar versions of more complex biologics like the majority of monoclonal antibodies may well require larger applications. To date, only one (infliximab) has been approved in Europe, and the approval was very recent. Until others have been approved, and some postmarket data have accumulated, it is reasonable to expect FDA to be more cautious about application size and scope. Furthermore, FDA is likely to maintain a very high bar with respect to establishing interchangeability for these products, with correspondingly high investment costs. Perhaps eventually, scientific advances might permit a fingerprinting-type analytical characterization approach for establishing biosimilarity and even interchangeability for these products. But a purely analytical approach seems unlikely even in the medium term.

The value of an interchangeability designation may be weakened further by the fact that most state pharmacy laws must be amended to permit substitution of biosimilars (whether or not found interchangeable) for their reference products (Parker 2013). The prospects for widespread and prompt adoption of legislation enabling biosimilar substitution remain uncertain. Current state pharmacy laws typically result in immediate substitution of therapeutically equivalent (A-rated) generic chemical drugs for their reference products. This automatic substitution, without the prescriber's involvement, drives rapid share loss by the branded reference product. But these laws were written for chemical drugs, and the vast majority will not—as drafted—result in substitution of a biosimilar (or interchangeable biosimilar) for its reference product. While several states (such as Virginia) have enacted legislation to govern substitution of interchangeable biosimilars, one state governor vetoed such legislation, and the bills have generated controversy (particularly with respect to notifying physicians that substitution has occurred), which could slow their introduction and enactment in the rest of the country (VA. CODE ANN. §§ 54.3408.04 (2013); Letter from California Governor Edmund G. Brown, Jr. to Members of the California State Senate (Oct. 12, 2013)). No state has considered authorizing substitution of noninterchangeable biosimilars.

There is also a question whether the decisions made by the World Health Organization (WHO) and FDA about the nonproprietary names of biologics and biosimilars could complicate substitution. Both the WHO and FDA and the United States Adopted Names (USAN) Council—in which FDA and others participate—play a role in assigning names to drugs including biological drugs. The BPCIA was silent on the issue of names, leaving this framework untouched. Consequently, since enactment, stakeholders have been discussing whether biosimilars need nonproprietary names that are distinguishable from the names of their reference products—neither the same, nor entirely different, but instead somehow distinguishable—to ensure that adverse events are appropriately attributed to the correct product. Some have suggested that identical names will facilitate either substitution or, perhaps, simply market uptake (for instance, if decision makers take the view that identical names are an additional sign of therapeutic equivalence). Others believe that the name will have no impact on payer and formulary decisions and that state pharmacy laws are likely to focus on whether FDA has designated the products as interchangeable—not whether the nonproprietary name is the same. The outcome of this debate remains uncertain (FTC 2013).

Elsewhere, one of us has considered how the decisions of various stakeholders—physicians, payers, and patients—are likely to affect biosimilar market evolution if they compete as noninterchangeable therapeutic alternatives. The rate of biosimilar penetration is expected to vary by disease indication, patient type, and physician specialty, and to incorporate both price and nonprice considerations (Grabowski, Guha, and Salgado 2014b). In addition, rates of biosimilar acceptance may vary according to physician-focused and patient-focused factors, such as whether the physician specialty is price sensitive or instead demonstrates brand loyalty in therapy choice; whether the biosimilar will be used long term as maintenance therapy or only once or twice (particularly if long-term clinical data are not available); whether the disease or condition is life-threatening (and the implications of therapeutic nonresponse consequently very serious); whether adverse reactions are perceived to be very serious; and whether the difference in ease-of-use or out-of-pocket cost to the patient between the reference product and biosimilar is expected to be high.

From a payer perspective, hospitals are expected to be most cost sensitive and receptive to initial biosimilar usage given the fact that biological products are generally subject to fixed payment bundling reimbursement schemes by Medicare and private payers. For biologics dispensed by physicians in clinics, the evaluation by pharmacy and therapeutics (P&T) committees of comparative quality and price considerations between biosimilars and their reference product will be critical in the case of deciding about other therapeutic alternatives. Interchangeability is likely to be a particularly important factor for the utilization of biosimilars in the case of self-administered biologicals dispensed by retail pharmacies, whether reimbursed through Medicare Part D or private health insurers.

In short, achieving an interchangeability designation will be associated with far greater development costs than simply obtaining approval as a biosimilar biologic. There may be diminished economic incentives for companies to pursue interchangeability, particularly if there are multiple biosimilars already in the market that have provided satisfactory outcomes for providers. The one-year exclusivity right related to interchangeability could provide only marginal economic value because it is not true market exclusivity, only the right to be identified as the only interchangeable product. More generally, the economic value associated with interchangeability will depend on the timing of approval, the competitive situation in the relevant therapeutic class, and whether a product is administered by physicians in clinics and hospitals or dispensed by retail pharmacies and self-administered by patients. This stands in marked contrast to therapeutic equivalence determinations in the Hatch-Waxman setting, which are automatic for the conventional generic product (upon approval and with no additional showing needed). This, in turn, leads to their rapid uptake as a result of state automatic substitution laws and corresponding economic incentives by insurers such as tiered co-payment formularies that apply in the retail pharmacy setting where most generics are dispensed.

IV. EXPECTED COSTS OF POSTMARKET ACTIVITIES AND REGULATORY COMPLIANCE

Biosimilar firms likely will need to have postmarket organizational infrastructures in place to deal with competitive and regulatory issues, at least if their products are not interchangeable. Specifically, unlike generic drug manufacturers who rely on FDA designation of therapeutic equivalence to the reference product and the operation of automatic substitution laws to drive their market share, manufacturers of biosimilars will need to invest in sales and marketing divisions to obtain market share. This would typically include a market research group and a trained sales force to call on prescribers. Biosimilar firms will also need to develop a brand strategy and promotional materials and to arrange for presentations to prescribers, payers, and formulary committees. Depending on the therapeutic category, it may also be desirable to create a patient-oriented website and direct-to-consumer promotional materials. While a firm without such an existing infrastructure can obtain these services from a contract sales organization, launching a biosimilar product on this more brand-to-brand style of commercial marketplace will require a substantial investment of upfront resources whether the work is done internally or externally.

A parallel investment will need to be made in a compliance infrastructure to ensure adherence to federal laws governing advertising and promotion. Firms will need to interact with FDA regulatory officials to ensure that advertising and promotion adhere to the approved labeling and to satisfy other legal and regulatory requirements related to promotional activity. Biosimilar firms also will need to ensure compliance with laws governing sample distribution, assuming that they choose to promote their products in this manner. Firms will also need to monitor compliance with federal and state sunshine laws, state lobbying laws that may apply to sales representatives and possibly even medical affairs personnel, and state laws that require reporting of expenses for advertising and promotion. A biosimilar company would likely also need to establish a medical affairs group to respond to outside requests for product information, and it may want that group to commission scientific publications as well as educational and research grants, and to coordinate advisory boards and consulting arrangements with medical professionals.

Beyond this infrastructure for promotion of their products and engaging in scientific exchange about those products and the relevant disease areas, biosimilar firms will need to establish other organizational groups and systems to deal with postmarket regulatory requirements to which generic firms do not typically devote significant resources. For instance, biosimilar firms will need to establish pharmacovigilance systems in order to trace, analyze, and report adverse events. Generic companies also have adverse event reporting requirements, but their pharmacovigilance obligations are not identical to those of innovators, and in any case they do not typically receive significant numbers of adverse event reports. Additional postmarket requirements for biosimilar firms (which generic firms may not face) could include mandated risk management strategies—including patient registries or even use and distribution restrictions—as well as performance of postmarket studies regarding immunogenicity or even verification of clinical effectiveness where indications have been extrapolated.

Like other biologics, biosimilar products will likely be the subject of manufacturing process improvements and changes over time, a topic discussed at greater length in chapter 18. These will require comparability protocols—including, in some cases, clinical testing—and may also require labeling changes. Dealing with unintentional drift in the product attributes over time could also require postmarket resources that generic small molecule drug applicants do not typically need to invest. These efforts will require medical and regulatory staff within the company, although likely the same manufacturing and regulatory personnel that were required for the premarket application process.

Further, in contrast to generic products, there is no statutory requirement for biosimilar labeling to be the same as the reference product's labeling. Although FDA policy on biosimilar labeling is presumably still under development, biosimilar firms will likely negotiate with the agency and implement decisions on any proposed labeling changes that emerge as a result of postmarket pharmacovigilance and other postmarket developments. This too will require a medical officer and a regulatory staff to handle FDA submissions and implement approvals.

Finally, unlike generic drug companies, biosimilar companies may well face significant exposure to tort liability for failure to update their labeling to reflect emerging safety data. Generic drug companies are currently unable to make unilateral changes to their product labeling, which must be the same as the innovator's labeling. Under a recent Supreme Court case, they therefore benefit from preemption of state tort law claims for failure to warn (*Pliva v. Mensing* 2011:2567). In response, FDA has proposed to change the rules for generic drugs, permitting safety-related changes without prior approval, thereby eliminating preemption and exposing generic companies to tort liability (78 Fed. Reg. 67985 (Nov. 13, 2013)). Although the agency has yet to announce its approach to safety labeling changes for biosimilars, it seems likely the agency will similarly permit unilateral changes without prior approval.

V. FULL BLAS AND BIOBETTERS AS STRATEGIC OPTIONS

Another factor that will affect the timing and extent of biosimilar market entry is the attractiveness, and therefore extent of use, of the full BLA option *as an alternative to* the biosimilar application. After passage of the BPCIA, several firms developing biosimilar-like products—including Teva—publicly stated that they might not file their applications under section 351(k) (FDA Week, 2010). They cited "drawbacks" in the biosimilar pathway, such as the need to wait for expiry of the reference product's twelve-year exclusivity, the likely substantial data requirements, the likely lack of interchangeability designations, and the "onerous" patent litigation provisions. For these and other reasons, they may choose to submit full marketing applications under section 351(a) rather than biosimilar applications under the BPCIA. The question has been raised whether these applications might be less substantial in scope and size than would ordinarily be required for a full BLA. As a regulatory matter, it is plausible that the applications could be less substantial than would be required for a first-in-class biologic, but the question has been raised whether FDA will (or lawfully may) permit biosimilar-like applications under the full BLA provision, which would allow those firms to avoid the burdens of the BPCIA. Some consider the August 2012 approval of Sicor's (now Teva's) section 351(a) application for tbo-filgrastim to illustrate the viability of this strategy. Although the application was submitted before enactment of the BPCIA, European authorities had previously approved tbo-filgrastim as biosimilar to Amgen's Neupogen, and the U.S. BLA largely relied on the same clinical data that supported European approval.

Whether these firms will file their applications or seek approval under section 351(a) will affect when—if at all—biosimilars of certain reference products enter the market. On the one hand, a firm that uses the full BLA pathway will likely seek to differentiate its product (and may permissibly do so). Indeed, that firm will not be constrained by the requirements of biosimilarity, so its product could incorporate deliberate differences from, or improvements to, the reference product (a so-called biobetter strategy). On the other hand, a firm that wishes to extrapolate indications or obtain an interchangeability designation will need to file its application under section 351(k). Even where interchangeability is of uncertain value, for complex biologics with multiple indications (such as the monoclonal antibodies), the cost savings from extrapolation may drive firms to the BPCIA.

Another factor that might affect sponsors' willingness to choose the section 351(k) pathway is whether the innovator has introduced a next-generation product that will divert sales from the first-generation biosimilar. The innovative process in biopharmaceuticals is dynamic in character, with the development of improved formulations, dosing, delivery methods that provide better clinical outcomes, and convenience/quality of life advantages for patients. As seen from the experience in Europe, the introduction of longer-lasting, next-generation PEGylated versions of Neupogen (filgrastim) and Procrit (epoetin alfa) significantly extended the duration of treatment and provided cost savings elsewhere in the health care system and quality of life benefits for patients. From an economic perspective, these next-generation products captured a large share of the overall market in their respective therapeutic classes (ESAs and G-CSFs) in many European countries (Grabowski, Guha, and Salgado 2014a). This in turn constrained the potential market for the first-generation products that are currently subject to biosimilar competition.

Complicating the decision for innovator firms considering a second-generation strategy is the significant discretion afforded to FDA with respect to exclusivity for these products. As noted, the law contains a provision that precludes a separate twelve-year exclusivity period for subsequent BLAs—filed by the same sponsor or manufacturer, a licensor, or a related entity—that involves a "modification to the structure" of an approved biologic that does not result in a "change in safety, purity, or potency" (42 USC § 262(k)(7)(C)(ii)(II)). FDA must therefore decide whether each particular next-generation product (e.g., a pegylated version of the original molecule) is entitled to its own twelve-year exclusivity period as opposed to falling under the umbrella of the first-generation product's exclusivity. A very high bar for manufacturers to obtain a twelve-year exclusivity period for next-generation products could have an adverse effect on investment incentives in the research and development of these products, at least in instances where the patent protection for these products is uncertain or narrow in scope. Given the importance of these decisions to innovators, FDA may well face litigation as it fleshes out its policy on biologics exclusivity. To the extent FDA's decisions turn on matters of science (what constitutes a "modification to the structure," for example, and what constitutes a "change" in safety), the courts are likely to defer to the agency. The area of exclusivity policy will be particularly important to watch, given its likely effect on innovative research and development incentives.

VI. CONCLUSION

The primary objective of the BPCIA, like the Hatch-Waxman Act before it, is to encourage increased price competition and consumer cost savings while maintaining a favorable environment for innovation. Although FDA has approved no applications to date, more than four years after enactment, it is clear that the premarket process for biosimilars entails the submission of data from multiple comparative clinical trials as well as overall a significantly greater investment of time and resources than are required of generic firms. Given this fact, as well as the likelihood that most biosimilars will not (for now) be rated as interchangeable with their reference products, one can expect fewer entrants and a pattern of biosimilar competition that is more likely to resemble branded competition than generic competition for the foreseeable future. Over a longer time frame, scientific advances could reduce the regulatory costs of developing biosimilars and allow demonstration of interchangeability through analytical characterization (e.g., "fingerprinting" structural analysis).

For the immediate future, a number of thorny regulatory issues remain to be sorted out, including permissible extrapolation of safety and effectiveness from one indication (demonstrated in clinical trials) to other indications (not tested), naming requirements, and the use of bridging studies to justify use of global marketing dossiers. These and related issues may need to be resolved on a class-by-class basis. It may take several years for this process to complete, during which time analytical science will improve and important new medical treatments will become available. FDA regulatory actions and decisions, as well as those of associated health care providers and payers, will need to evolve continually to take advantage of the opportunities of the BPCIA while ensuring that patients are protected and continued innovation is encouraged.

REFERENCES

Brown, Jr., E. California Governor Edmund G. Brown, Jr. to Members of the California State Senate, CA, October 12, 2013.

DiMasi, J. A. and H. G. Grabowski. 2007. "The Cost of Pharmaceutical R&D: Is Biotech Different?" *Managerial and Decision Economics* 28:469–79.

FDA Draft Guidances for Industry. 2013. "Scientific Considerations in Demonstrating Biosimilarity to a Reference Product; Quality Considerations in Demonstrating Biosimilarity to a Reference Protein Product; Q & As Regarding Implementation of the BPCI Act of 2009 (2012). http://www.fda.gov/drugs/developmentapprovalprocess/howdrugsaredevelopedandapproved/approvalapplications/therapeuticbiologicapplications/biosimilars/default.htm.

FDA Week. 2010. "FDA Says It Prefers, But Can't Mandate Firms Utilize Biosimilar Pathway." (December 7).

Federal Trade Commission. 2009. "Health Care Issues: Follow-On Biological Drug Competition." *FTC Report* (June).

Federal Trade Commission. 2013. "Public Workshop: Follow-on Biologics: Impact of Recent Legislative and Regulatory Naming Proposals on Competition." (November 15).

Grabowski, H., R. Guha, and M. Salgado. 2014a. "Biosimilar Competition: Lessons from Europe." *Nature Reviews: Drug Discovery* 13(February):99–100.

Grabowski, H., R. Guha, and M. Salgado. 2014b. "Regulatory and Cost Barriers Are Likely to Limit Biosimilar Development and Expected Savings in the Near Future" *Health Affairs*, June, 33(6) 1048–57.

Kozlowski, S., et al. 2011. "Developing the Nation's Biosimilar Progam." *New England Journal of Medicine* 365(5): 385–88.

Parker, N. 2013. "Do Pending State Biosimilar Substitution Bills Violate State Constitutional Law Principles of Non-Delegation and Incorporation by Reference?" *FDLI's Food and Drug Policy Forum* 3 (September).

Pliva v. Mensing, 564 U.S. ____ (2011).

Schneider, C. K., et al. 2012. Setting the Stage for Biosimilar Monoclonal Antibodies." *Nature Biotechnology* 30(December):1179–85.

U.S. Food and Drug Administration (FDA). 2013. "Biosimilar Authorization Performance Goals and Procedures Fiscal Years 2013 through 2017." http://www.fda.gov/downloads/Drugs/DevelopmentApproval Process/HowDrugsareDevelopedandApproved/ApprovalApplications /TherapeuticBiologicApplications/Biosimilars/UCM281991.pd.

———. 2013a. "Biosimilar User Fee Act."

PART SEVEN

New Wine in Old Bottles
FDA's Role in Regulating New Technologies

Introduction

FRANCES H. MILLER

THE BEST word to describe the U.S. Food and Drug Administration's (FDA) role in regulating new technologies is "complicated." Perhaps the words "very complicated" would be better, modified by terms like "difficult," sometimes "controversial," and on occasion even "fraught." Many of the regulatory stakeholders are organized, powerful, and politically sophisticated, and can constitute a potential impediment to the scientific integrity on which the agency prides itself. On a bad day, contemplating stumbling blocks like congressional gridlock, one could think the best descriptive word for FDA's regulatory role vis-à-vis some innovations might be "impossible." But on a good day, you could say that emerging technologies are simply challenging FDA to come up with creatively forward-thinking processes that fit the demands and opportunities presented by ground-breaking twenty-first century medicine.

The five chapters in this part, dealing with such disparate subjects as computerized medical devices, regenerative medicine, direct-to-consumer genetic testing, e-prescribing, and the appropriate use of racial and ethnic categories in biomedical contexts, share a common theme: FDA's current regulatory approach to medical technology leaves something to be desired. The first three chapters examine technological advances already in play, where the agency is trying to adapt existing legislation and rulemaking to innovations undreamed of when they were enacted—new wine in old bottles, as it were. The last two chapters, suggesting improvements in FDA's stance regarding pharmacovigilance and the role of race in research, are more sui generis.

Computerized medical devices have been around in one form or another for decades but lack an explicit regulatory categorization. FDA's method for regulating them has been quite different from its recent attempts to rein in direct-to-consumer (DTC) genetic testing and regenerative medicine, the new kids on the innovation block. As Cortez points out in his chapter, the agency has long been a "reluctant regulator" with regard to electronic device oversight, even though almost half of all medical interventions now involve computers. It has proceeded cautiously and deferentially on unfamiliar technical terrain, using nonbinding guidance documents to guide enforcement rather than straightforwardly adopting regulatory standards. With regard to regenerative medicine and DTC genetic testing, however, FDA has come out much more aggressively in what looks like an attempt to shut down at least some purveyors of these technologies.

Specifically, in 2010 FDA moved to enjoin Colorado physicians and their company from performing regenerative medical procedures wherein patients' stem cells are removed, manipulated, combined with an antibiotic, and then returned to their bodies. Although the defendants claimed their procedure, Regennex-C, was merely the practice of medicine and therefore beyond the agency's reach, FDA maintained that these treatments in fact constituted unapproved and misbranded drugs under the FDCA. The D.C. Circuit agreed with the agency's position in its 2014 *United States v. Regenerative Sciences* opinion, and permanently enjoined the defendants from utilizing the procedure. The court found that the agency also had regulatory authority under the Public Health Service Act to enjoin it because the defendants were manipulating the cells in question more than minimally. By giving FDA a clear win in this case, however, the D.C. Circuit left it with a big problem: how *should* the agency go about regulating "personalized medicine," which is destined to become far more widespread as stem cell science progresses?

Similarly, in 2013 the agency declared personal genome services to be Class III devices (i.e., those medical devices raising the most serious potential patient safety problems). It ordered 23andMe, a high-profile DTC genetic testing company, to discontinue marketing genomic information because it had neither applied for nor received premarket approval (PMA) as required by the FDCA. Distinguishing raw genomic data from genetic information, which can be used to predict propensity for illness and thus fits the definition of a medical device, FDA came down hard on the company; 23andMe can no longer provide what it still advertises as "health results," such as whether the purchaser carries the gene for Parkinson's disease or BRCA, although the company's website predicts that those results soon will be available. That will depend on whether it comes up with a strategy to safeguard patient safety (i.e., diminish the risks of inappropriate or unnecessary treatment, and of depression and/or suicide accompanying potentially upsetting medical information) that will be sufficient to satisfy FDA's concerns.

Why the markedly more aggressive FDA conduct toward these two recent medical innovations than toward computerized medical devices? In the case of 23andMe, the direct-to-consumer nature of the product is undoubtedly a large factor. With no medical professionals involved to interpret potentially worrisome health results for lay purchasers, the possibilities for psychological harm and inappropriate treatment are apparent. Moreover, the high-profile company, a "darling of the tech industry" and recipient of the 2008 *Time Magazine* Invention of the Year award, has never been shy in its marketing efforts. Even today, it uses Groupon coupons to pump up the sales of "ancestry information" and raw genomic data from the vials of spit that purchasers send the company—all that the firm now provides, pending further FDA action. Questions have also been raised about the accuracy of the company's testing methods and about other uses to which consumers' genetic information might be put, which the agency must consider in any application for approval. But categorizing genetic information as a Class III device is probably overkill as a long-run regulatory strategy. As Pike and Spector-Bagdady suggest in their chapter, FDA and the public will be better served by an approach more narrowly tailored to the risks presented by specific types of genetic vulnerability.

With respect to the Regenerative Sciences stem cell procedure, FDA's original 2008 warning was spawned by the company's website claims, accompanied by patient testimonials, that it employed the patient's own enhanced mesenchymal stem cells to regenerate bone and cartilage. The defendants were apparently the only practitioners of this untested therapy, and not only was patient safety an issue but a whiff of the snake-oil salesman permeated FDA's concern. (The defendant's website sells an "Advanced Stem Cell Support Formula" dietary supplement at $97.99 for a one-month supply.) In one sense, Regenerative was an easy target because it presented as a one-off situation. But the issue it raised about FDA's jurisdiction over regenerative medicine has far-reaching implications, as now buttressed by the D.C. Circuit's decision. In her chapter, Riley sensibly suggests that the agency consider adopting "responsible innovation" as an abbreviated pathway to approval for stem cell therapies, which are sure to proliferate as the science develops.

The fourth chapter in this section departs from the focus on the present and looks to the future in offering an ambitious plan to advance patient safety by combining e-prescribing and electronic medical records technology to improve pharmaceutical use and FDA oversight. Lamenting the agency's less-than-stellar record on pharmacovigilance, Rosenberg et al. advocate better pre- and postmarketing oversight of the way physicians actually employ approved products, including off-label use. They want to make doctors specify *why* they are e-prescribing particular medical interventions and then require them to report the medical outcomes of their treatment choices. FDA would then evaluate these data for further evidence about safety and efficacy. Their proposal is easy to state but gargantuan in its implications. Can doctors really be expected to track down the results of every prescription they write, let alone justify off-label use when it may have become the standard of care?

This far-reaching and nontrivial undertaking would require extensive and expensive physician buy-in to be effective, and doctors' opposition to taking on regulatory responsibilities they consider intrusive and onerous will be a given. Perhaps a better idea would be to start slowly by implementing a pilot program in a field like oncology, where the patient risks associated with inappropriate or ineffective therapy are higher, treatment standards are still evolving, and follow-up about the results of treatment is part of the process. The opportunity to link e-prescribing to electronic medical records to evaluate treatment efficacy and improve patient safety should not be squandered, but trying to do too much too fast seems doomed to failure.

Finally, in the last chapter Kahn uses a broad lens to examine FDA's current approach to the use of racial and ethnic categories across all of biomedicine. Pointing out that "[r]ace is not a coherent genetic concept; rather it is best understood as a complex and dynamic social construct," he compares FDA guidelines for collecting data on race and ethnicity in clinical trials with the International Conference of Harmonization's (ICH) "Ethnic Factors in the Acceptability of Foreign Clinical Data." FDA's guidelines are designed "to ensure consistency in evaluating potential differences in drug response among [U.S.] racial and ethnic groups" but explicitly warn that the categories they establish for purposes of data collection are neither genetic nor biological.

The ICH, by way of contrast, is focused on eliminating national regulatory differences in order to open up global pharmaceutical markets and "constructs race as some sort of component of a larger category of ethnicity." ICH views the more nuanced FDA guidelines as an impediment to harmonization, but Kahn's warning against having the United States adopt a "harmonized conception of race as genetic" simply to facilitate globalization is sound. He cites a memorable editorial in *Nature Biotechnology* to the effect that "[p]ooling people in race silos is akin to grouping raccoons, tigers and okapis on the basis that they are all stripey." Point made. In black and white.

Few knowledgeable people would argue that FDA is not an overworked and underfunded government agency, given its vast regulatory responsibilities encompassing at least a quarter of the U.S. economy. New medical technologies are constantly pushing the regulatory envelope, creating new agency challenges for effective oversight. The five provocative chapters in this part illuminate some of those challenges and offer intriguing perspectives on potential regulatory routes forward in a dynamic and difficult era.

CHAPTER TWENTY-THREE

Analog Agency in a Digital World

NATHAN CORTEZ

SINCE ITS inception, the U.S. Food and Drug Administration (FDA) has had to confront novel products from early pharmaceuticals to sophisticated biotechnology drugs. But perhaps nothing has stretched the agency's comfort zone more than computerized medical devices. When Congress granted FDA gatekeeping authority over devices in 1976, it did not anticipate software. By then, the germinations of computerized medicine were well under way. In 1967, for example, the *New York Times* reported that a computer in Washington, D.C., received electrocardiogram signals from France and returned its analysis by satellite within seconds.

Of course, the intervening decades would bring profound advances in computing. But FDA has always struggled with how to regulate computerized devices. From primitive diagnostic software and radiation machines to modern software that can turn smartphones into cardiac-event recorders, an old agency continues to apply an old regulatory framework to novel products.

This chapter reviews over three decades of federal documents, finding striking parallels between yesterday's concerns and today's. The same questions raised in the 1970s and 1980s remain today, largely untouched by Congress and FDA. But it is well past time to address them. The variety, complexity, and ubiquity of medical software grows relentlessly. A study by Fu (2011) found that beginning in 2006, more than half of all medical devices on the U.S. market relied on software in some way. Indeed, the market is now being saturated with software used on phones and other mobile devices, known as "mobile health" or "mHealth" (Cortez 2014a). As the use of software in medicine grows almost exponentially, it is time to reconsider FDA's old framework.

I. COMPUTERIZED MEDICINE

FDA jurisdiction over certain software derives from the broad definition of "device" in the Food, Drug, and Cosmetic Act (FDCA). Section 201(h) defines a "device" as any "instrument, apparatus, implement, machine, contrivance" that is intended to diagnose, cure, mitigate, treat, or prevent disease or other conditions, or that is "intended to affect the structure or any function of the body," including any component, part, or accessory. Broad as this definition is, many computer products are not FDA regulated. But this line is not always clear.

A. Beginnings

With a foundation laid in the 1940s and 1950s, computerized medicine emerged in the 1960s with university research creating promising-sounding programs like HELP, AIM, and PROMIS. "At the beginning of the 1960s," a doctor testified to Congress, "the computer was essentially unknown in the biomedical world" (*Computers in Health Care: Hearing Before the Subcomm. on Domestic and International Scientific Planning, Analysis, and Cooperation of the H. Comm. on Science and Technology*, 95th Cong. (1978)). Early support from the National Institutes of Health (NIH) and its National Library of Medicine were instrumental (November 2012). A 1963 NIH program funded computers at forty-five universities and six hospitals, which pioneered the first computers for health care diagnosis and delivery, including clinical decision support (*Information Technologies in the Health Care System:*

Hearing Before the Subcomm. on Investigations and Oversight of the H. Comm. on Science and Technology, 99th Cong. (1986)).

Later, NIH funding helped establish a network called SUMEX-AIM (Stanford University Medical Experimental Computer—Artificial Intelligence in Medicine), which included MYCIN, Stanford software that outperformed experts in identifying therapies for bacterial infections, and CASNET, Rutgers University software that helped diagnose glaucoma. Private sector research complemented these programs. For example, in the late 1970s, a drug company created software for diagnosing pediatric diseases, allowing physicians to "dial up to the remote computer, enter three or four findings to get back a list of possible causes, including the hard to remember congenital syndromes" (1986 Hearing).

Of course, many of these programs were not medical "devices" and were not even on FDA's radar. But they helped lay the foundation for today's sophisticated devices.

B. Aspirations

From the beginning, computers promised to reduce health spending. The first congressional hearings in 1978 focused on "whether and how computer science might help to stem the uncontrolled growth of medical costs." At hearings in 1981, a congressional representative lamented that "we are missing a great bet in the involvement of the computer" in controlling "escalating health care costs" (*Health Information Systems: Hearing Before the Subcomm. on Science, Research, and Technology and Subcomm. on Natural Resources, Agricultural Research, and Environment of the H. Comm. on Science and Technology*, 97th Cong. (1981)). This idea persists today. Almost every recent congressional hearing or agency announcement continues to broadcast the idea that computers will cut health spending.

Another longstanding notion is that computers can help reduce medical errors and improve the quality of care. Even early computers had enough memory and processing power to capture data and compute probabilities better than humans. The near-exponential growth of memory and processing power has obvious medical applications, like analyzing large quantities of clinical data or distinguishing standard diagnoses from outliers. The emergence of artificial intelligence

(AI) could refine medical algorithms, and communications networks could better share this data. Software was also used in medical devices to better calibrate diagnostic and therapeutic equipment.

A third longtime aspiration of computerized medicine is to expand access to care. A spokesperson for the U.S. Public Health Service said that the 1967 electrocardiogram in France showed how "long distance electrocardiogram analysis could be made available almost anywhere on earth" (*New York Times* 1967). Early hearings contemplated using computers to expand access to care in urban and rural communities. These ambitions surged recently with medical software on mobile devices (Cortez 2014a), which is already being deployed to expand access to care in developing economies (Kahn et al. 2010).

Science fiction has long embraced these ambitions. The idea that computers can process the complexities of the human body better than even expert physicians may trace back to the television show *Star Trek*, which featured a handheld "Tricorder" device that could instantaneously diagnose crew members. The 1986 congressional hearings included a *Stark Trek* reference, and in 2011 the X Prize Foundation announced a $10 million competition for the first group to create a real-life Tricorder (Cortez 2014a).

But these ambitions are also dismissed as mere science fiction. A doctor at the 1986 congressional hearing testified that "the notion that physicians will turn to a computer's advice exactly as they would turn to a human consultant, I think, is science fiction."

C. Hazards

There are well-documented hazards in relying on software to diagnose and treat patients (Leveson 1995, 2011; Fu 2011). Unlike physical devices, small errors in software can be disastrous. But software code is difficult, if not impossible, to test completely (Fu 2011). User interfaces can be awkward. And software is frequently operated in less than ideal environments, when the user is fatigued, distracted, or frustrated (Cortez 2014a). Repeated error alerts can numb users, who learn to bypass or ignore them. Users are also susceptible to "automation bias," the belief that computers are infallible (Citron 2008; Fu 2011). As the *New York Post* speculated in 1959, "The day will come sooner than you think when improper diagnosis and treatment of human ills will be virtually impossible" (November 2012). Finally, hospitals and physicians can be quick to adopt unproven new technologies without investing the resources to use them properly.

FDA is well aware of these hazards. In 1996, FDA detailed a litany of software device errors, including problems with cardiac devices, radiation therapy machines, and infusion pumps, observing that "software-related errors can be subtle" and that "seemingly small design flaws can result in serious problems" (FDA 1996). The agency is aware of adverse events caused by human–computer interaction, including cumbersome controls, counterintuitive displays, unsignaled default settings, and inadequate user manuals.

Recalls involving software have increased steadily over the last 30 years (Fu 2011). In fact, FDA's first public statement on software followed a series of deaths caused by computer software. Between 1985 and 1987, the Therac-25 (short for "therapeutic radiation computer") caused multiple deaths in the United States and Canada. The Therac-25 was the first radiation therapy machine controlled primarily by software, but it was plagued with bugs, a confusing interface, incomplete manuals, and repeated malfunctions (Leveson 1995). Decades later, the *New York Times* documented frighteningly similar problems caused by the latest generation of radiation therapy machines (Bogdanich 2010).

These bookends demonstrate that old problems persist. Unfortunately, FDA scrutiny has not been commensurate with how ubiquitous and critical software devices have become.

II. CONGRESS, AN INTERESTED BYSTANDER

Congress has long had an interest in computerized medicine. But this interest has generated very little legislation, particularly laws that would update FDA's old statutory framework. Early attention focused mostly on the benefits of computerized medicine. In 1978, the House Committee on Science and Technology held hearings on *Computers in Health Care*, with a sequel in 1981 on *Health Information Systems*. FDA officials did not testify at either hearing. An official from the U.S. Department of Health and Human Services testified at the 1981 hearing but mentioned FDA only in passing. In 1986, the same House committee held hearings on *Information Technologies in the Health*

Care System, this time asking FDA for a draft of its early, unpublished software policy.

After years of sporadic attention, Congress has shown a renewed interest in FDA regulation of software, prompted by massive federal investments in health information technology and the proliferation of mobile devices. In 2013, the House held several hearings to consider FDA regulation of mobile software (*Health Information Technologies: Hearing Before the Subcomm. on Health and Subcomm. on Communications and Technology of the H. Comm. on Energy & Commerce*, 113th Cong. (2013); *Examining Federal Regulation of Mobile Medical Apps and Other Health Software: Hearing Before the Subcomm. on Health of the H. Comm. on Energy & Commerce*, 113th Cong. (2013)).

Nevertheless, this attention has led to few bills and even fewer laws. In 2012, Congress did pass a law, but only to ask for recommendations on how to regulate newer health information technologies (Food and Drug Administration Safety and Innovation Act of 2012 (FDASIA), Pub. L. No. 112-144, § 618), which were published recently (FDA 2014). Although such bills seem to be gaining momentum, these bills generally try to taper FDA oversight, not strengthen it (Cortez et al. 2014). For example, bills introduced in the 113th Congress would limit FDA jurisdiction over clinical decision support software (The Sensible Oversight for Technology which Advances Regulatory Efficiency (SOFTWARE) Act, H.R. 3303, 113th Cong. (2013); The Preventing Regulatory Overreach To Enhance Care Technology (PROTECT) Act, S.2007, 113th Cong. (2014); The Medical Electronic Data Technology Enhancement for Consumers' Health (MEDTECH) Act, S.2977, 113th Cong. (2014)). Congress has not considered bills that would update FDA's statutory framework for devices, enacted in 1976 when few could imagine today's computerized products. (Note, however, that a bill not yet introduced as of February 2015, the 21st Century Cures Act, would represent more ambitious reforms to FDA regulation of software.)

III. FDA, THE RELUCTANT REGULATOR

Although FDA has spent almost four decades regulating computerized devices, its oversight remains piecemeal and sporadic, relying heavily on nonbinding guidance. FDA has never written comprehensive rules tailored to software. In the 1990s, FDA hinted that it was contemplating such a rule but never proposed one. In 2011, the agency explained that it never created an "overarching software policy" because "the use of computer and software products ... grew exponentially and the types of products diversified and grew more complex" (FDA 2011a).

A. Early Consideration

FDA's experience with computerized devices dates back to the 1970s, when it "completed premarket approval (PMA) reviews of computer-related products such as cardiac pacemaker programmers, patient monitoring equipment, and magnetic resonance imaging machines" (1986 Hearing). In the early 1980s, FDA began studying software more systematically. In 1981, it created a Task Force on Computers and Software as Medical Devices, which wrote an unpublished report. In 1984, a Program Management Committee on Software and Computerized Devices issued another nonpublic report.

By 1985, when the first reports emerged of patient deaths from the Therac-25, the need for an FDA software policy became apparent. In 1986, FDA made its first public statement, before the House Committee on Science and Technology. In 1987, FDA announced its first software policy in the *Federal Register* (Draft Policy for the Regulation of Computer Products, 52 Fed. Reg. 36,104 (Sep. 25, 1987)), beginning a quarter century of addressing software by guidance. But the 1987 draft was not particularly ambitious, simply explaining the scope of FDA jurisdiction and existing requirements that might apply. In 1989, FDA published an updated *Draft Policy for the Regulation of Computer Products*, which it withdrew 16 years later (70 Fed. Reg. 824, 890 (Jan. 5, 2005)).

B. Organization

Responsibility for software resides with the FDA Center for Devices and Radiological Health (CDRH), which has pockets of software expertise. For example, the Office of Science and Engineering Laboratories includes a Division of Electrical and Software Engineering, a Division of Physics, and Division of Imaging and Applied Mathematics, each of which covers software devices. For example, the divisions include a Laboratory of Software Engineering and a Laboratory of Medical Electronics, among others. These units support premarket product reviews, policy development, and other activities.

Yet, as important as these activities are, FDA's software expertise is still not commensurate with the volume of software devices on the market. Fu (2011) observed that "seldom does an FDA inspector assigned to review a 510(k) application have experience in software engineering even though the majority of medical devices today rely on software." Fu also notes that "software experts are notably underrepresented" in FDA's fellowship programs that seek to cultivate technical expertise. Unfortunately, recent proposals in Congress would not dedicate substantially more in-house resources or expertise to software oversight.

C. Regulation

Although dozens of FDA rules refer to software, very few establish broadly applicable requirements distinct from physical devices. Most software references appear in 21 CFR Parts 862–892, which classify more than 1,700 types of devices. Software also appears frequently in FDA rules for radiology products (21 CFR Parts 1000–1050). Beyond this, the only broad rules tailored to software devices appear in the Quality Systems Regulation (QSR), which specifies that software is subject to manufacturing design controls, must be validated, and must document its specifications (21 CFR Part 820).

Of course, many broad FDA rules apply to software devices because they are "devices." For example, like physical devices, software devices can be classified as Class I (low risk), Class II (moderate risk), or Class III (high risk) (21 USC §§ 360c, 360e), subject to the rules that apply to each class. But by and large, FDA has not written broadly applicable, legally binding rules tailored to software, relying instead on case-by-case adjudication and nonbinding guidance.

D. Adjudication

Since the 1970s, FDA has evaluated software devices case by case in both the premarket and postmarket stages. In the premarket stage, FDA can approve novel or high-risk devices through a premarket approval

(PMA) application, which requires clinical data that the device is safe and effective (21 USC §§ 360c, 360e; 21 CFR Part 814). But more commonly, FDA clears devices through its 510(k) notification process, which declares that the device is "substantially equivalent" to a device already on the market (21 USC § 360(k)). FDA then places devices into one of roughly 1,700 product categories (21 CFR Parts 862–892). For example, 21 CFR § 876.1300 describes an "ingestible telemetric gastrointestinal capsule imaging system," equipped with a camera, light, transmitter, and battery to take pictures of the small bowel and transmit them to a receiver. Sometimes, FDA will clear products without substantial equivalence through a de novo 510(k). For example in 2014, FDA cleared the PillCam, an ingestible capsule that videotapes the small bowel and colon before the capsule is excreted. FDA thus created a new classification for ingestible capsules that record video (21 CFR § 876.1330) as opposed to take pictures (§ 876.1300).

FDA has been criticized for clearing the vast majority of devices through its 510(k) process, sometimes with very little scrutiny (IOM 2011). Such critiques often highlight software devices as a particular concern. Yet, FDA continues to clear computerized devices as being substantially equivalent to noncomputerized predicates (Fu 2011; Cortez 2014a). Other software products may escape FDA scrutiny altogether if they qualify as general purpose articles, if they are designed and used by individual physicians or facilities, or if they are made "solely for use in research, teaching, or analysis" (21 CFR § 807.65).

FDA also regulates software devices in the postmarket stage, for example, by regulating manufacturing, adverse event reporting, and recalls. Some of FDA's early software activities included postmarketing enforcement (1986 Hearing). Today, as might be suspected, software failures are responsible for a growing proportion of product recalls (Fu 2011).

Though there are certainly benefits to careful product-by-product consideration, this approach can lead to regulation that is piecemeal and inconsistent. Software does not stand on terra firma with FDA, lacking broader rules that bind prospectively.

E. Guidance

The most prominent feature of FDA's approach to software is its heavy reliance on nonbinding guidance. My review found twenty-six separate guidance-type documents for software devices, including fifteen original documents and eleven updated versions. The guidances cover a range of topics, including premarket submissions, manufacturing controls, and cybersecurity. There are perhaps dozens more that specify "special controls" for Class II devices. For example, the ingestible camera above is governed by a *Class II Special Controls Guidance Document: Ingestible Telemetric Gastrointestinal Capsule Imaging Systems*. FDA guidance often incorporates even more guidance from standard-setting bodies like the International Electrotechnical Commission (IEC) and the International Organization for Standardization (ISO) (Cortez 2014a). Thus, FDA's framework for software relies on a loose scaffolding of de facto but not de jure rules.

IV. ENDURING CONCERNS

This review demonstrates that FDA still lacks a tailored regulatory framework for software devices. Despite profound advances in computing, there are striking parallels between concerns raised in the 1970s and 1980s and in recent years.

A. Innovation vs. Regulation?

Early hearings expressed concern that FDA regulation would "delay," "stagnate," or "stifle" software innovation (1986 Hearing). A witness in 1986 said that regulating medical software at that point would be like creating the Federal Aviation Administration after the Wright Brothers' inaugural flight. Vincent Brannigan, perhaps the leading legal expert on medical software at the time, emphasized that "the FDA proposal has the potential to simply wipe the whole industry out" (1986 Hearing).

Participants at recent hearings repeat this refrain (FDA 2011b; 2013 Hearings). Rare are the voices that argue that FDA regulation is not necessarily incompatible with technological advancement. In fact, every recent bill seems more concerned that FDA promote software innovation rather than regulate it (FDASIA 2012; Health Care Innovation and Marketplace Technologies Act, H.R. 2363, 113th Cong. (2013); SOFTWARE Act 2013; PROTECT Act 2014, MEDTECH Act 2014).

This sentiment is well meaning but blinkered. Software devices are in great need of more predictable, tailored oversight. The quantity, complexity, and variety of software devices are accelerating, not abating. Consider just one type of software, mobile health applications. There are more than 97,000 different mobile health applications on the market (research2guidance 2013). Hundreds, if not thousands, fall under FDA jurisdiction. Yet, despite studies showing that many do not work as claimed or lack scientific support, FDA relies on guidance and case-by-case scrutiny (Cortez 2014a). The market risks being flooded by apps that are ineffective or unsafe, which can undermine consumer confidence (Carpenter 2009).

B. Paralyzed by Change?

A second recurring notion is that software evolves too quickly for FDA. Agency personnel testified in 1986 that the "burgeoning growth of computers in medicine and its more pivotal roles poses new challenges to FDA" (1986 Hearing), a concern that has been repeated (FDA 1992). In fact, FDA explained that it did not publish "an overarching software policy" in part because "the use of computer and software products as medical devices grew exponentially and the types of products diversified and grew more complex" (FDA 2011a).

The pace of software innovation may partly explain why FDA relies heavily on guidance rather than rulemaking. Issuing guidance typically takes far less time than notice-and-comment rulemaking, and gives the agency more flexibility to change approaches. In fact, many scholars assume that guidance is appropriate for specifying technical or scientific standards that need more frequent updating. However, long-term reliance on guidance can become a crutch for agencies (Cortez 2014b), and there are signs of this in FDA's approach to software.

C. Is Software Different?

A third unanswered question is whether software differs enough from physical devices to warrant its own tailored regulations. On one hand, Brannigan testified to Congress that FDA personnel treated software like "some kind of new bedpan" (1986 Hearing). On the other hand,

FDA recognized early on that software differs in key ways from physical devices on issues like quality assurance and user errors. Even Brannigan observed that FDA personnel, "people of immense goodwill," were "wrestling" with software being very different from traditional devices (1986 Hearing).

Thus, it is surprising how few FDA regulations are tailored to software devices. The agency has long wavered on whether software deserves an overarching policy. In 1986, FDA declared, "No separate policy for computer software presently exists nor is one envisioned for the future" (1986 Hearing). Although the 1987 and 1989 draft policies contradicted that sentiment, FDA never finalized or codified them, despite labeling them as "prerulemaking" (54 Fed. Reg. 44,643).

Years later, software's place in the FDA universe continues to perplex. In a 2013 hearing, two congressional representatives asserted quite confidently (and quite incorrectly) that "software is not a medical device" (2013 Hearing; statements of Reps. Joe Pitts and John Shimkus). This lingering confusion suggests that it is well past time to update the FDCA to account for software.

D. Agency Expertise?

A well-recognized basis for agency authority is technical expertise. But FDA has long questioned its own expertise on software, recognizing its unusual complexity. In the agency's words, "The reliability of software systems and higher order integrated circuits are extremely difficult to assess because of their complexity," and it can be "impractical to test it for every possible input value, timing condition, environmental condition, logic error, coding error and other opportunity for failure" (1986 Hearing). Later FDA guidance repeated these concerns (FDA 1992). Brannigan observed that "even in the best of faith, with the best of will, the best of technology, the best of intentions," FDA could not adequately regulate software based on the 1976 Device Amendments (1986 Hearing). FDA's lack of confidence reveals itself during public hearings, when agency personnel frequently downplay FDA's expertise and defer to the software industry's. Again, this might reflect a combination of resource constraints and the challenge of applying an aging statutory framework to new technologies.

E. Regulating a Regulatory-Naïve Industry?

A related concern is that the software industry is not accustomed to federal regulation, particularly by FDA. Historically, the federal government has done more to incubate the software industry than regulate it (Katz & Phillips 1982). FDA recognized early on that "many manufacturers of new, computer-related medical technologies are not aware of their responsibilities under the law because they are not part of the traditional medical device industry" (1986 Hearing). Certainly, large modern software device manufacturers are well acquainted with FDA requirements. But the latest generation of software developers for mobile devices seems naïve to them.

A parallel theme is FDA's longstanding commitment to adopting the "least burdensome" approach to software. FDA's first statement on software promised to impose "the minimum level of regulatory control necessary" (1986 Hearing), a promise it would often repeat. Congress later codified this philosophy for all devices, requiring FDA to consider the "least burdensome" ways to evaluate device effectiveness and substantial equivalence (Food and Drug Administration Modernization Act, Pub. L. No. 105-115, 105th Cong. § 513 (1997)).

F. Regulating Medical Knowledge?

Finally, there has long been unease that FDA regulation of software, particularly clinical decision support, would allow it to regulate medical knowledge, contravening the limitation in FDCA § 1006 that the agency cannot regulate the practice of medicine. Perhaps accounting for this concern, FDA's early software guidances cite the exemption that allows practitioners and hospitals to use custom devices without prior FDA clearance (21 CFR § 807.65).

Software regulation also can implicate First Amendment free speech rights, as early commentators noted. At the 1986 hearings, a physician testified, "Computerized decision support programs are not much different than a physician taking the newest literature on a subject and applying it to their patient." Indeed, FDA's 1987 draft policy acknowledged that the First Amendment might limit its ability to regulate medical advice encapsulated in software. Scholars continue to consider this question (Candeub 2014).

FDA policy reflects these concerns in two ways. First, the agency is careful not to regulate software that replicates the functions of textbooks, articles, or reference sources, distinguishing software intended for "educational purposes" from that intended to "diagnose or treat patients" (FDA 1989). Second, early FDA policy exempted software devices that allow time for "competent human intervention before any impact on human health occurs," defined as a situation in which "clinical judgment and experience can be used to check and interpret a system's output" (FDA 1987). Such software would include, for example, most "expert" or "knowledge based systems," including "artificial intelligence and other types of decision support systems" (FDA 1989).

The conventional wisdom is that medical professionals will use such technologies wisely, without FDA oversight: "In the absence of any relevant regulations, the medical profession has been very cautious and critical in the acceptance of this kind of intelligent computing" (1986 Hearing). Yet, the more we learn about automation bias and errors from human–computer interaction, the more this conventional wisdom seems suspect.

V. TOWARD A NEW REGULATORY FRAMEWORK

Given these longstanding concerns, no longer should Congress be an interested bystander nor FDA a reluctant regulator. Though momentum seems to be building toward congressional action, proposals to date have been modest, if not regressive. On the modest end, the much-anticipated *FDASIA Health IT Report* does not propose any meaningful reforms to FDA oversight. On the regressive end, recent bills like the PROTECT Act and SAFETY Act would remove FDA jurisdiction over "clinical software," including most clinical decision support. Thus, this moment of renewed interest in FDA software regulation threatens to pass without meaningful reform. FDA's framework, based on the 1976 Medical Device Amendments and decades of guidance documents, needs to be updated and tailored to software. Such a framework should consider four types of improvements.

First, as a definitional matter, Congress should clarify that software can satisfy the definition of "device" in FDCA § 201(h). As a corollary, Congress should reject recent proposals to exclude from FDA oversight clinical decision support and other types of "clinical" software. Clinical software will become more ambitious, more sophisticated, and more likely to be relied upon by patients and providers alike. Research on automation bias and human–computer interaction suggests that many users will rely on computer advice without second guessing it, even if there is an opportunity for "competent human intervention."

Second, Congress should recognize that software devices differ enough from physical ones and push FDA to create tailored requirements, preferably through rulemaking rather than guidance. For example, device approval pathways, quality systems, labeling, and postmarket surveillance all could stand to be updated for software products. In particular, Congress might consider a premarket pathway better suited to software devices that can easily change and that are often based on common, off-the-shelf software modules used by multiple products. Concerns about deterring innovation might be addressed by experimenting with conditional approvals based on clear, binding postmarketing requirements. A new round of rulemaking for software would also allow FDA to update decades of policies enunciated in nonbinding guidance.

Third, Congress might create an Office of Software Devices to better focus FDA's attention and bolster its in-house expertise on software. Recent proposals stop short of doing so. For example, the *FDASIA Health IT Report* recommends a new "Health IT Safety Center," but it would not have regulatory authority and would reside within the Office of the National Coordinator for Health Information Technology (ONC), not FDA. Recent bills also propose new entities, but their function is informational, not regulatory. Most observers recognize that there are simply too many software products for FDA to review. Certification by private entities has been suggested (Powell et al. 2014). But asking certifiers to oversee fee-generating applicants can create conflicts of interest. Presumably, user fees to fund a new FDA Office of Software Devices would present less of a conflict. A new office would need to coordinate with the ONC, the Federal Communications Commission (FCC), and Federal Trade Commission (FTC), to be sure. But FDA is uniquely situated to protect public health.

The final component of successful software regulation is consistent enforcement, given the bewildering number of software devices on the market. Without real enforcement, we risk having a lemons market like the dietary supplement industry, in which most products are ineffective, unsafe, or both. Consistent enforcement can encourage high-value innovation in the long run. FDA regulation can even be "market-constituting," in that it sustains consumer confidence that would otherwise erode if flooded with substandard products (Carpenter 2009).

VI. CONCLUSION

This chapter tells the story of an old agency applying an old regulatory framework to very new technologies. Because neither Congress nor FDA has updated this framework, the same concerns over software regulation continue to linger. To be certain, other questions, beyond the scope of this chapter, also persist, including questions about legal liability, patient privacy, and regulation by other federal agencies like the ONC, FTC, and FCC. Although the federal government is beginning to address some of these questions, it insists on doing so under a very old statutory framework. It is well past time for Congress to consider a twenty-first-century framework for software devices.

REFERENCES

Bogdanich, W. 2010. "Radiation Offers New Cures, and Ways to Do Harm." *New York Times* (January 2013).

Candeub, A. 2014. "Digital Medicine, FDA, and the First Amendment." *Georgia Law Review* (forthcoming).

Carpenter, D. 2009. "Confidence Games: How Does Regulation Constitute Markets?" In *Government and Markets: Toward a New Theory of Regulation*, ed. Edward J. Balleisen and David A. Moss, 164–190. New York: Cambridge University Press.

Citron, D. K. 2008. "Technological Due Process." *Washington University Law Review* 85:1249.

Cortez, N. 2014a. "The Mobile Health Revolution?" *U.C. Davis Law Review* 47:1173.

———. 2014b. "Regulating Disruptive Innovation." *Berkeley Technology Law Journal* 29:173.

Cortez, N.G., I. G. Cohen, and A. S. Kesselheim. 2014. "FDA Regulation of Mobile Health Technologies." *New England Journal of Medicine* 371:372–379.

Food, Drug, and Cosmetic Act of 1938, Public Law 75-717 (codified as amended at 21 USC §§ 301-99).

Fu, K. 2011. "Trustworthy Medical Device Software." *Institute of Medicine Workshop on The FDA's 510(k) Clearance Process at 35 Years.*

Institute of Medicine (IOM). 2011. *Medical Devices and the Public's Health: The FDA 510(k) Clearance Process at 35 Years.* Washington: National Academies Press.

Kahn, J. G., J. S. Yang, and J. S. Kahn. 2010. "'Mobile' Health Needs and Opportunities in Developing Countries." *Health Affairs* 29(2):254.

Katz, Barbara G. and Almarin Phillips. 1982. "The Computer Industry." In *Government and Technical Progress: A Cross-Industry Analysis*, ed. Richard Nelson, 162–232. New York: Pergamon Press.

Leveson, Nancy G. 1995. *Safeware: System Safety and Computers.* New York: Addison-Wesley Professional.

Leveson, Nancy G. 2011. *Engineering a Safer World: Systems Thinking Applied to Safety.* Cambridge: MIT Press.

New York Times. 1967 (July 6). "Hearts in France Analyzed in U.S. in a Satellite Test."

November, Joseph. 2012. *Biomedical Computing: Digitizing Life in the United States.* Baltimore: Johns Hopkins University Press.

Powell, A.C., A. B. Landman, and D. W. Bates. 2014. "In Search of a Few Good Apps." *Journal of the American Medical Association* 311(18):1851.

research2guidance. 2013. *Mobile Health Market Report 2013–2017.*

U.S. Food and Drug Administration (FDA). 2014. "FDASIA Health IT Report: Proposed Strategy and Recommendations for a Risk-Based Framework."

———. 2011a. "Draft Guidance for Industry and FDA Staff: Mobile Medical Applications."

———. 2011b. "Public Workshop – Mobile Medical Applications Draft Guidance."

———. 1996. "Do It by Design: An Introduction to Human Factors in Medical Devices."

———. 1992. "Application of the Medical Device GMP to Computerized Devices and Manufacturing Processes: Medical Device GMP Guidance for FDA Investigators."

CHAPTER TWENTY-FOUR

Twenty-First-Century Technology with Twentieth-Century Baggage

FDA Regulation of Regenerative Medicine

MARGARET FOSTER RILEY

REGENERATIVE MEDICINE is perhaps *the* technology of the twenty-first century. It holds the promise of eliminating organ shortages, rebuilding limbs, and curing disease. Much of this has until now only existed in science fiction. But regenerative medicine has also inherited the twentieth-century ethical issues and political baggage associated with stem cell research. And in the United States, the primary regulator, the Food and Drug Administration (FDA), must regulate under statutes that are decidedly mid-twentieth century.

These are new technologies, and the scientific aspects of how they function are still being worked out. This raises safety issues in their application. In addition, because these are living cells, this is likely to be more complicated and more individualized than the chemical technologies FDA has a hundred years of experience regulating. Moreover, regenerative medicine, or more precisely the technologies comprised under that name, progenitor and stem cell technologies and therapies, also seems to be a battleground where many political, legal, and philosophical conflicts are waged.

I. WHAT IS REGENERATIVE MEDICINE?

Put simply, regenerative medicine is "use of natural human substances, such as genes, proteins, cells and biomaterials to regenerate diseased or damaged human tissue" (Orlando et al. 2011). But the central focus of regenerative medicine is pluripotent cells (stem cells that are capable of differentiating into more than one cell type). Regenerative medicine may be amazingly complex; it is a multidisciplinary field involving biology, engineering, and medicine. It involves stem cells: embryonic, fetal, adult, and induced pluripotent stem cells. It may require biomechanical design: biomaterials and biomolecules, including nanomaterials, designed to support the organization, differentiation, and proliferation of cells into functional tissues. It may require informatics to do things like support gene expression and protein expression and interaction analysis as well as tissue and cell modeling and manufacturing. Or, it may be as simple as removing stem cells from the body of a person, storing them, and then returning them to the body of the same person.

A. The Cell Sources

Stem cells used for regenerative medicine may be autologous, using cells from the intended recipient; allogenic, using human cells from a different donor than the recipient; or xenographic, using cells from other species (Tigerstrom 2008). All stem cells have the ability to reform and support a complete tissue from a single cell and the ability to self-renew. A number of different stem cell types may be involved.

So-called adult stem cells are found among differentiated cells in tissues or organs but are more plastic and can renew themselves and differentiate into some or all of the specialized cell types (NIH 2012). Hematopoietic stem cells (HSCs) in the bone marrow were discovered in the 1950s. Mesenchymal stromal cells (MSCs) have received the most recent attention (Lalu et al. 2012). Because adult stem cells, relative to embryonic stem cells, have been relatively uncontroversial, there is a tendency to assume that adult stem cells are well understood. However, there is little experience with in vivo processes of self-renewal, differentiation, and function and how the context of their use may affect their function (NIH 2012).

Embryonic stem cells (hESCs) are totipotent, meaning that they can differentiate into any cell type. In addition, hESC cell lines are easier to create and may remain useful for longer periods and provide greater research and therapeutic benefits (NIH 2012). The behavior of hESCs may differ considerably from MSCs and other adult stem cells, and this may have implications for therapeutic uses.

Of course, hESCs must be derived from an embryo, and for now, that usually means destroying the embryo, which is ethically problematic for some people (NIH 2012). Federal funding for derivation of hESCs, the part of the process that involves destroying the embryo, is now prohibited, although federal funding can be used for cells derived using existing or nonfederal sources so long as NIH guidelines are followed (Executive Order 2009). Induced pluripotent stem (IPS) cells were created, at least partly, to avoid the ethical issues raised by hESCs. Because they were created using a viral vector and they are not yet well understood, their clinical utility has been limited although potentially important (Takahashi et al. 2007; Chin 2009).

B. What Are the Safety Issues with Regenerative Medicine?

Generally, all biological products pose additional safety issues compared to traditional drugs and devices. Unlike drugs that are made up of comparatively simple chemical molecules, biological drugs are made up of far more complex proteins, cells, or even whole organisms. Since they need to be produced in precise conditions, they introduce additional safety in manufacturing concerns. They are harder to characterize, reproduce, and sometimes may be inherently unstable.

In addition, regenerative medicine poses unique safety issues. Because at its core it involves pluripotent cells, it has safety issues that are unique to those cells. These include increased possibilities of tumorigenicity, the potential for genetic abnormalities, and additional issues of immunogenicity (Goldring et al. 2011). Many of these issues are more prevalent with hESCs and iPSCs than therapies involving MSCs, but they hypothetically exist for all stem cell types (Lalu et al. 2012). Whether the full effects of these cells can be predicted, tracked, and controlled is still often an open question. As with any new technology, the biggest safety issue may not be the known risks but the unknown risks.

II. THE REGULATION OF REGENERATIVE MEDICINE

It is possible to argue that regenerative medicine is both under- and overregulated. But, overall, stem cell technology may be one of the most regulated technologies in the world. FDA finalized comprehensive regulations for cell-based therapies in 2004. Because of the ethical issues raised by hESCs, voluntary guidelines were issued by the National Academy of Sciences. Many academic institutions that conduct stem cell research have institutional stem cell research oversight committees. The very influential International Society for Stem Cell Research (ISSCR) has issued voluntary guidelines for hESC research and for Clinical Translation of Stem Cell research. NIH has a registry for hESC lines used with federal funding. Europe recently developed an overarching legislative policy tailored to new technologies including regenerative medicine (*EC,* Regulation (EC) No. 1394). Australia, Canada, and a number of Asian countries have laws on stem cell research and clinical translation. And in the United States, many states also have laws that affect stem cell use and research. Worldwide regulation of these products is not harmonized; the same product may be regulated as a device in one country, a drug in another, and completely unregulated in another (Kemp 2006).

The most important regulator, however, is FDA, which controls the approval process for new therapeutic products in the United States. Without that approval, none of these products can be used in the United States—and the United States is still the most important market for this type of technology. In an ideal world, FDA could construct regulations that would perfectly fit the issues raised by the technology. Such regulation would effectively protect patients from the potential safety issues but allow sufficient flexibility for quicker introduction into the market as FDA and industry gain experience. This, of course, is not an ideal world. FDA's authority is constrained by the statutes from which its authority emanates, the Food Drug and Cosmetic Act (FDCA) (21 USC § 301 et seq.) and the Public Health Service Act (PHSA) (42 USC § 262 (Sec. 351) and § 264 (Sec. 361)), neither of which could have anticipated this technology. Moreover, most of this technology has developed at a time when Congress was in no hurry to expand FDA's authority.

A. How Does FDA Regulate/Propose to Regulate Regenerative Medicine?

Because of the diversity of potential products, regenerative medical products do not fit squarely within any of the three main classifications: drug, biological drug (biologic), or device. Many of these products involve a combination of features that could make them eligible for two or even three categories. FDA typically classifies such a product according to its primary mode of action (21 USC § 353(g)). Because the major characteristic of regenerative medicine is the use of pluripotent cells, most products are subject to FDA's regulations regarding "good tissue practice" and many are classified as biologic drugs or devices (21 USC § 321 (g) and (h)).

FDA did not formally regulate most tissue or cellular products until the 1990s. First, it was unclear whether tissues should be regulated as drugs or medical devices or biologics. Second, it was uncertain whether all tissue products required the full oversight and testing that the approval process for those products required (Merrill 2002). FDA's first forays into tissue regulation were with heart valve allografts, which were treated as devices. FDA then turned in a different direction, seeking comprehensive regulation of tissue intended for human transplantation. FDA proposed a tiered risk-based approach for regulation of *all* "human cellular and tissue based products" (HCT/Ps) (FDA 1997). A major goal of the proposed regulations was to more clearly answer the question of which therapies merited the full treatment of the biologic/drug/device approval process and which could be regulated under the more limited regulatory controls designed to control communicable disease (Merrill 2002). The complete HCT/P rubric became final in 2004 and applies to all regenerative medicine's clinical technologies (Current Good Tissue Practice for Human Cell, Tissues, and Cellular- and Tissue-Based Product Establishments, 21 CFR 1271).

Under the regulation, HCT/Ps are articles containing or consisting of human cells or tissues that are intended for implantation, transplantation, infusion, or transfer into a human recipient. These include cellular therapies and combination products consisting of cells/tissue with a device and/or drug. They do not include vascularized organs for transplant, whole blood or blood products, secreted or extracted human products (except semen) (21 CFR 1271). All HCT/Ps are subject to Current Good Tissue Practices and registration under Section 361 of the PHSA, but tissues that are only minimally manipulated and that are used for their normal (homologous) function in the person from whom they were obtained do not require premarket approval. Cells and tissues that are extensively manipulated, combined with non-cell/non-tissue components, or used for nonhomologous functions are regulated under Section 351 of the PHSA and the FDCA as biological drugs or devices requiring premarket approval (21 CFR 1271.20).

B. Challenges to FDA Regulation

Most of the regenerative-medicine industry has not seriously questioned FDA's authority to regulate in this manner, and most concerns have been focused on speeding the approval process, clarifying scientific standards, and harmonizing FDA regulation with global regulation (Alliance for Regenerative Medicine). However, a number of clinics using autologous MSCs for clinical treatments of diverse conditions have vehemently opposed FDA authority. These laboratories extract adult stem cells (usually from blood, marrow, or fat) from patients, select and expand them, and then implant them into the patients to treat arthritis, multiple sclerosis, and many other conditions (Berfield 2013). Some of these clinics are engaged in outright fraud (60 Minutes 2010), others appear to be more legitimate, but they share a common refusal to acknowledge FDA's authority to regulate their activities as clinical experimentation. These companies believe that they are offering a treatment and therefore a service within the normal practice of medicine. As such, they believe that they are regulated only by state law and their activity is outside FDA's authority.

In 2012, FDA moved to stop a number of these stem cell clinics from selling their treatments, claiming that the "treatments" constituted unapproved drugs. One company, Regenerative Sciences, mounted what was ultimately an unsuccessful challenge. In 2014, FDA obtained an opinion in the *Regenerative Sciences* case that provides FDA significant, but not ironclad, support for its regulatory rubric for the use of such cells (*U.S. v. Regenerative Sciences* 2014:1314). The question before the court was whether Regenerative Sciences' autologous stem cell "mixture," "Regenexx," constituted a drug or biologic product subject to FDA regulation or whether it was an intrastate method of medical practice subject only to state law. Regenerative Sciences operated only in Colorado. Regenexx involved taking MSCs from a patient's bone marrow and expanding them using growth factors from the patient's blood. The expanded cells were then combined with a drug shipped in interstate commerce and then injected into the patient at the clinic.

In a strongly worded opinion, the Court of Appeals affirmed the lower court ruling that Regenexx was an adulterated and misbranded drug under the FDCA (*U.S. v. Regenerative Sciences* 2014:1314). The court also found that both the PHSA Sec. 361 HCT/P regulations and the PHSA Sec. 351/FDCA requirements applied because Regenexx was more than "minimally manipulated" (*U.S. v. Regenerative Sciences* 2014:1322). Finally, the court held that the interstate commerce requirement was also met (*U.S. v. Regenerative Sciences* 2014:1320).

There is insufficient room here to fully analyze the claims raised in the *Regenerative Sciences* case. However, it is worthwhile to look at a few issues as they have implications for FDA's authority with this technology going forward.

1. Are Autologous Treatments Different?

There is a common presumption among proponents for minimally regulated stem cell clinics that "your own cells can't hurt you." It is certainly true that risks of infection and potential immune reactions are significantly reduced by the use of autologous cells, but it is worth remembering that the very nature of cancer is having one's own cells become the engine of destruction. And all stem cells hypothetically carry risks of increased malignancy (Goldring et al. 2011). Thus, FDA's reluctance to assume such treatments are safe without actual clinical proof seems appropriate. The real questions should be how much proof is necessary—and what standards will FDA use?

There are also some regulatory oddities that are raised by autologous cells. FDA regulates all HCT/Ps under Section 361 of the PHSA whose focus is the prevention of communicable disease. In *Regenerative Sciences*, the court was persuaded by FDA's argument that removal of the cells created the risk of contamination, which in turn carries with it a potential for communicable disease. That may not withstand scrutiny with another court. Autologous cells that are properly handled and combined only with autologous growth factors bear little risk of contamination. FDA might logically counter, however, that proper handling requirements are the purpose of the regulation. It is nonetheless possible that another court might reasonably find that autologous cells, used under the oversight of a physician, that do not meet the full FDCA definitions of drugs, should escape basic regulation under Section 361 of the PHSA as well. That would make those products largely unregulated.

2. When Is Cell Therapy a Drug?

Since the first experiments with autologous cells manipulated ex vivo and reimplanted for structural repair in the mid-1990s, companies have argued that the technique should not be regulated as a product but rather as a service (Hyman, Phelps, and McNamara 1996). FDA has never acceded to that view. Of course, just because FDA says it is so does not necessarily make it so. But the FDCA drug definition was deliberately written broadly to accommodate future scientific developments, and the courts have backed up FDA's claims that products lacking many drug characteristics may still qualify as drugs (*United States v. An Article of Drug . . . Bacto-Unidisk*, 1969:798).

"More than minimal manipulation," the standard under which a product moves from Section 361 controls to being considered a drug, is an ambiguous term. The question of what constitutes "minimal manipulation" has always been controversial. FDA defines it as changing the "biological characteristics" of the material (21 CFR 1271.3(f)). While these words do not appear in any of the relevant statutes, they can be implied from the drug and device definitions. A process that alters the relevant biological characteristics affects the "structure or function" of the cells or tissue involved (21 USC § 321 (g)(1) and 21 USC § 321 (h) (3)).

Based on FDA's 1996 discussions with industry, it appears that FDA intended this to be an evolving standard so that if experience showed that the biological characteristics of the cells were not altered by the process, they could later be considered "minimally" manipulated (FDA 1997). FDA did "downregulate" cell selection as not constituting more-than-minimal manipulation under the 1997 proposed rules (FDA 1997). However, no other significant changes have been made since then. Moreover, this standard does not allow for easy downregulation when a process may be found to change the biological characteristics of a cell or tissue but where it is still found to be safe. FDA might assert authority to regulate the product as a drug but decline enforcement authority. But the standards may not be clear.

3. Practice of Medicine

The concept that FDA does not interfere with the "practice of medicine" has become a bit of a sacred cow concerning FDA's authority to regulate drugs and therapies. FDA itself is wont to repeat it as a mantra. The doctrine has its roots in the fierce protection of professional authority that was sought by physicians throughout the twentieth century even as government was extending aid to the health care enterprise and otherwise increasing regulation of industry players (Starr 1982). Thus, physicians are primarily regulated through state licensure requirements, a process that is itself significantly under the control of physicians. For FDA specifically, the principle is focused on the physician exemption from drug and device approval requirements. Physicians are not required to seek approval for new unlabeled uses of existing approved drugs or devices that they choose to use in the treatment of their patients (*Chaney v. Heckler* 1983:1180; 21 USC § 396 (2000)). There is no doubt that this off-label use has usually allowed for better medicine and potentially lowers costs. But the practice of medicine exemption is often presented as a much broader doctrine that means FDA cannot touch any aspect of the practice of medicine, often with little analysis as to why it exists, what its limits are, and whether it is worth protecting in all situations. It is perhaps not surprising that many of the stem cell clinics are operated by surgeons. At least until recently, most innovative surgery did not involve drugs and devices and was therefore conducted without FDA oversight—and is often conducted without *any* oversight (Reitsma 2005).

In any event, as the court found in *Regenerative Sciences*, the question here is actually easier (*U.S. v. Regenerative Sciences* 2014:1319). The approval requirements of the FDCA have always represented a limit on the practice of medicine. Although it is often appropriate for physicians to prescribe approved drugs off label, it is never acceptable for physicians to prescribe unapproved drugs. Therefore, the first question is not whether FDA's requirements limit the practice of medicine but rather whether the "treatment" constitutes a new drug or biological drug. Since the court answered that in the affirmative, it found that FDA has the authority to limit the practice of medicine to that extent (*U.S. v. Regenerative Sciences* 2014:1319).

4. Commerce Clause

One of the central questions posed by *Regenerative Sciences* was whether FDA had jurisdiction to regulate "the mixture" since almost all of the activity occurred intrastate. The Court of Appeals applied a traditional interpretation of the Commerce Clause, finding that since such cell therapy substantially affects commerce (*U.S. v. Regenerative Sciences* 2014:1320; citing *Wickard v. Filburn* 1942), there was a sufficient nexus with interstate commerce. That determination is somewhat controversial. First, the FDCA became law before the more expansive interpretation of *Wickard* was decided. Second, in the Supreme Court's Obamacare decision (*NFIB v. Sebelius* 2012:2566), four of the justices seemed predisposed to a narrower interpretation of the breadth of the commerce clause. In *NFIB*, however, Justice Roberts affirmed that Congress's power to regulate interstate commerce includes intrastate activity that substantially affects interstate commerce and also affirmed the continued vitality of the Court's opinion in *Raich* (*NFIB v. Sebelius* 2012:2589–93; *Gonzales v. Raich* 2005:1). The FDCA's power to regulate the interstate market for stem cell and other therapeutic products would be substantially undercut if it could not regulate products that stem cell clinics offer for sale in a solely intrastate market. This would seem also support the court's determination in *Regenerative Sciences*; it is also possible though for another court to interpret it quite differently. And we of course don't know what the Supreme Court would do in this context.

Under the facts of *Regenerative Sciences*, the Court of Appeals' Commerce Clause determination was significantly bolstered by the fact that a component of the mixture, doxycycline, was shipped in interstate commerce (*U.S. v. Regenerative Sciences* 2014:1320–21). It is likely, however, that many autologous cell products could be produced without a component shipped in interstate commerce. If a narrower interpretation of Commerce Clause authority were to be imposed, many of those products might escape FDA regulation under the FDCA. FDA might still retain authority under the PHSA because the focus of that statute is on the prevention and control of communicable disease. Because of that focus, the PHSA gives FDA even broader Commerce Clause authority than does the FDCA, and arguably supersedes conflicting state powers, and allows even more expansive incursions into the traditional spheres of the practice of medicine. But it also has limitations in that the statute arguably does not confer authority related to efficacy of a product, just safety.

III. WHAT SHOULD FDA DO GOING FORWARD?

As the foregoing discussion demonstrates, the D.C. Circuit, certainly an influential court, has found that FDA has authority for the positions it takes on the regulation of regenerative medicine generally and clinics offering autologous MSC treatments specifically. This judicial support for the HCT/P rubric, a creative but certainly not statutorily mandated regulatory scheme, is extremely valuable for FDA going forward. It means that FDA's enforcement efforts in requiring compliance with its regulations have real teeth. This should aid the agency's efforts to eliminate potentially unsafe and unproven stem cell treatments in the United States.

But FDA's regulations may be so onerous that the result is the "off-shoring" of many of these treatments. They will still be offered to patients, just not in the United States. Several of these companies have left the United States and there is considerable "stem cell tourism" involving these types of technologies (Cyronoski 2013). Indeed, Regenexx, the expanded stem cell product at issue in the *Regenerative Sciences* case, is now offered only in the Cayman Islands. Ironically, in demanding a more rigorous pathway to market, FDA may actually be exacerbating potential safety issues since these offshore products are subject to little significant regulation. Although some patients will not travel for the treatments, many will. Moreover, it is unlikely that that is the end of the story; the issue has become politicized. There has been considerable backlash against the perception that FDA is overstepping its authority in regulating these stem cell procedures. Even

Texas Governor Perry questioned FDA's authority to regulate these companies and promised to make Texas "the stem cell capital of the United States" (Berfield 2013).

It also means that there may be significant delays in bringing legitimately less risky stem cell products to market. FDA should find some "middle way" to speed and shortcut pathways for certain stem cell therapies even if it has the authority to demand full clinical study. When FDA chose not to treat HCT/Ps as devices but rather as drugs, it gained broader and more specific oversight authority. But it lost flexibility. There is no abbreviated pathway like the device 510(k) notification pathway for drugs. In addition, unlike devices, there is no clear pathway to "down classify" stem cell products to less risky classifications as the agency and industry gain experience. It is not just the stem cell clinics that are complaining. There is evidence that the costs of large clinical trials are a roadblock to much clinical translation of many of the technologies that are part of regenerative medicine (*Nature* [editorial] 2013). This is so widespread that the ISSCR guidelines offer a "responsible innovation" approach as an alternative to clinical research (ISSCR 2008). Proponents of that approach argue that such innovation constitutes a "middle way" between research and treatment appropriate for a small number of seriously ill patients (ISSCR 2008). Under certain circumstances, innovative but not-yet-approved treatments would be available for patients who have no alternatives, especially where funding opportunities for traditional clinical trials is insufficient. However, such an approach has some considerable risks in that it may lack appropriate oversight and may also expose large numbers of patients to ineffective or even dangerous therapy (Sugarman 2012). At a minimum, FDA would need to determine what level of preliminary safety and efficacy standards is appropriate. It is not clear that this pathway in practice would be much different from some of the fast-track pathways that are already available (*Nature* (editorial) 2013).

The best way may be to alter the "minimal manipulation" and homologous use standards. Those standards are easy to apply but may be overinclusive. There may be situations where the biological characteristics of the cells or tissues are altered but their behavior is predictable, safe, and effective. FDA needs to seek alternatives and find ways to achieve the downregulation for certain therapies that was envisioned in the 1990s. The current body of science may be insufficient for FDA to establish alternative standards now. FDA missed an opportunity to gather useful data from the rampant use of stem cells in animals by failing to provide guidance until that use was widespread. There are few reports of injuries in such animals but little data that supports efficacy (Cyronoski 2013a). Industry and FDA also need to partner more effectively to build that data in humans. Industry has had little incentive to publish studies of techniques that have not succeeded, but that knowledge is crucial to developing alternative standards.

IV. FUTURE ISSUES

While it is by no means a perfect fit, the statutory structure of the FDCA and PHSA has provided sufficient authority for FDA to regulate regenerative medicine. But that may end as the technologies evolve. FDA is ill suited to make determinations that go beyond safety and efficacy but instead focus on whether it should be ethically permissible to do something. This has been a question lurking in the background in both gene therapy and potential technologies like human cloning and so-called "three-parent" embryos. FDA's authority to regulate those areas is certainly questionable (Merrill 2002). Thus far, ethical questions are on the periphery of FDA's regulation of regenerative medicine. Nevertheless, FDA could find itself grappling with questions of ethics in the not-too-distant future. So far, the field of regenerative medicine is characterized by replacing diseased or damaged tissue. But the same technologies may be used for enhancement as well as replacement. The decision about whether such enhancement is ethically acceptable is outside the statutory safety and efficacy rubric in which FDA functions. FDA may be wise to avoid wading into that debate.

V. CONCLUSION

Regenerative medicine is a field that holds great promise but will require significant investment and creativity—more than scientific creativity—to succeed. It will also require significant luck in avoiding political quagmires that could hamstring the development of the technology. FDA's role will be essential. It must try to develop responsive and nimble regulation, thus helping the technology—and itself—avoid some of those political pitfalls.

REFERENCES

60 Minutes. 2010. "21st Century Snake Oil." http://www.cbsnews.com/news/21st-century-snake-oil-16-04-2010/.

Alliance for Regenerative Medicine, http://alliancerm.org/industry.

Berfield, Susan. 2013. "Stem Cell Showdown: Celltex vs. the FDA." *Business-Week* (January 3).

Chaney v. Heckler, 718 F.2d 744 (D.C. Cir. 1983).

Chin, Mark H., Mike J. Mason, Wei Xie, Stefano Volinia, Mike Singer, Cory Peterson, et al. 2009. "Induced Pluripotent Stem Cells and Embryonic Stem Cells are Distinguished by Gene Expression Signatures." *Cell Stem Cell* 5:111–23.

Cyronoski, David. 2013. "Cowboy Culture." *Nature* 494:166.

Cyronoski, David. 2013a. "Stem Cells Boom in Vet Clinics." *Nature* 496:148.

Editorial. 2013. "Preventive Therapy." *Nature* 494:147.

Executive Order 13505 of March 9, 2009. Removing Barriers to Responsible Scientific Research Involving Human Stem Cells. *Federal Register*, Vol. 74, No. 46 (2001):10667–68. http://www.gpo.gov/fdsys/pkg/FR-2009-03-11/pdf/E9-5441.pdf.

FDA, Proposed Approach to Regulation of Cellular and Tissue-based Products, February 28, 1997 (announced but not published in the *Federal Register*); available at http://www.fda.gov/downloads/BiologicsBloodVaccines/GuidanceComplianceRegulatoryInformation/Guidances/Tissue/UCM062601.pdf.

Goldring, Chris E. P., Paul A. Duffy, Nissim Benvenisty, Peter W. Andrews, Uri Ben-David, Rowena Eakins, et al. 2011. "Assessing the Safety of Stem Cell Therapeutics." *Cell Stem Cell* 8:618–28.

Gonzales v. Raich, 545 U.S. 1, 17 (2005).

Hyman, Phelps & McNamara, P.C. letter to Dockets Mgmt. Branch, FDA (Feb. 16, 1996), available at http://www.regulations.gov/#!documentDetail;D=FDA-1995-N-0200-0026 (commenting on Products Comprised of Living Autologous Cells Manipulated ex vivo and Intended for Implantation for Structural Repair or Reconstruction, FDA Docket No. 95N-0200, 60 Fed. Reg. 36808 (July 18, 1995)).

International Society for Stem Cell Research (ISSCR). 2008. "Guidelines for the Clinical Translation of Stem Cells." (December 3). http://www.isscr.org/docs/default-source/clin-trans-guidelines/isscrglclinicaltrans.pdf.

Kemp, Paul. 2006. "History of Regenerative Medicine: Looking Backwards to Move Forwards." *Regenerative Medicine* 1:653–69.

Lalu, M. Manoj, Lauralyn McIntyre, Christina Pugliese, Dean Fergusson, Brent W. Winston, John C. Marshall, et al. 2012. "Safety of Cell Therapy with Mesenchymal Stromal Cells (SafeCell): A Systematic Review and Meta-Analysis of Clinical Trials." *PLOS One* 7:347559.

Merrill, Richard A. 2002. "Human Tissues and Reproductive Cloning: New Technologies Challenge FDA." *Houston Journal of Health Law & Policy* 3:1–82.

National Federation of Independent Business v. Sebelius, 132 S. Ct. 603 (2012).

National Institutes of Health (NIH), U.S. Department of Health and Human Services. 2012. The National Institutes of Health resource for stem cell research: Stem Cell Basics. http://stemcells.nih.gov/info/basics/defaultpage.

Orlando, Guiseppe, Kathryn J. Wood, Robert J. Stratta, James J. Yoo, Anthony Atala, and Shay Soker. 2011. "Regenerative Medicine and Organ Transplantation: Past, Present and Future." *Transplantation* 91:1310–17.

Reitsma, Angela and Jonathan Moreno. 2005. "Ethics of Innovative Surgery: US Surgeons' Definitions, Knowledge and Attitudes." *Journal of American College of Surgery* 200:103–10.

Starr, Paul. 1982. *The Social Transformation of American Medicine*. New York: Basic Books.

Sugarman, Jeremy. 2012. "Questions Concerning the Clinical Translations of Cell-Based Interventions Under an Innovation Pathway." *Journal of Law Medicine & Ethics* 40:945.

Takahashi, Kazutoshi, Koji Tanabe, Mari Ohnuki, Megumi Narita, Tomoko Ichisaka, Kiichiro Tomoda, and Shinya Yamanaka. 2007. "Induction of Pluripotent Stem Cells from Adult Human Fibroblasts by Defined Factors." *Cell* 131(5):861–72.

Tigerstrom, Barbara J. 2008. "The Challenges of Regulating Stem Cell-Based Products." *Trends in Biotechnology* 26:653–58.

United States v. An Article of Drug . . . Bacto-Unidisk, 394 U.S. 784, 798 (1969).

United States v. Regenerative Sciences 741 F. 3d 1314 (D.C. Cir. 2014).

Wickard v. Filburn, 317 U.S. 111, 128–29 (1942).

CHAPTER TWENTY-FIVE

Device-ive Maneuvers

FDA's Risk Assessment of Bifurcated Direct-to-Consumer Genetic Testing

ELIZABETH R. PIKE AND KAYTE SPECTOR-BAGDADY

THE 2007 launches of 23andMe and deCODEme heralded the beginning of direct-to-consumer (DTC) access to personalized genetic medical information (Kalf et al. 2014). Although DTC companies originally provided targeted testing for discrete disorders such as Huntington's disease, DTC companies now offer large-scale and whole genome (genomic) sequencing that provide information about genetic predispositions to disease and pharmacogenomic responses to particular drugs. This broader inquiry raises novel concerns about how best to regulate the collected information.

The U.S. Food and Drug Administration (FDA) began direct engagement with the DTC genetic testing industry in 2010, issuing 23 "Untitled Letters" to companies that suggested that DTC offerings were medical devices that needed to be registered, cleared, or approved by FDA (2014b). In response, many companies began requiring a physician intermediary while some exited the market entirely. In 2013, FDA sent a more formal "Warning Letter" to industry leader 23andMe detailing its violations of the Federal Food, Drug, and Cosmetic Act and demanding that 23andMe cease marketing its health-related genomic information (Gutierrez to Wojcicki 2013).

In the wake of FDA's enforcement actions, some DTC companies bifurcated into entities that produce a *data*-only product—a file of As, Ts, Cs, and Gs without interpretation—and entities that produce an *information*-only product, interpreting data to produce analysis regarding genetic predispositions such as to particular diseases. Because data-only offerings fall outside FDA's definition of a device, they are unlikely to be regulated by FDA. Producing genomic information, however, falls squarely within FDA's definition of a device and raises novel questions about how this product should be regulated.

This chapter describes the challenges that FDA faces in regulating genomic information as a medical device, including classifying these products according to levels of risk and implementing standard risk-mitigation strategies. Ensuring the safety and effectiveness of genomic information requires ensuring analytic and clinical validity, as well as minimizing risks of disclosure. This chapter proposes a path forward for satisfying these requirements.

I. EVOLUTION OF DIRECT-TO-CONSUMER GENETIC TESTING

Since the start of the Human Genome Project in 1990, knowledge of the relationship between the human genome and physical traits and predispositions to diseases has greatly expanded. In 1990, genetic tests were available for only 100 genes specifically associated with disease (Brower 2010). By April 2012, there were tests for 2,600 disease-causing variants (Palmer 2012).

Although once the exclusive province of clinicians and researchers, companies increasingly offer genetic testing directly to consumers without clinician involvement (Hudson et al. 2007). Comparatively inexpensive DTC tests predict carrier status for diseases like cystic fibrosis; the probability of developing disease in the future, such as breast cancer or Alzheimer's disease; the ability to metabolize certain drugs, such as warfarin (a blood thinner); or traits as mundane as excessive earwax (Bair 2012; Beaudet and Javitt 2010; McGuire et al. 2010). Increasingly, however, DTC companies are moving from discrete genetic tests like the examples above to larger-scale genomic sequencing. In 2008, for example, Knome first offered DTC genomic sequencing for $350,000 per person (Singer 2008).[1] The cost of genomic sequencing has since come down considerably with Illumina announcing the first "$1,000 genome" in January 2014 (Winnick 2014).

Although the expanded scope of DTC services offers benefits to consumers—potentially including greater autonomy and empowerment (Brower 2010; Frueh et al. 2011; Hudson et al. 2007)—expanded DTC access also raises concerns. These concerns include that consumers might get tested "without adequate context or counseling, will receive tests from laboratories of dubious quality, and will be misled by unproven claims of benefit" (Hudson et al. 2007:1392). Moreover, consumers might make "unwarranted, and even irrevocable, decisions on the basis of test results and associated information, such as the decision to terminate a pregnancy, to forgo needed treatment, or to pursue unproven therapies" (Hudson et al. 2007:1393-4).

Some of these concerns have merit. Reviews conducted by the U.S. Government and Accountability Office (GAO) in 2006 (GAO 2006) and 2010 (GAO 2010) raised serious questions about the accuracy and reliability of information being returned to consumers. A 2008 report by the Secretary's Advisory Committee on Genetics, Health, and Society (SACGHS) also noted gaps in the regulatory oversight of these products (SACGHS 2008). Addressing some of these concerns requires additional regulatory oversight by FDA.

II. CURRENT REGULATION OF DIRECT-TO-CONSUMER GENETIC TESTING

FDA protects public health by assuring the "safety, effectiveness, quality, and security" of medical interventions such as drugs and medical devices (FDA 2014). FDA defines medical devices as any "instrument, apparatus, implement, [or] machine" that is "intended for use in the diagnosis of disease or other conditions, or in the cure, mitigation, treatment, or prevention of disease . . ." (Federal Food, Drug, and Cosmetic Act (FDCA), 21 USC § 321h). Medical devices that pose risk to patients or consumers generally must be cleared or approved by FDA before they can be marketed or distributed (FDCA, 21 USC § 360e; 21 USC § 807.81). For potential violations, FDA often begins enforcement with either a Warning Letter, which highlights violations that may lead to enforcement action such as a recall or seizure of products, or an Untitled Letter for less significant violations (FDA 2014c).

In 2010, FDA sent 23 Untitled Letters to companies in the DTC genetic-testing industry, notifying them that their products were considered medical devices and therefore needed to be classified and registered, cleared, or approved, as appropriate (Spector-Bagdady and Pike 2014). For example, the letter to deCODE Genetics argued that despite deCODE never having submitted any analytic or clinical validity data for its tests, "your website states that the deCODEme Complete Scan identified twelve common genetic variants and provides interpretation of their associated risk for developing breast cancer . . . [and that] consumers may make medical decisions in reliance on this information" (Gutierrez to Collier 2010). These Untitled Letters resulted in many recipients exiting the DTC market. One exception, 23andMe, responded by beginning the FDA clearance process for seven of its then-254 tests (Perrone 2012).

But on November 22, 2013, FDA issued a Warning Letter to 23andMe noting that it "failed to provide adequate information" in support of its devices and was therefore ordered to "immediately discontinue marketing" its product. In addition, FDA stated that the entire personal genome service would be considered a Class III device—warranting the highest level of regulatory oversight (Gutierrez to Wojcicki 2013). As a result, 23andMe began selling individuals their genetic data (the file of As, Ts, Cs, and Gs) and ancestry information but stopped providing health-related genetic information ("Status of our health-related genetic reports"). In February 2015, FDA authorized 23andMe to market a DTC test to determine whether a consumer is a carrier for Bloom Syndrome, a rare autosomal recessive disorder. In its press release, FDA also announced that it intends to exempt similar carrier screening tests from premarket review in the future (FDA 2015). In response, 23andMe noted that it would not immediately begin returning Bloom carrier status testing results or other health results, instead preferring to wait until it could offer a more comprehensive product under the new FDA guidance ("23andMe Granted Authorization by FDA to Market First Direct-to-Consumer Genetic Test Under Regulatory Pathway for Novel Devices.").

III. BIFURCATED DIRECT-TO-CONSUMER GENETIC TESTING ENTITIES

In response to the changing regulatory landscape, some DTC companies bifurcated into entities that produce either a data-only or information-only product (Vorhaus 2012). For example, in November 2012, Gene By Gene announced "DNA DTC" ("23andMe, Inc. Provides Update on FDA Regulatory Review"; "Gene By Gene Launches DNA DTC"), a DTC sequencing service that provides raw genomic data without interpretation. DNA DTC marked the first "truly direct-to-consumer" offering of such a data-only product (Vorhaus 2012).

Other companies have made the parallel offering of information-only services. For example, openSNP, an open-source project, allows consumers to input raw genomic data and receive medical information; openSNP is compatible with raw data from 23andMe, deCODEme, or FamilyTreeDNA ("Welcome to openSNP"). Promethease offers another free basic report and an enhanced report for five dollars ("Promethease"; "Promethease/Features").

FDA's approach to regulating these bifurcated entities is still somewhat unsettled. For data-only products—the files of As, Ts, Cs, and Gs—FDA's approach is straightforward: because genomic data does not itself diagnose, cure, mitigate, treat, or prevent disease, to the extent it is marketed and sold as a stand-alone product, it falls outside FDA's definition of a medical device and therefore outside its regulatory jurisdiction (Carmichael 2010). Officials at FDA have acknowledged that for companies that produce a data-only product, "[i]f they don't make any medical claims about that data, then they're free to provide information [without approval] as far as we're concerned" (Carmichael 2010).

To be used in the diagnosis, cure, mitigation, treatment, or prevention of disease, data-only products require a parallel information offering—which *is* of concern to FDA. Indeed, FDA sent several 2010 Untitled Letters to "software program[s] that analyz[e] genetic test results created by an external laboratory" (FDA 2014b) because even though these companies were not producing genetic data from biological samples, they were producing the medical information marketed for the diagnosis, cure, mitigation, treatment, or prevention of disease. This reasoning would similarly apply to information-only products such as Promethease (Palmer 2012).

Attempting to regulate information-only services raises novel questions about FDA's regulatory authority that are beyond the scope of this chapter but have been discussed elsewhere (Spector-Bagdady and Pike 2014).[2] The remainder of this article therefore focuses on the challenges FDA faces in regulating genomic information as a medical device, in particular, the challenges of classifying genomic information on the basis of perceived risk and the limited applicability of standard risk-mitigation strategies.

IV. REGULATING GENOMIC INFORMATION AS A MEDICAL DEVICE

As discussed previously, a medical device is a product "intended for use in the diagnosis of disease or other conditions, or in the cure, mitigation, treatment, or prevention of disease" (Federal Food, Drug, and Cosmetic Act (FDCA), 21 USC § 321h). Because data-only products that return the list of As, Ts, Cs, and Gs do not themselves cure, mitigate, treat, or prevent disease, they generally fall outside FDA's definition of a device. Genomic information, by contrast, *can* be used in the diagnosis, mitigation, or perhaps even prevention of disease, and so the process of producing it falls squarely within FDA's jurisdiction. Regulating genomic information as a medical device, however, poses serious challenges for FDA. In particular, FDA will have difficulty classifying genomic information on the basis of perceived risk and successfully implementing common risk mitigation strategies.

A. Risk Classification of Genomic Information

Risk classification is unwieldy as applied to bifurcated genomic information. FDA classifies medical devices into three categories—Class I, II, and III—on the basis of risk to the consumer. The greater the risk to the consumer, and the more control presumed necessary to assure the safety and effectiveness of the device, the higher the classification a device receives (FDA 2012).

Class I devices are assumed to pose low to moderate risk and are subject to "General Controls," including registration and labeling requirements. Devices are classified as Class II if general controls alone are perceived to be "insufficient to provide reasonable assurance of . . . safety and effectiveness" (21 CFR § 860.3(c)). Class II devices are subject to both premarket notification and "Special Controls," including stricter labeling requirements and/or postmarket surveillance (FDA 2012; FDA 2013). If there is insufficient evidence that Special Controls provide assurance of safety and effectiveness, FDA will consider a device Class III, thereby requiring rigorous "premarket approval," including "sufficient valid scientific evidence to assure that the device is safe and effective for its intended use(s)" (21 CFR § 860.3(c); FDA 2012a). If FDA is particularly concerned about a device, the Secretary of Health and Human Services may restrict a Class III device to sale, distribution, or use "only upon written or oral authorization of a practitioner licensed by law to administer or use such device" (FDCA, 21 USC § 360j(e)(1)).

Assessing the risks posed by genomic information is challenging because there are a number of different types of risks to consider. In addition to the risk inherent to the potential underlying disease, risks also include that a sample might be inaccurately sequenced (thereby lacking analytic validity); the sequence might be inaccurately interpreted (thereby lacking clinical validity); the interpretation might be inaccurately communicated; and that recipients might take actions in response to the information that are detrimental to their health (Gniady 2008). In the 1980s, similar concerns arose regarding DTC HIV tests. FDA initially concluded that the risks of DTC HIV tests were so great that they would be approved only for use by professionals within a health care environment and with appropriate counseling (54 FR 7279). FDA later concluded that DTC HIV tests "may be approvable" provided that all technical aspects of the test were validated, that instructions and educational materials were comprehensible to a lay audience, that adequate pre- and post-test counseling were provided, and that consumers would be able to take appropriate follow-up action (60 FR10087).

Like DTC HIV tests, one risk of genomic information stems from actions a consumer might take in response to the information received (Gniady 2008). Controlling genomic information on the basis of potential actions is unusual, although not unprecedented. With DTC HIV tests, the concern was that positive results would trigger widespread suicide and panic, which could not have been mediated by clinical professionals who had previously been the bearers of such news (Gniady 2008). By contrast, the most concerning medical actions discussed in the context of DTC genomic information—such as a mastectomy upon learning about a predisposition to breast cancer—require clinician involvement (Frueh et al. 2011; Palmer 2012). Some responses, however, such as a decision to self-medicate or dose in response to pharmacogenomic predispositions, could fall outside the influence of clinicians thereby raising similar concerns (Gutierrez to Wojcicki 2013).

Assessing the risk of returning genomic information is also more complex than assessing the risk of DTC HIV testing because many different types of results can arise from one test. In 2010, an FDA official stated that a test for a benign trait, such as baldness, would be considered Class I—if considered a device at all (Rubin 2010). Meanwhile, 23andMe's Warning Letter stated that "FDA has spent significant time evaluating . . . whether certain uses might be appropriately classified into [C]lass II" before ultimately concluding that, without additional information, FDA was going to consider 23andMe's entire product Class III (Gutierrez to Wojcicki 2013).

For information-only products that return a range of assessments from breast cancer to earwax type, assessing the level of perceived risk is difficult. FDA could classify an entire information-only product on the basis of the riskiest piece of information returned—a "weakest-link" approach to classification. In this scenario, information-only products that return genomic information about both baldness *and* BRCA mutations could potentially be considered Class III. With this approach, if at least one Class III piece of information is returned, the entire information-only product would be considered Class III.

Alternatively, FDA could assign each type of genomic information being returned to a class. Bundled products would therefore require independent assessments of each variant being returned. For example, information about BRCA mutations could be Class III (with its attendant premarket approval requirements and risk mitigation strategies), while information provided by the same device about baldness could be considered Class I. But an approach requiring individualized assessments of each potential piece of genomic information being returned—including information about analytic and clinical validity, along with information about potential responses to this information—would be time consuming to implement (Beaudet and Javitt 2010) and might result in a stark disparity between the most advanced genetic information available clinically versus DTC.

B. Risk Mitigation Strategies for Genomic Information

Should genomic information be classified as a Class II or Class III device, as appears likely, FDA faces additional challenges in implementing risk mitigation strategies sufficient to ensure its safety and effectiveness.

For Class II devices, pertinent special controls include labeling and postmarket review. If genomic information is classified as Class III, FDA can implement "[s]uch other requirements as FDA determines are necessary . . ." (21 CFR § 814.82(a)(9)), which may include counseling or, if a product is a restricted device, mandated clinician involvement. But each of these controls or conditions has particular challenges when it comes to providing additional assurances of the safety and effectiveness of genomic information.

One control FDA has over regulated medical devices is restrictions regarding the device's "labeling," which cannot be "false or misleading in any particular" (FDA 2009). Moreover, the "labeling must bear adequate directions for use and any warnings needed to ensure the safe and effective [ ] use of the device" (FDA 2009). In vitro diagnostics—a category of devices into which DTC genetic testing falls—must include "any warnings . . . appropriate to the hazard presented by the product" (21 CFR § 809.10(a)(4)). This special control seems particularly challenging to implement for information-only products, in part, because it would be impossible to fully disclose the risks of all possible results being returned given the large number of potentially returnable results and the rapidity with which the science is advancing (when combined with the time-consuming nature of the clearance/approval process). The alternative—making a general disclosure about the range of risks that can arise from any of the tests—might be too general to be effective.

Should genomic information be classified as at least a Class II medical device, FDA can mandate postmarket review, generally used when premarket review cannot adequately assure FDA of safety and effectiveness (FDA 2012; 2013). However it would be difficult to craft postmarket review that provides additional assurances of safety and effectiveness. This is true in part because the effectiveness of genomic information is difficult to measure (Palmer 2012); recent research has, however, found that "neither the health benefits envisioned by DTC-[genetic testing] proponents . . . nor the worst fears expressed by its critics . . . have materialized to date" (Roberts and Ostergren 2013:182).

Should genomic information be classified as a Class III device, FDA could require counseling for tests that "'have the potential to cause distress'" (Brower 2010:1612). FDA took this approach for DTC home HIV tests, which were finally approved as a Class III device requiring anonymous over-the-phone counseling (Palmer 2012). Some DTC genetic companies already offer counseling, but this service is generally available only after an individual purchases the service. Counseling therefore might be unavailable to assist in the decision of whether to purchase testing in the first place (Bair 2012). A requirement that DTC companies provide counseling would likely make DTC genetic testing more expensive and time consuming—both of which will limit its accessibility. And consumers might balk at mandated counseling, particularly if they choose DTC services specifically to learn medical information independently from clinical professionals.

FDA could also determine that some information—for example, mutations associated with risk for breast cancer or Huntington's disease status—has such potential to cause distress that it should be a restricted Class III device, which must be prescribed through a clinician intermediary. For example, in its Warning Letter to 23andMe, FDA pointed out that some of its concerns were "typically mitigated by . . . management under a physician's care" (Gutierrez to Wojcicki 2013). A requirement that clinicians be involved in the communication of genomic information would effectively end the DTC genetic testing industry. It is not clear, however, that clinician intermediaries make the communication of genomic information more effective or resulting patient response more beneficial to their health. Genomic interpretation is complicated, and most clinicians lack the training to feel comfortable returning genomic information (Beaudet and Javitt 2010; Mardis 2010; Palmer 2012). Moreover, a clinician's involvement is often already required for the most concerning types of therapies resulting from genetic test results. For example, while FDA has cited unfounded prophylactic mastectomies as a driving concern for regulating the DTC genetic testing industry (Gutierrez to Wojcicki 2013), such surgical intervention would require physician involvement (Palmer 2012).

C. A Path Forward for Regulating Genomic Information

Three steps are necessary to ensure the safety and effectiveness of DTC genomic information. First, the underlying data must be analytically valid—that is, the genomic data sequence must be accurate and precise. Second, the information must be clinically valid—the findings must be causally associated with clinical outcomes. And third, the risks of disclosing the genomic information must be minimized. FDA's ability to effectively regulate genomic information hinges upon the approach taken to each of these challenges.

The first step, ensuring the analytic validity of the underlying data, applies primarily to data-producing products. Although not directly applicable to information-only products, analytic validity is a necessary predicate to ensuring that information-only products are safe and effective. FDA could rely on the Clinical Laboratory Improvements Amendments (CLIA) of 1988, which help "ensure the accuracy, reliability, and timeliness of patient test results . . ." (CLIA). Some DTC genetic testing companies, such as 23andMe, already produce data in a CLIA-certified laboratory. If genomic information is classified as a Class III device, FDA could require as a condition of premarket approval that information-only products rely on data sequenced in a CLIA-compliant manner.

The second step, ensuring clinical validity, requires regular review of evolving scientific literature. Reviewing the scientific literature and assessing the clinical validity of hundreds of variants being returned for each submission is incredibly labor intensive and, in a fast-moving scientific field, is likely to evolve as rapidly as the review can progress. Companies seeking approval can shoulder the burden of providing assurance of clinical validity, but FDA would still ultimately have to clear or approve these claims. FDA could therefore focus on classifying and clearing tests of particular concern—including those for breast and ovarian cancer, Huntington's disease, Alzheimer's disease, and critical drug responses—while leaving less-risky interpretation accessible DTC in the interim—ensuring continued access to this more innocuous information.[3]

The third step, ensuring the safety and effectiveness of genomic information by minimizing the risks of disclosure, can be accomplished through the use of special controls and conditions of premarket approval. Although several possible controls become conceptually unwieldy when applied to DTC genomic information (e.g., seizure of product), effective labeling requirements could help ensure the safety and effectiveness of information-only products. Even if labeling requirements fall short in ensuring safety and effectiveness, there is some reassurance in the fact that many of the most concerning actions taken in response to genomic information require clinician involvement, providing an additional safety mechanism.

V. CONCLUSION

FDA's recent engagement with DTC genetic testing companies has brought major changes to this nascent industry. A result of this uncertain regulatory environment is bifurcated DTC genetic testing entities: entities that provide either a data-only product—a file of As, Ts, Cs, and Gs without any interpretation—or an information-only product, interpreting this genomic data and providing information about genetic risks and pharmacogenomic interactions. Although data-only products generally fall outside FDA's regulatory purview, regulating genomic *information* as a medical device will pose serious challenges.

FDA's device-regulation structure is based on perceived risk, an assessment that is particularly difficult for the wide array of genomic information that can be returned. In addition, typical risk mitigation strategies—including labeling, postmarket review, counseling, and mandated clinician involvement—will be more complicated to implement for information-only products.

Ensuring the safety and effectiveness of genomic information as a medical device requires ensuring its analytic and clinical validity and minimizing risks of disclosure. To do so, FDA could rely on CLIA to ensure that genomic information is based on analytically valid genomic data. FDA could also focus on validating tests for the riskiest types of information. FDA could minimize the risks of disclosure through labeling requirements while also being reassured that the most concerning therapeutic actions taken in response to genomic information often require clinician involvement. This approach allows continued access to important medical information while enabling FDA to meet its public health mission of ensuring the safety and effectiveness of this new technology.

NOTES

The findings and conclusions in this chapter are those of the authors and do not necessarily represent the official position of the Presidential Commission for the Study of Bioethical Issues or the Department of Health and Human Services. The use of trade names and commercial sources in this report is for identification only and does not imply endorsement by the Presidential Commission for the Study of Bioethical Issues or the Department of Health and Human Services. The authors would like to thank Lisa M. Lee and Paul Lombardo for their thoughtful review of this article and Tenny R. Zhang for her editorial assistance.

1. Knome no longer offers this service. In 2010, Knome changed its model and currently markets to medical institutions, researchers, and pharmaceutical companies (Palmer 2012).

2. For example, FDA's ability to regulate the labeling of medical devices derives from the government's ability to regulate commercial speech, but some of the newest genomic interpretation platforms neither sell their services nor advertise their websites. Without any commercial transaction, it is not clear that FDA has the authority to regulate these services. Moreover, unlike other more typical devices, the device here is entirely speech—or information. These services therefore touch upon FDA's ability to regulate true speech, an issue that has recently made its way through the courts (Spector-Bagdady and Pike 2014).

3. This argument was originally made in Spector-Bagdady, K. and E. R. Pike. 2014. "Consuming Genomics: Regulating Direct-to-Consumer Genetic and Genomic Information." *Nebraska Law Review* 92:677–745, but this article offers an expanded analysis and assessment of risk classification and risk mitigation for genomic information.

REFERENCES

23andMe. "Status of our health-related genetic reports." Last accessed April 22, 2014. https://www.23andme.com/health/.

23andMe. 2013. "23andMe, Inc. Provides Update on FDA Regulatory Review." Last modified December 5, 2013. http://mediacenter.23andme.com/press-releases/23andme-inc-provides-update-on-fda-regulatory-review/.

23andMe. 2015. "23andMe Granted Authorization by FDA to Market First Direct-to-Consumer Genetic Test Under Regulatory Pathway for Novel Devices." February 19, 2015. http://mediacenter.23andme.com/blog/2015/02/19/fdabloomupdate/.

Bair, S. 2012. "Direct-to-Consumer Genetic Testing: Learning from the Past and Looking Toward the Future." *Food and Drug Law Journal* 67:413–33.

Beaudet, A. L. and G. Javitt. 2010. "Which Way for Genetic-Test Regulation?" *Nature* 466:816–8.

Brower, V. 2010. "FDA to Regulate Direct-to-Consumer Genetic Tests." *Journal of the National Cancer Institute* 102:1610–12, 1617.

Carmichael, Mary. 2010. "DNA Dilemma: The Full Interview with the FDA on DTC Genetic Tests." *Newsweek*. (August 17) http://www.newsweek.com/dna-dilemma-full-interview-fda-dtc-genetic-tests-222812.

Centers for Medicare and Medicaid Services. "Clinical Laboratory Improvement Amendments (CLIA)." Last accessed April 23, 2014. https://www.cms.gov/Outreach-and-Education/Medicare-Learning-Network-MLN/MLNProducts/downloads/CLIABrochure.pdf.

Frueh, F. W. et al. 2011. "The Future of Direct-to-Consumer Clinical Genetic Tests." *Nature Review Genetics* 12:511–15.

Gene By Gene. "Gene By Gene Launches DNA DTC." Last modified November 29, 2012. http://www.prnewswire.com/news-releases/gene-by-gene-launches-dna-dtc-181360271.html.

Gniady, J.A. 2008. "Regulating Direct-to-Consumer Genetic Testing: Protecting the Consumer Without Quashing a Medical Revolution." *Fordham Law Review* 76:2429–75.

Government Accountability Office (GAO). 2010. *Direct-to-Consumer Genetic Tests: Misleading Test Results Are Further Complicated by Deceptive Marketing and Other Questionable Practices*. Washington, DC: GAO.

———. 2006. *Nutrigenetic Testing: Tests Purchased from Four Web Sites Mislead Consumers*. Washington, DC: GAO.

Gutierrez, Alberto. Alberto Gutierrez to Earl M. Collier, Jr., June 10, 2010. http://www.fda.gov/downloads/MedicalDevices/ResourcesforYou/Industry/UCM215241.pdf.

Gutierrez, Alberto. Alberto Gutierrez to Ann[e] Wojcicki, November 22, 2013. http://www.fda.gov/iceci/enforcementactions/warningletters/2013/ucm376296.htm.

Hudson, K. et al. 2007. "ASHG Statement on Direct-to-Consumer Genetic Testing in the United States." *Obstetrics and Gynecology* 110:1392–5.

Kalf, R. R. J. et al. 2014. "Variations in Predicted Risks in Personal Genome Testing for Common Complex Diseases." *Genetics in Medicine* 16:85–91.

Mardis, E. R. 2010. "The $1,000 Genome, the $100,000 Analysis?" *Genome Medicine* 2:84.

McGuire, A. L. et al. 2010. "Regulating Direct-to-Consumer Personal Genome Testing." *Science* 330:181–2.

openSNP. "Welcome to openSNP." Last accessed April 22, 2014. http://opensnp.org/.

Palmer, J. E. 2012. "Genetic Gatekeepers: Regulating Direct-to-Consumer Genomic Services in an Era of Participatory Medicine." *Food and Drug Law Journal* 67:475–522.

Perrone, Matthew. 2012. "23andMe Seeks FDA Approval for Personal DNA Test." *Associated Press*. (July 30). http://www.businessweek.com/ap/2012-07-30/23andme-seeks-fda-approval-for-personal-dna-test.

Roberts, J. S. and J. Ostergren. 2013. "Direct-to-Consumer Genetic Testing and Personal Genomics Services: A Review of Recent Empirical Studies." *Current Genetic Medicine Reports* 1:182–200.

Rubin, Rita. 2010. "FDA Groups Genetic Tests with Medical Devices, Requiring Approval." *USA Today* (June 14). http://usatoday30.usatoday.com/news/health/2010-06-15-genetictesting15_ST_N.htm.

Secretary's Advisory Committee on Genetics, Health, and Society (SACGHS). 2008. *U.S. System of Oversight of Genetic Testing: A Response to the Charge of the Secretary of Health and Human Services*. Washington, D.C.: Department of Health and Human Services. http://oba.od.nih.gov/oba/SACGHS/reports/SACGHS_oversight_report.pdf.

Singer, Emily. "The Genetic Early Adopters." *MIT Technology Review*. Last modified September 8, 2008. http://www.technologyreview.com/news/410787/the-genetic-early-adopters/.

SNPedia. "Promethease." Last accessed April 22, 2014. http://www.snpedia.com/index.php/Promethease.

———. "Promethease/Features." Last accessed April 22, 2014. http://www.snpedia.com/index.php/Promethease/Features.

Spector-Bagdady, K. and E. R. Pike. 2014. "Consuming Genomics: Regulating Direct-to-Consumer Genetic and Genomic Information." *Nebraska Law Review* 92:677–745.

U.S. Food and Drug Administration (FDA). 2015. "FDA Permits Marketing of First Direct-to-Consumer Genetic Carrier Test for Bloom Syndrome" (February 19). http://www.fda.gov/NewsEvents/Newsroom/PressAnnouncements/ucm435003.htm.

———. 2014. "FDA Fundamentals." (February 12). http://www.fda.gov/AboutFDA/Transparency/Basics/ucm192695.htm.

———. 2014b. "Letters to Industry." (April 22). http://www.fda.gov/medicaldevices/resourcesforyou/industry/ucm111104.htm.

———. 2014c. "Warning and Untitled Letters." (April 22). http://www.fda.gov/downloads/AboutFDA/Transparency/PublicDisclosure/GlossaryofAcronymsandAbbreviations/UCM212064.pdf.

———. 2013. "Regulatory Controls." (April 11). http://www.fda.gov/Medical Devices/DeviceRegulationandGuidance/Overview/GeneralandSpecial Controls/default.htm.

———. 2012. "Classify Your Medical Device." (December 3). http://www.fda.gov/MedicalDevices/DeviceRegulationandGuidance/Overview /ClassifyYourDevice/default.htm.

———. 2012a. "Premarket Approval (PMA)." (January 24). http://www.fda.gov/medicaldevices/deviceregulationandguidance/howtomarket yourdevice/premarketsubmissions/premarketapprovalpma/default.htm.

———. 2009. "General Controls for Medical Devices." Last modified May 13, 2009. http://www.fda.gov/MedicalDevices/DeviceRegulationandGuidance /Overview/GeneralandSpecialControls/ucm055910.htm.

Vorhaus, Dan. 2012. "DNA DTC: The Return of Direct to Consumer Whole Genome Sequencing." *Genomics Law Report.* Last modified November 29, 2012. http://www.genomicslawreport.com/index.php/2012/11/29/dna -dtc-the-return-of-direct-to-consumer-whole-genome-sequencing/.

Winnick, Edward. 2014. "Illumina Launches Two New Platforms at JP Morgan Conference; Claims $1,000 Genome." *Genomeweb.* Last modified January 14, 2014. http://www.genomeweb.com/sequencing/illumina-launches-two -new-platforms-jp-morgan-conference-claims-1000-genome.

CHAPTER TWENTY-SIX

A New Regulatory Function for E-Prescriptions

Linking FDA to Physicians and Patient Records

ANDREW ENGLISH, DAVID ROSENBERG, AND HUAOU YAN

I. INTRODUCTION

In this chapter, we propose broadening the conventional use of electronic-prescription (e-prescription) technology to include a new regulatory function. Currently, e-prescriptions mainly provide digital means for more efficiently accomplishing the time-honored purchase-order function of paper prescriptions[1] (Kannry 2011; Stross 2012). We propose deploying this technology to also serve FDA regulatory objectives in overseeing the postmarket safety and efficacy of medical products.[2]

FDA faces significant difficulties in monitoring the postmarket safety and efficacy of medical products. A recent example is the diabetes drug Avandia. Approved in 1999, Avandia reportedly caused more than 47,000 people to suffer heart attacks, strokes, or death before FDA recognized the potential risk and intervened in 2010 to curtail the drug's marketing (Harris 2010). Indeed, the risk was first uncovered not by FDA but by plaintiffs' lawyers prosecuting product liability claims against the manufacturer, with medical studies subsequently confirming the drug's dangers based on data publicly divulged as a result of the litigation's settlement (ibid.). The case of all-metal hip replacements demonstrates that inadequate monitoring also hobbles FDA's oversight of medical devices. As early as 2008, DePuy Orthopaedics knew that its hip replacements were failing prematurely; even earlier, the company was alerted that the metallic implants were leaking chromium and cobalt into surrounding tissue, resulting in damage to vital organs, muscle, and bone (Editorial 2013; Jaslow 2012). Yet, FDA took no action until 2011, despite the 300 complaints it received during the preceding three years (Meier 2013; Meier 2010a).[3]

Achieving physician compliance with its warnings, advisories, and practice guidelines is another major problem for FDA. This regulatory failure stems both from FDA difficulties in communicating content in a timely, informative, and attention-getting manner and from physicians lacking adequate time, training, and willingness to keep abreast with and adhere to the agency's prescriptions regarding the safe and efficacious use of medical products. The sedative triazolam provides an example. After a protracted debate over its safety, FDA directed the manufacturer to issue a new warning that the drug should only be prescribed for seven to ten days (Moore et al. 1998). Yet follow-up studies found that 85 percent of triazolam prescriptions were for significantly longer periods (ibid.). Similarly, despite a two-year informational campaign about the addictive properties of propoxyphene conducted by FDA and the drug's manufacturer, there was no material effect on prescription volume or on the number of overdose deaths (Soumerai et al. 1987). Furthermore, concern about its advice being ignored prompted FDA originally to severely restrict Avandia's use rather than rely on additional warnings (Harris 2010; Tavernise 2013).

The regulatory function we advance for e-prescriptions can simultaneously and substantially improve FDA's accomplishment of its two objectives of compliance and monitoring by interfacing the agency with physicians and computerized patient records.

In particular, the e-prescription could be used to encourage physicians' compliance with FDA advisories and more generally with the norms of best medical practice by increasing their awareness of and focusing their attention on up-to-date FDA warnings, notices, and actions at the critical point of issuing the purchase order, together with access to the underlying studies and reports concerning the safety and efficacy of medical products. Simultaneously, the e-prescription system will dramatically improve FDA monitoring of medical product usage for risk signals and patterns and hence for early regulatory intervention. This system would represent a significant advance over the agency's current reliance on voluntary, relatively circumscribed modes of oversight. Our proposal would require physicians to specify on the e-prescription their treatment purpose for the prescribed medical product and would be called upon to report their knowledge of the medical outcomes, both favorable and unfavorable. By collecting and evaluating these data, e-prescription technology will provide FDA with a long sought experiential and evidentiary basis for reliable and expeditious formulation, implementation, and review of regulatory decisions.

The major FDA regulatory role we are proposing has not previously been considered in the literature or by policy makers. Most safety-related studies concern the technology's purchase-order function, for example, in eliminating mistakes due to illegible handwriting or incorrect specification of the dose level or frequency (Shamliyan et al. 2008; Nanji et al. 2011). Researchers have also investigated linking e-prescriptions to patient records to provide physicians and pharmacists with computer programmed alerts, dosage defaults, and other checks against potentially dangerous interactions between the newly prescribed treatment and a patient's ongoing medical conditions and product usage (Kaushal et al. 2010; Denekamp 2007; Gandhi et al. 2005; Nanji et al. 2011). Beyond the purchase-order function, e-prescriptions are increasingly used to provide doctors with clinical guidelines on recommended protocols and practices for treating certain diseases or conditions (Lang et al. 2007; Kawamoto et al. 2005), but critically, they do not involve a regulatory linkage between e-prescriptions and FDA designed to facilitate the agency in accomplishing its compliance and monitoring objectives. A small-scale e-prescription system initiated by McGill University goes the furthest of any we have found in tracking indications for which physicians prescribe drugs as well as providing them with current information regarding product uses, risks, generic alternatives, pricing, and best practice protocols (Tamblyn et al. 2006; Buckeridge et al. 2007; Eguale et al. 2010). This program, however, stops well short of our proposal, which, in contrast to McGill's, is primarily aimed at promoting regulatory objectives. To that end, the e-prescription system we propose would comprehensively collect data for continuous agency assessment, algorithmically and otherwise, relating not only to the diagnostic, therapeutic, and other indications prompting the physicians to prescribe a given medical product but also, to a matter of utmost regulatory importance: the safety and efficacy of the prescribed course and outcome of treatment.

In part 2, we outline the operational mechanics of an e-prescription system tasked to serve FDA regulatory objectives of compliance and monitoring. In part 3, we discuss the system's regulatory benefits, including augmentation of existing regulatory oversight. In part 4, we assess potential costs of our proposal.

II. OPERATIONAL OVERVIEW OF PROPOSED E-PRESCRIPTION SYSTEM

Our aim is to introduce the concept of an e-prescription technology that would provide as capacious, directive, and wide-ranging an interface between physicians and patient records on the one hand and FDA on the other as the agency needs to achieve its regulatory goals. As such, we will not consider how FDA implements our proposal, including its determination of how much and how insistently information should be communicated and elicited through e-prescriptions. Similarly, we do not address technical matters of software and content design.

A. Compliance

On opening an e-prescription form, the physician would be prompted to enter a user-identification and password, patient medical identification number, and then the name (brand, generic, chemical) or class, type, and function of the particular drug or device in question. The program would offer the option of viewing (or in response to a search request, provide) a list of and comparative information on safety, efficacy, and pricing of alternative products.

Having entered a particular product, the physician will be asked to select from a dropdown menu of FDA-approved medical indications or conditions for the given product. The physician can also enter another, "unapproved" (e.g., "off-label," warned against) use. Choosing an off-label or other unapproved course of treatment at this juncture will trigger an "override" feature of the e-prescription program, which we discuss below.

The heart of the compliance function will be notifying physicians of FDA warnings, cautions, best practice guidance, and other safety and efficacy information (including a link to the product insert) regarding a drug's or device's proposed use. Crucially, this advisory will be delivered at the moment of its greatest relevance and effectiveness, when the physician decides to prescribe the product (Kawamoto et al. 2005). The particular notices would be ordered and presented to optimally effectuate their clear and concise communication consistent with FDA priorities, with each soliciting a physician response indicating either understanding of the content or, if not, the difficulty involved. The program will indicate whether, when, and the extent to which notices have been upgraded (or downgraded) in seriousness or supplemented by new adverse event data and studies. If desirable, the system could be designed to allow physicians who have recently submitted e-prescriptions for the given product to skip all but contraindications and major warnings contained in the previously sent advisories.

Furthermore, to foster the best treatment outcome, the program will search the patient's computerized records for problematic factors in his/her medical history and for potentially harmful interactions from concomitant drug or device uses. The program will seek to identify such hazards by comparing the entire array of FDA warnings and cautions against the patient's records for available information relating to prior or ongoing drug and device use, medical conditions, treatment experiences, family background, and other risk-related factors and behaviors.

A physician response or search of patient records indicating use of the medical product for an FDA unapproved course of treatment or one that might pose concomitant drug or other usage risks of a patient-specific nature would trigger an "override" feature. Consistent with the agency's predominantly informational approach that affords physicians latitude in treatment, the override feature will condition but not preclude submission of the e-prescription for the unapproved use. To satisfy the condition, the physician will be required to expressly select from a menu of unapproved uses or to specify the unlisted, unapproved use. At that point, FDA contraindications, warnings, cautions, guidelines, and other contrary recommendations would appear sequentially, each with a request for the doctor to acknowledge having read and understood the advisory. This interrogatory process will not only serve the compliance function but also will be critical to promoting the monitoring function.

B. Monitoring

Our proposal's monitoring feature is designed to identify how medical products are being used and how safely and effectively they are performing in practice. The information sought includes not only adverse but also positive results. Although outcomes may often be described simply in terms of the use, for example, that the patient showed no evidence of malady X after use of product Y, the goal of our proposal is to secure more detailed accounts. Thus, for devices, it would be most helpful to know how long and under what type of circumstances the particular product was in use, and if and when the need arose to repair or replace it. As far as practicable, e-prescription monitoring would produce calibrated appraisals.

The primary monitoring feature of the proposal would be activated upon completion and submission of the e-prescription. At that point, not only will the purchase order be transmitted to the pharmacy but all entries on the e-prescription and all physician acknowledgments will automatically be conveyed to both FDA (appropriately anonymized) and to the patient's computerized records.

Although completed, the e-prescription would remain open until the physician's final report of the treatment outcome. Frequently, the outcome report will be based on information supplied by the patient in the normal course, for example, during a follow-up examination. In cases determined by FDA to present particular regulatory concerns, such as those involving use of especially risky products or investigation of a newly emergent danger, the physician may be prompted by the program to contact the patient for relevant information. The program will permanently close the e-prescription when the physician submits a report designated as final.

Many other cases may involve long-term courses of treatment, the outcome of which may be delayed and the prescribing physician's connection with the patient may lapse. Here, the program may be designed to allow the physician to reassign the reporting obligation directly or via an existing medical record to the patient's subsequent health care providers, enlisting them as coadministrators of the e-prescription and responsible for entering updated treatment results. The e-prescription technology can also respond to the "open" prescription case by searching pertinent entries in computerized patient records for information about the outcome of the treatment in question that would suffice to effect closure.

Despite these features, there will be cases where the follow-up effort required to close a prescription is not practical. When obstacles prevent meaningful follow-up, the physician may close the e-prescription by entering a final report of "outcome unascertainable."

Reporting gaps are likely to be random and surmountable through appropriate statistical techniques. Thus the e-prescription system should provide an unprecedented wealth of information about drugs and devices, their medical effects, both positive and negative, and how and for what condition they are being used; and it will do so in real-time.

III. EXPECTED BENEFITS

E-prescription technology will improve compliance by providing easy and timely access to complete, up-to-date information through a directive and interactive transmission medium—physicians must acknowledge receipt and appreciation of FDA prescriptions. Moreover, to authorize off-label and other unapproved uses, physicians must explain their choice.

Furthermore, the two regulatory functions of our proposal reinforce each other. Facilitating compliance with timely risk and practice guidance should diminish the need and cost for monitoring; and monitoring will encourage compliance as physicians realize that their responses to e-prescription queries are being recorded and scrutinized by FDA.

With respect to monitoring, reporting would be an automatic part of every prescription, and the data accumulated would be subject to FDA's direct unmediated access. This mode of surveillance contrasts markedly with existing FDA systems that are voluntary and fragmented in nature, such as MedWatch, which creates its database of adverse events from physicians, patients, and others who choose to report them, and disseminates alerts to physicians who choose to receive them (FDA 2012; Craigle 2007). FDA established MedSun to monitor adverse events in the use of medical devices based on reports from a network of voluntarily participating health care facilities, and upon investigation, to share the findings with the members of the network (FDA 2012a). However, as of 2010, only about four hundred hospitals had signed up (FDA 2010), limiting the information collected and disseminated. Similarly, Sentinel involves medical insurers/providers volunteering to search their proprietary databases for relevant information upon FDA request, but crucially, the system depends on the agency having otherwise already acquired evidence of adverse reactions or inefficacy before requesting database mining (Mini-Sentinel Coordinating Center 2011). These FDA programs provide vital information but, as exemplified by performance of MedWatch, are prone to underreporting and unreliability due in part to their voluntariness (Meier 2010a; Meier 2010b).

There are evident regulatory benefits from combining a comprehensive and more dependable database with constant evaluation of incoming information. Naturally, the feed of information from every e-prescription generated represents an enormous increase in response rate over current, voluntary programs. The e-prescription generated databases can create real-time, continuous analysis by algorithms designed to catch problems quickly as well as provide researchers with information for developing causal hypotheses and testing them against the compiled evidence, indeed, even by using the postapproval data to run retroactive randomized clinical investigations of a given product's safety and efficacy.[4] The vast amount of data could be examined as a whole or parsed as needed to scale monitoring up to the national level or down to those served by a particular physician group or hospital.

Of particular importance, the e-prescription system would provide FDA with currently unavailable information on off-label medications, which accounts for as much as 21 percent of the prescriptions for some drugs (Radley 2006). Clinical trials run for an approved use may not capture the adverse effects of other, unapproved uses (Stafford 2008). And off-label prescriptions are often written notwithstanding the lack of evidence on safety and efficacy (ibid.). Our proposal enables FDA to evaluate the efficacy and safety of off-label and other unapproved medical product uses, set alerts for new or increased use of medical products for unapproved purposes, correlate such uses with adverse (or positive) patient outcomes, and adjust its policies and advisories accordingly.

Furthermore, our e-prescription proposal would promote FDA compliance objectives in the same manner as the agency pursues them. Physician adherence to rules and norms is encouraged mainly through information provided in approved product labels, inserts, and directions, and augmented as needed through specially issued warnings and updates. Medical malpractice suits and products liability deter nonconformity as well. However, physicians' compliance remains largely a matter of their conscious choice to become aware of and follow agency prescriptions. Although our proposal requires physicians to report their treatment decisions and outcomes, it leaves them free to choose the course of treatment, consistent with the general medical practice exemption from FDA regulation.[5]

The proposed e-prescription system could also be readily integrated with FDA's existing voluntary data reporting and sharing systems, incorporating and supplementing their databases. The e-prescription system can supplement Sentinel by providing a comprehensive database on treatment outcomes for all patients. It would also complement MedWatch. MedWatch would continue to allow consumers and require pharmaceutical companies to report adverse reactions, while our system would provide algorithmically generated warning signals from its vast database. Moreover, our proposed e-prescription system would render existing nongovernmental registries and databases, which are limited to specific types of medical devices or are proprietary, unnecessary as researchers could subject the e-prescription database to virtually unlimited types and combinations of investigative searches.

IV. POTENTIAL COSTS

The foregoing benefits are substantial. However, difficulties will arise in establishing an e-prescription system that can effectively serve FDA regulatory objectives. To begin with, building a fully integrated nationwide system will require federal funding and the expertise of FDA and other agencies with allied regulatory jurisdiction such as the Centers for Disease Control and Prevention. Next, we review the chief design problems, concluding that despite the challenges they present, none poses a significant obstacle to effectuating our proposal.

Development of the information technology component of the system is a nontrivial challenge. The system must be standardized and operate across multiple public and private entities. Given the many different preexisting systems, either the design of the overarching system must be able to encompass a wide variety of platforms or a major overhaul of infrastructure would need to occur.

The creation and editing of the content for the compliance function of the e-prescription system also poses significant design hurdles. Effectuation of that function hinges on informational prompts and questions whenever physicians fill out a prescription. The information being conveyed cannot regurgitate scientific studies, requiring significant effort by physicians to glean the meaning of the relevant reports. Nor should it simply provide links to lists of alerts that physicians can choose to access, as this would replicate the voluntary participation problems that undermine the reliability of MedWatch and other existing programs. Rather, FDA notices must be translated into straightforward prompts and questions that cogently and compellingly inform doctors of the relevant problems and side effects with the prescribed product. These cues would have to be carefully formulated, prioritized, and curated by FDA, and the agency would need to balance the value of conveying more information against not overloading the physician at this critical decision point. This process may be laborious and expensive in its initial setup as well as in continually updating the system. Moreover, these high capital investments will have to be undertaken despite relatively little empirical and experiential basis for anticipating the actual operating costs and benefits of the system.[6]

Our proposal will require continued expenditure from individual health care providers on necessary information technology and training. However, although the providers will bear most of these relatively fixed costs, they will reap a small amount of the benefits, which largely accrue to consumers and insurers in the form of better and less expensive treatment. Such a collective action problem could inhibit investment in the e-prescription system. FDA supervision as well as subsidization of participation by health care providers in system development would be required and costly.

Moreover, the system must not overtax the prescribing physician at the critical but typically brief moment of prescription. At least one study of e-prescription systems with clinical decision support suggested that clinicians override most medication alerts (Isaac 2009). Thus, physicians have to be trained on the software so that they can effectively and efficiently use the system and understand the importance of the alerts and questions generated rather than viewing the software as a burden (Drazen 2009). Beyond training costs, funding will be required for designing software and content to reduce "alert fatigue" and physicians' overall burden, for example, by minimizing the number and complexity of questions while at the same time countering their "clicking-through" and similar short-cutting behaviors by randomizing the sequence and varying the grammar of questions.

Any attempt to gather patient health information on the scale we propose raises privacy concerns. Our proposal must conform to the current federal patient privacy framework (DHHS 2002). Under DHHS privacy rules, protected health information that is individually identifiable cannot ordinarily be used without the person's consent. Thus, the collected data must be carefully stripped of its identifying characteristics while still maintaining to the greatest extent the research and monitoring value of the data set. If investigation of an FDA risk signal or pattern requires direct contact with patients, access to identifying information must be available, whether on some showing to a court or neutral FDA arbiter.[7] Furthermore, the number of breaches of patient privacy has corresponded with the increasing digitization of medical records (Perlroth 2011). To ensure privacy, this process will entail expenditures not only for training those entrusted with patient data but also for vigilance and oversight to prevent security lapses and assure compliance (McGilchrist 2007).

Finally, we note the possible costs of allowing a physician's e-prescription responses to be admitted into evidence to prove or disprove medical malpractice. This concern is speculative. A study of the McGill e-prescription system indicates physicians reported treatment indications with a high rate of completeness and accuracy (Eguale et al. 2010). Moreover, physician responses will add little to what can be readily gleaned from the treatment context, for example, from a patient's record showing that the doctor prescribed a drug approved only for adults to treat an infant. Although admission could motivate misreporting, the e-prescription system might also lower the rate and degree of liability findings by (1) reducing the number of medical errors and thus the number of lawsuits and (2) documenting physicians' answers to questions about potential side effects, thereby demonstrating reasonable care by the prescriber (Ransbotham 2011). While e-prescriptions might provide clear evidence of physician error, this kind of definitive evidence could reduce malpractice costs by encouraging settlement (ibid.). Moreover, the fact that these acknowledgements could later be used in litigation might encourage doctors to take the warnings seriously, thereby leading to fewer adverse effects, lower rates of defensive medicine, and stronger legal barriers against the filing of weak malpractice claims. And critically, regardless of the effects on liability, e-prescription evidence would render judicial determinations less prone to error.

V. CONCLUSION

It is critically important for the FDA to possess a robust, comprehensive postmarketing system of oversight to continuously surveil medical product usage, safety, and efficacy, and to ensure physician awareness of and compliance with its regulatory prescriptions. Leveraging e-prescription technology can provide FDA with such a system, the capability of which in acquiring monitoring and compliance data would far exceed its existing programs and those operated by other regulators and nongovernmental organizations. At the same time, the system could deliver critical warnings regarding drug and device use to practitioners at the very moment they need that information the most. Our proposal for developing e-prescription technology thus would not merely improve the purchase-order process but also would crucially and powerfully enhance FDA regulatory capacity to assure the safety and efficacy of medical products and delivery of medical services on which everyone's health and well-being depends.

NOTES

The authors express their thanks for comments from participants in the Conference on The Food and Drug Administration in the 21st Century held in May 2013 by the Petrie-Flom Center for Health Law Policy, Biotechnology, and Bioethics at Harvard Law School and research provided by Joshua Minix on an earlier draft. David Rosenberg also thanks the Harvard Law School Summer Research Program for funding our work.

1. For convenient reference, the e-prescription "purchase-order" function includes orders issued by health care providers to pharmacies or other private or public suppliers of FDA-regulated pharmaceuticals and medical devices.

2. "Medical products" encompasses the full range of drugs, devices, and other items, and related methods, means, and protocols for their use subject to FDA jurisdiction and control pursuant to 21 USC § 301 et seq.

3. The manufacturer issued a warning about the high early failure rate in 2010 (Meier 2010b).

4. Opening the database to non-FDA researchers would provide additional resources for monitoring the safe and efficacious use of medical products. It would also facilitate investigations related to the host of recommended medical procedures and therapies that are advanced by nongovernmental organizations, such as the recent study showing widespread failure to adopt improved cancer therapies (Editorial 2013a).

5. However, the admonitory approach is not a necessary feature. Thus, our proposal is also fully compatible with FDA initiatives, such as the Risk Evaluation and Mitigation Strategy (REMS) Program, that impose more control over the use of certain medical products involving particularly pronounced risk (National Comprehensive Cancer Network 2013).

6. Electronic medical-record systems have not produced the anticipated amount of benefits (Abelson 2013).

7. The precise mechanism to evaluate the need for anonymity will need to be determined later. A balance must be struck between society's need for the best medical products and practices and a patient's desire for privacy, but no patient should be allowed to absolutely veto such access at will.

REFERENCES

Abelson, Reed and Julie Creswell. 2013. "In Second Look, Few Savings from Digital Health Records." *New York Times* (January 10).

Buckeridge, David L., Aman Verma, and Robyn Tamblyn. 2007. "Ambulatory e-Prescribing: Evaluating a Novel Surveillance Data Source." *Intelligence and Security Informatics: Biosurveillance, Lecture Notes in Computer Science* 4506:190–195.

Craigle, Valeri. 2007. "MedWatch: the FDA Safety Information and Adverse Event Reporting Program." *Journal of the Medical Library Association* 95(2):224–225.

Denekamp, Yaron. 2007. "Clinical Decision Support Systems for Addressing Information Needs of Physicians." *Israel Medical Association Journal* 9(11):771–776.

Department of Health and Human Services (DHHS). 2002. "Standards for Privacy of Individually Identifiable Health Information," 45 CFR §§ 160, 164.

Drazen, Erica et al. 2009. *Saving Lives, Saving Money In Practice: Strategies for Computerized Physician Order Entry in Massachusetts Hospitals*, available at http://mehi.masstech.org/sites/mehi/files/documents/cpoe2009.pdf.

Editorial. 2013. "What a Company Knew About Its Metal Hips." *New York Times* (February 10).

Editorial. 2013a. "Inadequate Treatment of Ovarian Cancer." *New York Times* (March 13).

Eguale, Tewodros et al. 2010. "Enhancing Pharmacosurveillance with Systematic Collection of Treatment Indication in Electronic Prescribing." *Drug Safety* 33(7):559–567.

Gandhi, Tejal K., et al. 2005. "Outpatient Prescribing Errors and the Impact of Computerized Prescribing." *Journal of General Internal Medicine* 20(9):837–841.

Harris, Gardiner. 2010. "F.D.A. to Restrict Avandia, Citing Heart Risk." *New York Times* (September 23).

Isaac, Thomas et al. 2009. "Overrides of Medication Alerts in Ambulatory Care." *Archives of Internal Medicine* 169:305, 307–308.

Jaslow, Ryan. 2012. "Metals from Hip Replacement Present Toxic Risk for Millions, Investigation Warns." *CBS News* (February 29). http://www.cbsnews.com/news/metals-from-hip-replacements-present-toxic-risk-for-millions-investigation-warns/ (accessed February 9, 2014).

Kannry, Joseph. 2011. "Effect of E-Prescribing Systems on Patient Safety." *Mount Sinai Journal of Medicine* 78(6):827–833.

Kaushal, Rainu, Lisa M. Kern, Yolanda Barrón, Jill Quaresimo, and Erika L. Abramson. 2010. "Electronic Prescribing Improves Medication Safety in Community-Based Office Practices." *Journal of General Internal Medicine* 25(6):530–536.

Kawamoto, Kensaku, Caitlin A. Houlihan, and Andrew E. Balas. 2005. "Improving Clinical Practice Using Clinical Decision Support Systems: A Systematic Review of Trials to Identify Features Critical to Success." *British Medical Journal* 330:765–772.

Lang, Eddy S., Peter C. Wyer, and R. Brian Haynes. 2007. "Knowledge Translation: Closing the Evidence-to-Practice Gap." *Annals of Emergency Medicine* 49(3):355–363.

McGilchrist, Mark et al. 2007. "Assuring the Confidentiality of Shared Electronic Health Records." *British Medical Journal* 335(7632):1223–1224.

Meier, Barry. 2013. "F.D.A. to Tighten Regulation of All-Metal Hip Implants." *New York Times* (January 16).

———. 2010a. "Health Systems Bears Costs of Implants with No Warranties." *New York Times* (April 2).

———. 2010b. "With Warning, A Hip Device is Withdrawn." *New York Times* (March 9).

Mini-Sentinel Coordinating Center. 2011. *About Mini-Sentinel, Background.* http://mini-sentinel.org/about_us/ (accessed February 9, 2014).

Moore, Thomas J., Bruce M. Psaty, and Curt D. Furberg. 1998. "Time to Act on Drug Safety." *Journal of the American Medical Association* 279(19):1571–1573.

Nanji, Karen C., et al. 2011. "Errors Associated with Outpatient Computerized Prescribing Systems." *Journal of the American Medical Informatics Association* 18:767–773.

National Comprehensive Cancer Network. 2013. *Risk Evaluation and Mitigation Strategies.* http://www.nccn.org/rems/ (accessed April 26, 2015).

Perlroth, Nicole. 2011. "Digital Data on Patients Raises Risk of Breaches." *New York Times* (December 18).

Radley, David C., Stan N. Finkelstein, and Randall S. Stafford. 2006. "Off-Label Prescribing Among Office-Based Physicians." *Archives of Internal Medicine* 166:1021, 1023, 1025.

Ransbotham, Sam et. al. 2011. "Medical Malpractice Claims and Electronic Medical Records." *Social Science Research Network* (October 4).

Shamliyan, Tatyana A., Sue Duval, Jing Du, and Robert L. Kane. 2008. "Just What the Doctor Ordered." *Health Services Research* 43:32–53.

Soumerai, Stephen B., Jerry Avorn, Steven Gortmaker, and Sharon Hawley. 1987. "Effect of Government and Commercial Warnings on Reducing Prescription Misuse: The Case of Propoxyphene." *American Journal of Public Health* 77(12):1518–1523.

Stafford, Randall A. 2008. "Regulating Off-Label Drug Use—Rethinking the Role of the FDA." *New England Journal of Medicine* 358:1427.

Stross, Randall. 2012. "Chicken Scratches vs. Electronic Prescriptions." *New York Times* (April 28).

Tamblyn, Robyn et al. 2006. "The Development and Evaluation of an Integrated Electronic Prescribing and Drug Management System for Primary Care." *Journal of the American Medical Informatics Association* 13(2):148–159.

Tavernise, Sabrina. 2013. "F.D.A. Lifts Some Restrictions on Avandia." *New York Times* (November 25).

U.S. Food and Drug Administration (FDA). 2012. *FDA Adverse Event Reporting System* (FAERS) (formerly AERS) (September 10). http://www.fda.gov/Drugs/GuidanceComplianceRegulatoryInformation/Surveillance/AdverseDrugEffects/default.htm.

——— 2012a. *MedSun: Medical Product Safety Network* (December 11). http://www.fda.gov/medicaldevices/safety/medsunmedicalproductsafetynetwork/default.htm.

——— 2010. *Agency Information Collection Activities* (July 9). Federal Register Vol. 75, No. 131. http://www.gpo.gov/fdsys/pkg/FR-2010-07-09/pdf/2010-16807.pdf.

CHAPTER TWENTY-SEVEN

Race and the FDA

JONATHAN KAHN

RECENT DEBATES over the appropriate use of racial and ethnic categories in biomedical contexts have often concentrated on the practices of individual researchers with the aim, in part, to help researchers appreciate the nuances and complexities of the racial categories they use (e.g., Burchard et al. 2003; Cooper et al. 2003; Krimsky and Sloan 2011; Gomez and Lopez 2013). Since the inception of the Human Genome Project, much time and attention has been devoted to ensuring that biological knowledge emerging from advances in genetic research is not used inappropriately to make racial categories appear biologically given or "natural." Scientists may and do disagree on the utility of using particular racial or ethnic categories as *surrogates* for genetic groupings, arguing about different frequencies of particular genetic variations. But race is not a coherent genetic concept; rather, it is best understood as a complex and dynamic social construct (e.g., American Anthropological Association 1998; Collins 2004). Since Richard Lewontin's ground-breaking work on blood group polymorphisms in different groups and races in the 1970s, scientists have understood that race will statistically explain only a small portion of genetic variations (Lewontin 1972). As a 2005 editorial in *Nature Biotechnology* put it, "Race is simply a poor proxy for the environmental and genetic causes of disease or drug response. . . . Pooling people in race silos is akin to zoologists grouping raccoons, tigers and okapis on the basis that they are all stripey" (Editorial 2005).

A focus on individual practices, while necessary, overlooks the myriad structural forces that teach researchers and clinicians to see and use race in particular and often problematic ways. The U.S. Food and Drug Administration's (FDA) evolving practices with respect to the use of racial and ethnic categories in pharmaceutical research and development must be understood within a historical and institutional context that has been becoming increasingly concerned with the place of race in biomedicine. Federal initiatives that shape the production and use of racial categories in biomedical research provide the starting point for approaching the broader racialization of contemporary biomedicine. While there may be certain legitimate uses for racial classifications in the regulatory process (e.g., keeping track of health disparities), it is imperative that regulatory bodies be alert to preventing slippage into unwarranted and dangerous uses of race as genetic. The basic concern here, then, is to explore how particular mandates developed to serve bureaucratic needs of data management, political needs of inclusiveness, and commercial needs of market expansion may be shaping scientific research and medical practice in problematic ways that threaten to reinvigorate long-discredited notions of race as a fixed genetic construct.

Prominent among these forces is a wide array of federal mandates that dictate the characterization and application of genetically based biomedical interventions, such as pharmaceuticals and diagnostic tests, in relation to socially defined categories of race. Key federal mandates include: the NIH Revitalization Act of 1993 (Pub. L. No. 103-43, 107 Stat. 122 (1993), which directs the National Institutes of Health to establish guidelines for including women and minorities in clinical research; the Food and Drug Administration Modernization Act of 1997 (FDAMA), which directs that "the Secretary [of Health and Human Services] shall, in consultation with the Director of the National Institutes of Health and with representatives of the drug manufacturing industry, review and develop guidance, as appropriate, on the inclusion of women and minorities in clinical trials" (Food and Drug Modernization Act: § 115); and two subsequent Food and Drug

Administration Guidances for Industry. The first, a 1999 guidance entitled "Population Pharmacokinetics" (FDA 1999), made recommendations on the use of population pharmacokinetics in the drug development process to help identify differences in drug safety and efficacy among population subgroups, including race and ethnicity. The second, a 2005 guidance entitled "Collection of Race and Ethnicity Data in Clinical Trials" (FDA 2005), recommends a standardized approach based on federal census categories promulgated by the Office of Management and Budget (OMB) for collecting and reporting race and ethnicity information in clinical trials that produce data for applications to FDA for drug approval.

The NIH Revitalization Act requires researchers to certify that they have enrolled adequately diverse populations, have made sufficient efforts to enroll diverse populations, or could provide a biomedical justification for not enrolling diverse populations (for example, not enrolling women in a study of prostate cancer). At the time of the act's passage, Otis Brawley, then director of the National Cancer Institute's Office of Special Populations research, worried that the NIH Revitalization Act's "emphasis on potential racial difference" might "foster... the racism that its creators want to abrogate by establishing government sponsored research on the basis of the belief that there are significant biological differences among the races" (Epstein 2007:95–96).

Along with the NIH, FDA showed an increasing interest in issues of inclusion in the late 1980s and early 1990s. FDAMA and related Guidances for Industry, however, do not mandate the same sort of subgroup analyses as the NIH Revitalization Act. They simply urge that companies collect data by sex, race, and ethnicity and be on the lookout for "differences of clinically meaningful size" (Epstein 2007:129).

These federal mandates have had a profound effect upon the use of racial categories in biomedical research, clinical practice, product development, and health policy. Their construction and definition of racial categories is structured by Office of Management and Budget's Revised Directive 15 on "Standards for Maintaining, Collecting, and Presenting Federal Data on Race and Ethnicity." The standards were developed "to provide a common language for uniformity and comparability in the collection and use of data on race and ethnicity by Federal agencies" (Office of Management and Budget 1997). Directive 15 set forth the basic categories of race and ethnicity that are currently used in the census and inform data collection and monitoring across a wide array of federal programs: American Indian or Alaska Native, Asian, black or African American, Native Hawaiian or Other Pacific Islander, and white. In addition, there are two categories for data on ethnicity: "Hispanic or Latino" and "Not Hispanic or Latino" (Office of Management and Budget 1997). By conditioning grants and approvals on the collection of data according to the OMB categories, they provide powerful incentives to introduce race into biomedical contexts, regardless of its relevance.

I. FDA GUIDELINES ON RACE AND ETHNICITY

In January 2003, FDA announced the promulgation of a Draft Guidance for Industry on the Collection of Race and Ethnicity Data for Clinical Trials for FDA Regulated Products (68 Fed. Reg. 4788; FDA 2003a). After notice and comments, FDA issued the final guidance in September 2005, which recommends that individuals or corporations submitting drug approval applications use "a standardized approach for collecting and reporting race and ethnicity information in clinical trials conducted in the United States and abroad for certain FDA regulated products" (FDA 2005). As any federally funded researcher knows, these mandates impose significant requirements and provide incentives to identify and collect research data according to categories of race and ethnicity.

In a "Talk Paper" discussing the issuance of the Draft Guidance in 2003, FDA elaborated upon their nature and purpose: "FDA regulations require drug sponsors to present an analysis of data according to age, gender and race. An analysis of modifications of dose or dosage intervals for specific groups is also required when manufacturers submit a new drug application for approval by FDA. To accomplish this, FDA recommends that the drug manufacturers use the OMB race and ethnicity categories during clinical trial data collection to ensure consistency in evaluating potential differences in drug response among racial and ethnic groups" (FDA 2003b).

Consistency was a key theme throughout the Draft Guidance. It exhibited a general concern to regularize the collection and submission of data on race and ethnicity across the spectrum of clinical trials and the drug development process. The Draft Guidance specifically recommended the use of the OMB categories of race and ethnicity, first, to "help ensure consistency in demographic subset analyses across studies" and, second, to help evaluate "potential differences in the safety and efficacy of pharmaceutical products among population groups" (FDA 2003a). The Guidance elaborated the rationale for this concern by referencing some studies that show *on average* members of certain OMB racial groups *may* respond differently to certain drugs than members of other OMB racial groups. That is, the Guidelines connected race and physiology.

The Guidelines themselves were promulgated against a backdrop of a perceived need to harmonize the use of racial categories across national regulatory regimes to facilitate the globalization of pharmaceutical markets. A central initiative in this regard was led by the International Conference on Harmonization (ICH), formally known as the International Conference on Harmonization of Technical Requirements for Registration of Pharmaceuticals for Human Use. The ICH was formed in 1990 at a meeting in Brussels hosted by the European Federation of Pharmaceutical Industries and Associations (EFPIA). It is structured around the three regions of the United States, European Union (EU), and Japan and includes major regulatory agencies and pharmaceutical trade associations from each region.

The ICH was born of a concern to address the problems created by diverse regulatory standards imposed by the governments of the major pharmaceutical markets. Founding members believed that the harmonization of standards for product development and regulatory approval would greatly increase the efficiency and economy of drug development and pave the way for the creation of a truly global pharmaceuticals market (International Conference on Harmonization n.d.; Lee 2005).

A key consideration for global drug development and registration involves the acceptability of data in different regions. To address this issue, in 1998, the ICH adopted Guideline E5, "Ethnic Factors in the Acceptability of Foreign Clinical Data." ICH E5 is intended to facilitate drug registration in the different ICH regions by recommending a framework for evaluating the impact of ethnic factors on a drug's safety and efficacy in a manner that will enable appropriate evaluation of ethnic factors (International Conference on Harmonization 1998). Together with the NIH Revitalization Act, and the FDAMA, ICH E5 provides a critical backdrop to FDA guidance. ICH E5 defines "ethnic factors" as "factors relating to races or large populations grouped according to common traits and customs" (International Conference on Harmonization 1998:9). It makes a further key distinction between what it characterizes as "intrinsic" versus "extrinsic" ethnic factors. It defines "intrinsic ethnic factors" as "factors that help define and identify a subpopulation and may influence the ability to extrapolate clinical data between regions. Examples of intrinsic factors include genetic polymorphism, age, gender, height, weight, lean body mass, body composition, and organ dysfunction." In contrast, it defines "extrinsic ethnic factors" as "factors associated with the environment and culture in which a person resides. Extrinsic factors tend to be less genetically and more culturally and behaviorally determined" (International Conference on Harmonization 1998:9–10). At first blush, this distinction seems straightforward and relatively unproblematic. However, ICH E5 elaborates upon these definitions in a chart that situates race as an intrinsic *genetic* characteristic. That is, it constructs race as some sort of genetic component of a larger category of ethnicity (Kuo 2008). This is problematic for a number of reasons, not least of which being that whatever they are, race and ethnicity are not genetic.

In contrast to the ICH E5 Guideline, the FDA Guidance contains the OMB caveat that its racial and ethnic categories are not to be interpreted as biological or genetic. Nonetheless, its recommendations, which are based on physiological processes, exist in tension with the caveat. This tension was recognized and seized upon by many pharmaceutical companies in offering comments to the Draft Guidelines in 2003. The response of pharmaceutical companies, however, was not uniform. Large pharmaceutical companies with global marketing concerns focused in particular on inconsistencies between FDA mandated use of the distinctively American OMB categories of race and ethnicity (e.g., "African American," "Hispanic") and those used internationally in other ICH regions of Japan and Europe. Their comments tended to call for a more sophisticated use of population categories that could be more easily integrated with the structure of ICH E5 and more readily translatable across regions. Smaller biotechnology companies tended to be less concerned with the international ramifications of the Guidance but rather urged the adoption of new genetic technologies to provide more precise population categories for the collection of data.

Generally speaking, comments submitted by pharmaceutical corporations expressed concern over (1) inconsistent definitions of race and ethnicity; (2) questionable accuracy of definitions of race and ethnicity;

(3) the negative impact that using OMB categories of race and ethnicity might have on global trial recruitment; and (4) the creation of unnecessary and unscientific differences among populations through the use of inappropriate racial and ethnic categories. Underlying the concerns of large pharmaceutical corporations in particular was a perceived need to develop a globally applicable standard for the collection of racial and ethnic data—clearly more in line with the mandate of ICH E5. In short, where the ethnic categorizations of ICH E5 were perceived as opening up global markets, the OMB-based racial and ethnic classifications of the FDA Guidance were perceived as potential barriers to globalization of drug markets.

II. INCONSISTENT DEFINITIONS

The Pharmaceutical Research and Manufacturers of America (PhRMA) is a pharmaceutical trade association describing itself as representing "the country's leading pharmaceutical research and biotechnology companies," whose mission is "to conduct effective advocacy for public policies that encourage discovery of important new medicines for patients by pharmaceutical and biotechnology research companies" (PhRMA n.d.). It opened its comments to the FDA Draft Guidance with an admonition that "for these categories to be valuable globally and to permit identification of ethnic differences, there should be only one set of agreed ethnic/racial categories" (PhRMA 2003). It therefore recommended that the issue be brought to the ICH as a forum for the development of globally consistent categories. Comments submitted by Pharmacia (2003) (subsequently acquired by Pfizer) that largely replicated comments submitted by PhRMA focused in particular on the myriad ways in which the Guidance threatened to obstruct pharmaceutical globalization. With regard to inconsistent definitions of race and ethnicity, Pharmacia observed that

> The OMB race and ethnicity categories can be used only in the US, not in the EU or in Japan; this is especially true for the ethnicity questions (Hispanic/Latino vs. Not Hispanic/Latino). A definition of the ethnicity varies among the ICH countries, as well as non-ICH countries. There will be more opportunities for the US to utilize foreign clinical data in evaluating safety and efficacy of new drugs in

the future. Therefore, it is recommended that the race and ethnicity categories should be more scientific and globally accepted so that the data comparison becomes more meaningful and provides valuable information in evaluating potential differences or similarities in safety and efficacy of new drugs among population subgroups.

—(PHARMACIA 2003:1)

There was a clear recognition here that the OMB definitions of race and ethnicity were not static, scientifically objective categories. Pharmacia (and PhRMA) was concerned that imposing the U.S. regulatory definitions of race on the pharmaceutical industry would inhibit the globalization of pharmaceutical markets.

Comments submitted by Abbot Laboratories expressed a similar concern that the OMB categories were "oversimplified" and "vague" and urged that FDA "recommend a better definition of race and ethnicity that can be understood by a subject in a study and be consistent across the board" (Abbot 2003:3). Bristol Meyers Squibb also noted that the Guidance's "proposed ethnicity and racial categories may be understood differently in different parts of the world" and urged the development of "better defined categories" (Bristol Meyers Squibb 2003). Thus, for example, it proposed that the OMB category of "Black or African American" be revised to "Black, of African heritage or African American" (ibid.). Given the OMB's own caveat that its categories were not genetic or biological, the clear concern with such a proposal was not to present a more "scientific" definition of race but rather to produce a more globally acceptable definition. The two are not necessarily the same.

Of all the OMB categories incorporated in the Guidance's mandate, the ethnicity category of "Hispanic or Latino" caused particular concern in terms of consistent global application. Bristol Meyers Squibb noted that "the requirement that Hispanic or Latino versus not Hispanic or Latino ethnicity be collected even in trials that are conducted entirely outside the U.S. seems contradictory to the spirit of the ICH guidelines. If ethnicity designations, as per the guidance, are to reflect the sociocultural construct of the society, then the proposed category is generally inappropriate outside the United States" (ibid.). Again, there was an explicit concern for potential conflict with the ICH E5 guidelines and resulting barriers to the efficient globalization of markets.

III. QUESTIONABLE ACCURACY

Pharmacia also used the asserted inappropriateness of the Latino/Hispanic category in a global context to highlight the questionable accuracy of the Guidance's terminology, asserting that "[t]he terms Hispanic and Latino will not have the same meaning outside the U.S. as they do within the U.S. According to the definition, Spaniards are considered Hispanic, but they are both culturally and racially more similar to French than Mexicans" (Pharmacia 2003). Similarly, Bristol Meyers Squibb noted that "terminology—like 'Latino'—can be confusing outside the United States, while the medical relevance of such category is not demonstrated inside the US" (Bristol Myers Squib 2003). Pharmacia went on to make similar criticisms of the accuracy of other OMB categories, noting that "there is no distinction among the Asian group, which may be more genetically variable" (Pharmacia 2003). In mentioning similarity and difference, "medical relevance," and "genetic variability," these comments go beyond the earlier stated concern for globally consistent definitions to the basic scientific accuracy of the categories themselves.

Significantly, however, smaller biotechnology companies specializing in genetic research urged a different approach to overcoming similar problems of definitional accuracy. Biotech companies such as Genaissance Pharmaceuticals and DNAPrint Genomics did not have the global reach of corporations such as Pharmacia or Bristol Meyers Squibb. Genaissance, for example, marketed its technology to large pharmaceutical corporations rather than engaging directly in global drug development and marketing. The focus of their criticisms of FDA Guidance, therefore, was less on developing globally consistent categories of race and ethnicity and more on using their own proprietary genetic technologies to provide purportedly more scientifically objective and accurate definitions of race and ethnicity.

In its comments on the OMB Categories employed by the Draft Guidance, Genaissance focused on genetic accuracy, observing that

> Although these categories may be useful for national demographics, they are substandard with regard to the state-of-the-art in genetic analysis of ancestry. In a population such as the United States that increasingly is mixed, the boundaries between these classifications

are likely to be blurred further. For example, Genaissance has conducted genetic analysis of Hispanic populations from Florida and California. It is very clear that the label "Hispanic" encompasses individuals with African descent and Native American descent, as well as Caucasian descent.

—(GENAISSANCE 2003)

Genaissance here very subtly introduced the concept of genetic ancestry as a metric to assess the validity of the OMB categories of race and ethnicity. Nonetheless, Genaissance recognized the questionable accuracy of using *any* racial or ethnic categories in the context of pharmaceutical research and development. Commenting on the Draft Guidance's discussion of the relation between race and drug metabolism, Genaissance noted that "the link between these clinical outcomes and race is anecdotal at best and discriminatory at worst. New genetic technologies offer much more precise relationships between the genotype of an individual and the clinical management of disease" (Genaissance 2003).

Genaissance presented a solution to this problem in the form of its proprietary technology, which it asserted "would afford a high-resolution genetic identification of ancestry, consistent analysis of ethno-geographic backgrounds, and possible use directly to diagnostics for improvement of drug therapy" (Genaissance 2003). That is, it urged FDA to replace OMB categories of race with genetic categories of ancestry, recommending "the adoption of new genetic systems for ancestry determination rather than antiquated and potentially inaccurate racial denominations" (Genaissance 2003).

Unlike the suggestions from large pharmaceutical corporations, Genaissance here was less concerned with global uniformity per se and more with the purported scientific accuracy of the categories—accuracy to be provided by its own technology. One basic problem with Genaissance's claim to more rigorous scientifically based categories of ancestry was that its own discussion of them was premised on definitions of ancestral population that essentially replicated the OMB categories. Thus, in discussing its genetic analysis of samples from U.S. populations, it grouped the samples as "African American, Asian, Caucasian, and Hispanic/Latino" (Genaissance 2003).

In a similar vein, the comments from DNAPrint Genomics urged that "for the sake of science and the health of us all . . . it is time to incorporate molecular anthropological data metrics" to supplement the OMB categories in the collection of racial and ethnic data for clinical trials (DNAPrint 2003). It argued that its own proprietary genetic concept of "Biogeographical Ancestry" (BGA) was better suited to evaluating drug response than the OMB categories of race and ethnicity. Like Genaissance, DNAPrint emphasized the lack of accuracy inherent in self-reporting of race and suggested "that the FDA should pay more attention to molecular characterization of population structure when evaluating and assisting with the construction of clinical trials" (DNAPrint 2003).

Ironically, the genetic approaches taken by Genaissance and DNAPrint comport well with ICH E5's own categorization of race as an intrinsic genetic factor. However, the genetic approaches also suffer from the same dangers and inaccuracies as the ICH E5 definition despite the patina of scientific rigor layered upon them in the comments. In the end, where big pharmaceutical corporations simply wanted to regularize race and ethnic categories in whatever form they take so as to facilitate global drug development, small biotech companies wanted to take control of the actual process of racial and ethnic categorization and transform it into a function of genetics as applied through their proprietary technologies.

IV. NEGATIVE IMPACT ON GLOBAL TRIAL RECRUITMENT

Beyond inconsistency and inaccuracy, large pharmaceutical companies also expressed a pragmatic concern that being required to collect data according to the OMB categories could have a significant detrimental impact on their ability to recruit human subjects for clinical trials in a global environment. Here again, the category of Hispanic/Latino was of particular concern. Pharmacia addressed this issue most directly, noting that "Asking subjects about their race/ethnicity may be very sensitive in many circumstances and could be viewed as a bureaucratic burden. Conducting a study in Japan, e.g., and asking a subject whether they are Hispanic may result in patients taking questionnaires less seriously and compromising other data being collected" (Pharmacia 2003).

The specific reference to Japan echoes discussions surrounding the adoption of ICH E5 regarding the need to open up Japanese markets to Western pharmaceuticals. Companies wishing to conduct clinical studies in Japan in accordance with the ICH E5 guidelines certainly would not have wanted their efforts complicated or even subverted by the dictates of the FDA Guidance. Recruiting human subjects for clinical trials is difficult under the best of circumstances. In a commercial environment where ever-increasing numbers of clinical trials are being outsourced to countries around the world, Pharmacia recognized that the regulatory construction of race and ethnicity might pose a great barrier to the globalization of markets. This was a result of the medical construction of racial and ethnic difference that underlay Japanese barriers to the approval of Western pharmaceuticals.

V. THE CREATION OF UNNECESSARY AND UNSCIENTIFIC DIFFERENCE

Many of the comments challenged the accuracy and consistency of the OMB categories as a basis for collecting clinical data. An important subset of these concerns was a recognition by both Pharmacia and PhRMA that using the social categories of the OMB in the context of drug development might lead to the creation of the perception of relevant differences where in fact nonexistent differences would present unnecessary barriers to global drug development. Pharmacia noted that "The first paragraph [of the Draft Guidance] states that the categories are not based on scientific principles. It is understandable that the U.S. government wants to sort issues by various socio/cultural groups. However, if there is no scientific basis for examining the effects (either positive or negative) in these groups, doing so may provide an opportunity for identifying differences where none exist. Collecting the data by these definitions is one thing, using it to distinguish effects in different populations is another" (Pharmacia 2003).

Pharmacia recognized that racial and ethnic data was a double-edged sword. While it might be used to open up Japanese markets, it also might be misused and misinterpreted in a manner that obstructed markets. Once again, the OMB categories were criticized as barriers to globalization. Ironically, here commercial considerations to avoid having race be used as a barrier to commercial market expansion aligned with broader social and political concerns that race not be employed in a genetically reductivist manner.

VI. CONCLUSION

FDA responded to the comments and issued its final Guidance in September 2005. The final Guidance remained substantially the same as the draft. Among the significant revisions was added text that allowed the omission of the characterization of Hispanic or Latino for international clinical trials and a change in the characterization of "Black, of African Heritage" to "Black" for studies conducted abroad (70 Fed. Reg. 54946). The Guidance continued to recommend the use of the OMB categories when collecting data, even for studies conducted outside the United States, but recognizes that "these categories may not adequately describe racial and ethnic groups in foreign countries" (FDA 2005). FDA, therefore, did seem to have been at least somewhat responsive to the concerns expressed by large pharmaceutical companies that the categories not impede global research, development, and marketing. Nonetheless, the final Guidance made no concession toward the suggestions to adopt purportedly more genetically based classifications of ancestry proposed by Genaissance and DNAPrint.

The story of the FDA Guidelines points up the enduring conceptual power of race and ethnicity to shape understandings of human populations in diverse venues. Of greatest concern, perhaps, is that the drive to regulate race also threatens to geneticize race. Throughout this story, race and ethnicity were presented largely as barriers to globalization—differences that need to be somehow overcome in order for markets to grow. Harmonization, both as an explicit concern of the ICH and as the unavoidable backdrop to the FDA Guidance, provided the impetus to produce regular, standardized categories of race and ethnicity. In the discussions surrounding FDA Guidance, a prominent attribute of calls for arguments about the proper treatment of race was an appeal to genetics—whether as an "intrinsic" aspect of race or as a component of "Biogeographical Ancestry." These purportedly more objective or scientific understandings of race and ethnicity as a function of genetics were proposed as a means to stabilize the inconsistency of social categories and provide a basis for unifying global markets. In the drive to harmonize international drug development, we must be careful to avoid adopting a harmonized conception of race as genetic.

NOTE

This chapter draws extensively on the following article published previously by the author: "Harmonizing Race: Competing Regulatory Paradigms of Racial Categorization in International Drug Development." 5 *Santa Clara Journal of International Law* 35 (2006).

REFERENCES

68 *Fed. Reg.* 4788 (January 24, 2003).
70 *Fed. Reg.* 54,946 (September 19, 2005).
Abbott Laboratories. 2003. "Comments on the Draft Guidance for Industry on the Collection of Race and Ethnicity Data in Clinical Trials for FDA Regulated Products." http://www.fda.gov/ohrms/dockets/dailys/03/Mar03/032703/80059c93.pdf.
American Anthropological Association. 1998. "Statement on Race." *American Anthropological Association*, http://www.aaanet.org/stmts/racepp.htm (accessed March 3, 2014).
Bristol-Myers Squibb Pharmaceutical Research Institute. 2003. "RE: Docket No. 02D-0018." (March 24) http://www.fda.gov/ohrms/dockets/dailys/03/Mar03/032603/02d-0018-c000006-01-vol1.pdf.
Burchard, E. G., E. Ziv, N. Coyle, S. L. Gomez, H. Tang, A. J. Karter, et al. 2003. "The Importance of Race and Ethnic Background in Biomedical Research and Clinical Practice." *New England Journal of Medicine* 348:1170–75.
Collins, Francis. 2004. "What We Do and Don't Know About 'Race,' 'Ethnicity,' Genetics and Health at the Dawn of the Genome Era." *Nature Genetics* 36:S13–S15.
Cooper, R. S., J. S. Kaufman, and R. Ward. 2003. "Race and Genomics," *New England Journal of Medicine* 348:1166–70.
DNAPrint Genomics, Inc. 2003. "Collection of Race and Ethnicity Data in Clinical Trials for FDA Regulated Products; Draft Guidance, Comment Number: EC," Docket: 02D-0018 (February 4). http://www.fda.gov/ohrms/dockets/dailys/03/Feb03/020603/8004e14c.html.
Editorial. 2005. "Illuminating BiDil." *Nature Biotechnology* 23:903. http://www.nature.com/nbt/journal/v23/n8/full/nbt0805-903.html.
Epstein, Steven. 2007. *Inclusion: The Politics of Difference in Medical Research*. Chicago: University of Chicago Press.
Food and Drug Administration Modernization Act of 1997, Pub. L. No. 105-115, 111 Stat. 2296 (1997).

Genaissance Pharmaceuticals. 2003. "Commentary on the FDA Guidance for Industry"; "Collection of Race and Ethnicity Data in Clinical Trials." *Federal Register* Docket No. 02D-0018 (March 28). http://www.fda.gov/ohrms/dockets/dailys/03/Apr03/040103/8005b2de.pdf.

Gomez, Laura and Nancy Lopez, ed. 2013. *Mapping 'Race': Critical Approaches to Health Disparities Research*. New Brunswick: Rutgers University Press.

International Conference on Harmonization. n.d. "About ICH: History." http://www.ich.org/about/history.html (accessed March 3, 2014).

———. 1998. "Ethnic Factors in the Acceptability of Foreign Clinical Data." http://www.ich.org/fileadmin/Public_Web_Site/ICH_Products/Guidelines/Efficacy/E5_R1/Step4/E5_R1__Guideline.pdf.

Krimsky, Sheldon and Kathleen Sloan, ed. 2011. *Race and the Genetic Revolution*. New York: Columbia University Press.

Kuo W. H. 2008. "Understanding Race at the Frontier of Pharmaceutical Regulation: An Analysis of the Racial Difference Debate at the ICH." *Journal of Law, Medicine & Ethics* 36:498–505.

Lee, J. John. 2005. "Comment: What Is Past Is Prologue: The International Conference on Harmonization and Lessons Learned from European Drug Regulations Harmonization." *University of Pennsylvania Journal of International Economic Law* 26:155.

Lewontin, Richard. 1972. "The Apportionment of Human Diversity," *Evolutionary Biology* 6:381–84.

NIH Revitalization Act of 1993, Pub. L. No. 103-43, 107 Stat. 122 (1993).

Office of Management and Budget. 1997. "Revisions to the Standards for the Classification of Federal Data on Race and Ethnicity." http://www.whitehouse.gov/omb/fedreg_1997standards/.

Pharmaceutical Research and Manufacturers of America (PhRMA). n.d. "About PhRMA." http://www.phrma.org/about_phrma/ (last visited September 25, 2006).

———. 2003. "PhRMA Comments/Recommendations – Draft Guidance for Industry on the Collection of Race and Ethnicity Data in Clinical Trials for FDA Regulated Products." PhRMA. http://www.fda.gov/ohrms/dockets/dailys/03/Apr03/040103/8005b2dc.pdf.

Pharmacia. 2003. "Pharmacia's 5 March 2003 Comments re Guidance for Industry – Collection of Race and Ethnicity Data in Clinical Trials." Pharmacia. http://www.fda.gov/ohrms/dockets/dailys/03/Mar03/030703/02d-0018-c000001-01-vol1.pdf.

U.S. Food and Drug Administration (FDA). 1999. "Guidance for Industry: Population Pharmacokinetics." http://www.fda.gov/downloads/drugs/guidancecomplianceregulatoryinformation/guidances/ucm072137.pdf.

———. 2003a. "Draft Guidance for Industry: Collection of Race and Ethnicity Data in Clinical Trials." http://www.fda.gov/OHRMS/DOCKETS/98fr/5054dft.doc.

———. 2003b. "FDA Talk Paper: FDA Issues Guidance for Collection of Race and Ethnicity Data in Clinical Trials for FDA Regulated Products." http://aidsinfo.nih.gov/news/649/fda-issues-guidance-for-collection-of-race-and-ethnicity-data-in-clinical-trials-for-fda-regulated-products.

———. 2005. "Guidance for Industry: Collection of Race and Ethnicity Data in Clinical Trials." http://www.fda.gov/downloads/RegulatoryInformation/Guidances/ucm126396.pdf.

Contributors

Mark Barnes, JD. Harvard Law School; Multi-Regional Clinical Trials Center, Harvard University; Ropes & Gray, LLP
Barbara Bierer, MD. Harvard Medical School; Multi-Regional Clinical Trials Center, Harvard University; Brigham and Women's Hospital
Marie Boyd, JD. University of South Carolina School of Law
Daniel Carpenter, PhD, AM. Center for American Political Studies, Harvard University
R. Alta Charo, JD. University of Wisconsin Law School and School of Medicine & Public Health
I. Glenn Cohen, JD. The Petrie-Flom Center for Health Law Policy, Biotechnology, and Bioethics, Harvard Law School
Katrice Bridges Copeland, JD. Penn State Law
Nathan Cortez, JD. Southern Methodist University Dedman School of Law
Jennifer Devine, JD, LLM. U.S. Food and Drug Administration, Office of Global Regulatory Operations and Policy (when chapter was written)
Alla Digilova, JD. Multi-Regional Clinical Trials Center, Harvard University
Andrew English, JD. Willkie Farr & Gallagher, LLP
Barbara J. Evans, PhD, JD, LLM. University of Houston Law Center
Shannon Gibson, LLB, LLM. University of Toronto Faculty of Law
Henry Grabowski, PhD. Duke University
Jeremy Greene, MD, PhD. Johns Hopkins University
Kate Greenwood, JD. Seton Hall University School of Law
Lewis A. Grossman, PhD, JD. American University Washington College of Law; Covington & Burling, LLP
Peter Barton Hutt, LLB, LLM. Covington & Burling, LLP
Jonathan Kahn, JD, PhD. Hamline University Law School
Aaron S. Kesselheim, MD, MPH, JD. Program on Regulation, Therapeutics, and Law (PORTAL), Brigham and Women's Hospital; Harvard Medical School

Coleen Klasmeier, JD. Sidley Austin, LLP
Geoffrey Levitt, JD. Pfizer, Inc.
Rebecca Li, PhD. Multi-Regional Clinical Trials Center, Harvard University
Erika Lietzan, JD, MA. University of Missouri School of Law
Holly Fernandez Lynch, JD, M.Bioethics. The Petrie-Flom Center for Health Law Policy, Biotechnology, and Bioethics, Harvard Law School
Trudo Lemmens, LLL, LLM, DCL. University of Toronto Faculty of Law
Michelle M. Mello, PhD, M.Phil., JD. Stanford Law School; Stanford School of Medicine
Frances H. Miller, JD. University of Hawaii at Manoa; Boston University School of Law
Susan Moffitt, PhD, MPP. Brown University
Patrick O'Leary, JD. Sidley Austin, LLP
Efthimios Parasidis, JD, M.Bioethics. Moritz College of Law and College of Public Health, The Ohio State University
Genevieve Pham-Kanter, PhD. School of Public Health, Drexel University; Colorado School of Public Health; University of Colorado Denver; The Edmond J. Safra Center for Ethics, Harvard University
Elizabeth R. Pike, JD, LLM. Presidential Commission for the Study of Bioethical Issues, Department of Health and Human Services
W. Nicholson Price II, JD, PhD. University of New Hampshire School of Law
Arti Rai, JD. Duke University School of Law
Martin H. Redish, JD. Northwestern University School of Law; Sidley Austin LLP
Margaret Foster Riley, JD. University of Virginia School of Law
Christopher T. Robertson, PhD, JD. James E. Rogers College of Law, University of Arizona; The Petrie-Flom Center for Health Law Policy, Biotechnology and Bioethics, Harvard Law School; The Edmond J. Safra Center for Ethics, Harvard University
Benjamin N. Roin, JD. MIT Sloan School of Management
David Rosenberg, JD. Harvard Law School
Theodore W. Ruger, JD. University of Pennsylvania Law School
Howard Sklamberg, JD, MALD. U.S. Food and Drug Administration, Office of Global Regulatory Operations and Policy
Kayte Spector-Bagdady, JD, M.Bioethics. Presidential Commission for the Study of Bioethical Issues, Department of Health and Human Services
Huaou Yan, Law Clerk to Judge James Knoll Gardner, Eastern District of Pennsylvania (2015-17)

Index

Abbott Laboratories, 508
abbreviated new drug application (ANDA), 318, 319, 324*n*20, 385, 417
AbbVie, 124
Abigail Alliance for Better Access to Developmental Drugs, 180–81
Abigail Alliance v. Eschenbach, 180, 182, 183*n*1
abnormalities, 457
abuse: of drugs, 3; of *Park* doctrine, 169
ACA. *See* Affordable Care Act of 2010
academic-private partnership platforms, 128–29
Accelerated Approval rule, 67
accelerated approvals, 252, 255–56
acceptance, of biosimilar interchangeability, 422–23
accepted medical practice, off-label use and, 221–22
access: to drugs, 60, 369–71, 374–78; to information, 94; to markets, 275–78; to medicine, 85; to new therapies, 272–75; to participant-level clinical trials data, 118

accountability, 6, 110
accuracy, of racial and ethnic categories, 509–11, 512
Acme Markets, Inc., 149
active pharmaceutical ingredient (API), 40
active postmarket analysis: cost of, 295–97, 298; facilitation of, 292–95; health information technology and, 292–95; regulatory authority and, 292–95; role of, 286–99, 287–91; sponsors and, 10, 287, 288–89, 290, 291, 292, 293, 294, 295, 296, 297–98; tort claims relating to, 290–91
ACT UP, 65, 66
acute diseases, 2
Administrative Conference, 386
administrative law judge (ALJ), 151, 152
Administrative Procedure Act (APA), 23, 386
adulterated products, 42, 79, 82, 361
adult stem cells, 456, 460
adverse drug experience, 98, 99–100
advertising, 259–61

advocacy: for AIDS, 65–67; for patients, 35, 36, 59
Affordable Care Act of 2010 (ACA), 83, 92, 182, 402, 411, 414
African American, 504, 508, 510
African Medicines Regulatory Harmonization (AMRH), 54
agencies: FDA relating to, 18, 39; regulatory, 46–47, 52–53. *See also* digital world, analog agency in; European Medicines Agency; health agency communication; interagency coordination
aggregated results, of data, 116
AIDS, 35; advocacy for, 65–67; AZT for, 65, 66; food and drug activism rise, 62; ganciclovir for, 66; IND for, 65, 66, 67, 69; NDA for, 65–67, 68; successful movement for, 71; Treatment IND rule, 65; trimetrexate, 66
AIM, 439
Alaska Native, 504, 510
ALJ. *See* administrative law judge
Alzheimer's disease, 37, 254, 255, 471, 480
amendments. *See specific amendments*
America Invents Act, 354
American Cancer Society, 72
American Indian, 504, 510
American Medical Association, 62
Amgen, 427
AMRH. *See* African Medicines Regulatory Harmonization
amygdalin (Laetrile), 64–65
analog agency. *See* digital world, analog agency in
analysis: risk-benefit, 59, 98. *See also* postmarket analysis
analysis system, risk identification and, 95–96

Anatomical Therapeutic Chemical (ATC) Classification system, 314, *314–15*
ANDA. *See* abbreviated new drug application
animal products, 3
antibiotics, 97, 263
Antiviral Drugs Advisory Committee, 68
APA. *See* Administrative Procedure Act
API. *See* active pharmaceutical ingredient
approvals: accelerated, 252, 255–56; of biosimilars, 13, 404, 419; conditional, 258, 259, 262–63; of drugs, 24–26, 36, 37, 59, 67, 269–81; fast track, 251–52, 272; market, 85–86; premarket, 9, 95, 247, 435, 444, 476; of regulators, 127, 135. *See also* global marketing approval system; new drugs; staggered approval model
Arenella, Peter, 156–57
artificial intelligence, 440–41
Ashcroft v. Free Speech Coalition, 213
Asian, 504, 510
ATC. *See* Anatomical Therapeutic Chemical Classification system
authority: of FDASIA, 44; for user fees, 336, 337. *See also* legislative authorities; regulatory authority
authorizing statute, of FDA, 1
autologous stem cells, 456, 460; manipulation of, 462–63
autologous treatments, 461–62; MSC, 465
Avandia, 486. *See also* rosiglitazone
AZT, for AIDS, 65, 66

baldness, 477
ban. *See* off-label promotion; off-label use
Barnes, Mark, 6, 109, 115–33
barriers: procedural, 348–49; regulatory, 347–49
BCPIA. *See* Biologics Competition and Innovation Act of 2009
benefits: of data transparency, 116–19; of e-prescription, 492–97, 498*n*5; of negotiated rule making, 393–94; of products, 36, 37. *See also* risk-benefit analysis
beta blockers, 253
BGA. *See* Biographical Ancestry
Bierer, Barbara, 6, 109, 115–33
bifurcated DTC genetic testing, 474–75
Bill and Melinda Gates Foundation, 54, 55
BIO. *See* Biotechnology Industry Organization
Biographical Ancestry (BGA), 511, 513
biological macromolecules, 407
biological product regulation, 3
biologic license applications (BLAs), 416, 426–28
biologics, 345; follow-on, 402–13
Biologics Competition and Innovation Act of 2009 (BCPIA), 354
Biologics Price Competition and Innovation Act of 2010 (BPCIA), 12, 402; exclusivities of, 403–6, 407, 409, 410–11, 415–16; Hatch-Waxman compared to, 404–6, 416, 420, 424, 428; implementation of, 415; objective of, 414, 428; pathway of, 403–4
biomarkers, genetic, 271

biomedical industry, 7, 94, 112; broken system, 163–67; conclusion to, 173; enforcement in, 162–75; reform proposal, 167–72
biomedicine. *See* racial and ethnic categories, in biomedicine
biosimilarity, 415, 417
biosimilars, 12, 408; approval of, 13, 404, 419; clinical data requirements for, 419; costs of, 418–26; definition of, 414–15; disclosure for, 354; generations of, 427–28; manufacturers of, 418–20; molecules and, 414; premarket investments for, 418; regulation standards for, 417–20; user fees for, 26, 44. *See also* interchangeability designations, of biosimilars
biosimilars, FDA regulation of, 364–65; biosimilarity, 415; conclusion to, 428–29; costs associated with, 418–20, 424–26; innovation incentives and exclusivity provisions, 415–16; introduction to, 414–17; strategic options for, 426–28
biosimilars, firms for: comparability protocols of, 425; investment of, 424–25; labeling issues for, 425–26; postmarket organization infrastructure of, 424–26; product information from, 425; regulatory compliance of, 424–26; tort liability of, 426
Biosimilar User Fee Act of 2012 (BsUFA), 418
biospecimens, 102
Biotechnology Industry Organization (BIO), 406–7
Black, 504, 508

BLAs. *See* biologic license applications
blood group polymorphisms, 501–2
Bloom Syndrome, 473
Blue Cross-Blue Shield, 2
Blumberg, Eric, 165
Bolar, 318–19
books: on health information, 61–62. *See also* literature
Boyd, Marie, 12, 382–401
BPCIA. *See* Biologics Price Competition and Innovation Act of 2010
brand-name drugs, 88, 383, 384
Brannigan, Vincent, 447, 448, 449
BRCA mutations, 435, 477
breakthrough therapies, 69, 252, 258, 273
breast cancer, 62, 67, 471, 477, 479, 480
Bristol-Myers Squibb, 371, 508, 509
British Medical Journal policy, 125–29
broken system, of biomedical industry, 163–67
Brown & Williamson, 87
BsUFA. *See* Biosimilar User Fee Act of 2012
Buffalo Pharmacal Company, Inc., 148

cancer: breast, 62, 67, 471, 477, 479, 480; ovarian, 480
cancer-related therapies, 68
cancer representatives, 68, 72*n*1
Cannan, Keith, 309, 310
cap, on drug prices, 375–76
Caribbean regulatory system, 55
Caronia, Alfred, 184, 189–90
Carpenter, Daniel, 10, 77, 306–25
case selection, under *Park* doctrine, 170–71

categorical standard, 224–25, 229
Caucasian, 510
CBE. *See* changes-being-effected regulations
CDRH. *See* FDA Center for Devices and Radiological Health
CED. *See* coverage with evidence development
cells: MSCs, 456, 457, 460, 461; pluripotent, 457, 459; sources of, 456–57; therapies, 462–63; totipotent, 457. *See also* stem cells
Center for Medicare and Medicaid Services (CMS), 71, 88, 292
Centers for Disease Control and Prevention, 494
Central Hudson Gas & Electric Corp. v. Public Service Commission of New York, 188, 223
Central Hudson test, 190, 192, 193–94, 195–96; application of, 229–32; evolution of, 223–24, 225; false or misleading speech, 229–30; government's substantial interest, 231; regulation extent, 231–32
cGMP. *See* Good Manufacturing Processes
changes-being-effected (CBE) regulations, 384, 385–86
Charo, R. Alta, 9, 303–5
Chin, Denny, 190–91
Chin, Gabriel, 153–54
chromatography, 348
chronic diseases, 2
CIAs. *See* Corporate Integrity Agreements
Ciba Corp v. Weinberger, 316
civil death, 111, 153–54, 156, 159
claims: failure-to-warn, 382, 383. *See also* promotional claims; tort claims

Class I medical devices, 445, 475, 477
Class II medical devices, 445, 447, 475, 476, 477, 478
Class III medical devices, 27, 435, 473, 475–76, 477–80
CLIA. *See* Clinical Laboratory Improvements Amendments
clinical data requirements, for biosimilars, 419
Clinical Laboratory Improvements Amendments (CLIA), 480, 481
clinical software definition, 452
clinical studies, in Japan, 511–12
Clinical Translation of Stem Cell research, 458
clinical trials, 2; biosimilar costs of, 418; controlled, 80, 101–3, 253; design of, 84; disclosure of, 337, 340*n*5; foreign, 84; globalization of, 118; multinational, 38; negative, 336–37; premarket, 95, 247–48, 254–55; RCTs, 101–3, 330. *See also* data; participant-level clinical trials data
ClinicalTrials.gov, 120, 122
clinic treatments, of stem cells, 460–61
CMS. *See* Center for Medicare and Medicaid Services
coal-tar colors, user fees and, 24
Cohen, Glenn, 179–83
collateral consequences, 146–61
"Collection of Race and Ethnicity Data in Clinical Trials," 503
Commerce Clause, 79, 182, 240, 464–65
commercial speech: courts and, 187–89, 194–95, 206, 213; First Amendment and, 207–9, 213; jurisprudence of, 261–62; protected expression and, 233; Supreme Court relating to, 187–89, 194–95, 206, 213, 224–26
commercial speech, off-label promotion prohibition and doctrine of: categorical standard, 224–25, 229; Central Hudson test, 190, 192, 193–94, 195–96, 223–24, 225, 229–32; consumer protection, 228–29; doctrinal framework for, 223–25; justification for, 225–26; measurement of, 225–29; nonexpressive conduct regulation, 226–28
commissioners, 20–21, 26
committees: for negotiated rule making, 387, 389–96, 397*n*4. *See also specific committees*
communication: health agency, 334, 335, 336, 338, 340*n*2; regulators of, 329, 335, 340*n*4; about safety, 11. *See also* drugs
company-specific platforms, 126–27
comparability protocols, of biosimilar firms, 425
comparative efficacy and cost, 83, 84
comparative information, 37
comparison studies, of new drugs, 80
competition, for orphan drugs, 375
compliance. *See* regulatory compliance, of biosimilar firms
compliance, of e-prescription system, 489, 491; benefits, 490, 494; notifications with, 490; override feature with, 490; patient records searched with, 490
Compliance Policy Guides, 23
computerized medical devices, 433, 434
computerized medicine, 439–42

computers: health care access expanded with, 441; health care costs controlled by, 440; medical errors reduced with, 440
concrete issues, for negotiated rule making, 390–91, 397nn5–7
conditional approvals, 258, 259, 262–63
conduct and speech regulations, 237–39
confidence, in FDA, 60–61
confidentiality, 116, 123
conflicts of interest, 112–13, 118; conclusion to, 144; data and methods relating to, 140–41; introduction to, 134–37; multiple physician-industry ties, 138–40; previous literature about, 137–38, 142; results of, 141–43
Congress: digital world relating to, 442–43; hearings held by, 443; laws passed by, 443
Congressional intervention, about activism, 63
constitutional protection, 236–37
consumer-directed health care, 5
consumers: behavior of, 76, 81; protection of, 228–29; republic of, 81
consumption patterns, 76, 78, 79, 82, 86
contamination, with regenerative medicine, 461–62
contraceptive devices, 2
control, of drug safety communication: drug sponsor, 334–38, 339; health agency, 334, 335, 336, 338, 340n2; regulators, 335, 340n4; third parties, 334–35, 340n3; trust relating to, 338

controlled clinical trials, 80, 101–3, 253
convergence. See harmonization
cooperative enforcement, 146
Copeland, Katrice Bridges, 7, 109, 110–12
corporate integrity, 1, 7–8
Corporate Integrity Agreements (CIAs), 7–8, 112, 162, 163
corporate officers: accountability of, 110; ignorance of, 111; misbranding charges faced by, 149, 159nn1–2; misconduct of, 111; probation of, 155; targeting of, 165–67. See also responsibility
Cortez, Nathan, 13, 438–54
costs: of active postmarket analysis, 295–97, 298; comparative efficacy and, 83, 84; containment of, 85; of health care, 35; of negotiated rule making, 395–96, 398n9; of orphan drugs, 369–71; product, 76; of R&D, 210, 263, 280; safety and, 80. See also biosimilars
costs, of biosimilars: clinical trials, 418; for clinical trials, 418; for development, 419; interchangeability designations of, 423–24; for manufacturing plants, 419; for postmarket activities, 424–26; quality and, 420; regulations of, 418–20, 424–26
costs, of e-prescription system: with design changes, 495; with federal funding, 494; for individual health care providers, 495; with infrastructure overhaul, 495; for medical malpractice, 496–97; with medical records, 495, 498n6; for prescribing

physicians, 495–96; for privacy, 496, 498n7
countervailing power, for negotiated rule making, 393, 398n8
courts: commercial speech and, 187–89, 194–95, 206, 213; enforcement actions of, 25; ignorance of, 211–14. *See also* Supreme Court
coverage with evidence development (CED), 280
credibility, of data sharing, 117
criminal prosecutions, 171
criminal punishment, exclusions as, 154–56
Critical Path Institute, 68
criticism, of negotiated rule making, 395–96
cross-labeling, 97
CTRP. *See* International Clinical Trials Registry Platform
cultural foundation, for empowered patient: changing health information environment, 61–62; confidence in FDA, 60–61; decline of trust, 60–61; rights revolution, 61
Current Good Tissue Practices, 460
cystic fibrosis, 471

DAB. *See* Departmental Appeals Board
DARRTS. *See* Document Archiving, Reporting and Regulatory Tracking System
data: about drugs, 11, 15, 19, 24; genomic, 14; and methods, of conflicts of interest, 140–41; protection safeguards for, 117; for racial and ethnic categories, 504; regulators relating to, 119, 122; research study, 330, 340n1; unmediated, uncurated, 117
data, sharing of: credibility of, 117; data gathering and, 3, 7, 37, 38; environment for, 117; model for, 115, 117; of participant-level clinical trials, 115–16, 122–24; plan for, 121; with sponsors, 116, 117, 118, 119
data, transparency of: for aggregated results, 116; *British Medical Journal* policy and, 125–29; conclusion to, 130; data sharing model relating to, 115, 117; early registration requirements, 120–21; FDA's role in, 122–23; history and catalysts of, 119–24; for participant-level, 115–30; policy makers efforts with, 119; for pooled results, 116; risks, benefits, and issues of, 116–19; for summary level, 116; voluntary industry relating to, 125–29
database, for of e-prescription system, 493, 498n4
death: from Avandia, 486; civil, 111, 153–54, 156, 159; from generic drugs, 383
Declaration of Helsinki, 116, 121
decline, of trust, 60–61
deCODE Genetics, 473
deCODEme, 470, 473
defamation, 213
definitions: of biosimilars, 414–15; of clinical software, 452; by FDAAA, 97–98; of interest, 391; of regenerative medicine, 456. *See also* inconsistent definitions
Delalutin. *See* 17-alpha hydroxyprogesterone caproate
Departmental Appeals Board (DAB), 152

Department of Health and Human Services (HHS), 150, 165, 202, 442
Department of Justice (DOJ), 165, 171, 172
Department of Veterans Affairs, 296
DePuy Orthopaedics, 487, 498n3
deregulation, vitamin, 63
DES. *See* Drug Efficacy Studies
DESI. *See* Drug Efficacy Study Initiative
designation: of orphan drug, 372–73; Priority Review, 272
design change costs, of e-prescription system, 495
deterrence, 162, 166–67; of misconduct, 163, 164; penalties relating to, 164–65, 168
development: of biosimilars, 419; trade, 47–48. *See also* research and development
devices, 1, 3; Class II, 447; contraceptive, 2; under FDCA, 439; software, 442, 452
devices, medical: computerized, 433, 434; industry for, 18, 29n3
Devine, Jennifer, 5, 36, 39–58
dietary supplements promotional claims, 208–9
difference, in software devices, 452
difference creation: unnecessary, 512; unscientific, 512
Digilova, Alla, 6, 109, 112, 115–33
digital world, analog agency in, 438; conclusion to, 453; Congress and, 442–43; new regulatory framework for, 451–53
digital world, computerized medicine in: aspirations for, 440–41; beginning of, 439–40; hazards of, 441–42

digital world, enduring concerns of: agency expertise, 449; different software, 448–49; innovation, regulation vs, 447–48; medical knowledge regulation, 450–51; paralyzed by change, 448; regulatory-naïve industry, 450
digital world, FDA and, 443; adjudication of, 445–46; considerations of, 444; guidance of, 446–47; organization of, 444–45; regulation of, 445
direct-to-consumer advertising (DTCA), 259–61
direct-to-consumer (DTC) genetic testing, 433, 434; bifurcated, 474–75; challenges of, 471; concerns of, 472; conclusion to, 481; current regulation of, 472–73; evolution of, 471–72; expanded scope of, 472; genomic information, 14, 470, 475–81
discipline, for drug safety communication, 332
disclosure, 329; for biosimilars, 354; for clinical trials, 337, 340n5; of genomic information, 480–81; mandatory, 117, 123; public, 293
diseases: acute, 2; chronic, 2; food and drug activism for, 67–69; prognosis of, 93; risk of, 476; system of, 92. *See also specific diseases*
disease-specific platforms, 127–28
distributional equity, 85
DNA DTC, 474
DNA molecule, 2
DNAPrint Genomics, 509, 510, 511
doctrinal framework, 223–25
Document Archiving, Reporting and Regulatory Tracking System (DARRTS), 289

Index 527

DOJ. *See* Department of Justice
domestically focused agency, FDA as, 39
Dotterweich, Joseph H., 148–49
DQSA. *See* Drug Quality and Security Act
Draft Guidance for Industry on the Collection of Race and Ethnicity Data for Clinical Trials for FDA Regulated Products, 504–5, 509–10, 512, 513
drug activism. *See* food
Drug Amendments of 1962, 18, 19
drug application user fees, 331
Drug Efficacy Studies (DES), 307, 313, 317, 318
Drug Efficacy Study Initiative (DESI), 10–11, 307, 308, *310*, *311*; effectiveness with, 318, 324n18; experience of, 303, 305; final report of, *317*; impact of, 319–22; panels, *310*, *311*, 312; as research tool, 319–22; withdrawals of, 313–17, *314–15*, 321
Drug Price Competition and Patent Term Restoration Act of 1984, 362, 369, 385
Drug Quality and Security Act (DQSA), 45–46, 353
drugs: abuse of, 3; adverse experience of, 98; approval of, 24–26, 36, 37, 59, 67, 269–81; brand-name, 88, 383, 384; cell therapy as, 462–63; data about, 11, 15, 19, 24; effectiveness of, 3, 11, 309, 318; expenditures on, 343–44; expense of, 344–45; generic chemical, 420; grandfathered, 308–9, 318; as ineffective, 311, 312, 313, 317, 318, 319; lag in, 77; off-label, 8;

old, 317–18; pharmacogenomic, 276–77; postmarket analysis and review of, 287–91; prescription, 63, 83; price cap on, 375–76; recall of, 343; regulation of, 1–2, 10, 80, 383–85; RLD, 385, 397n2; shortages of, 343; sponsors of, 334–38, 339; unapproved, 7, 8, 60. *See also* generic drugs; new drugs; safety; *specific drugs*
drugs, efficacy of, 306–7; old drugs market, 317–18; regulating in retrospect, 308–13; unmaking markets, 313–17
drugs, safety communication about, 328; concluding thoughts about, 339–40; control of, 334–38; discipline about, 332; doubts about, 329; goal of, 339; governance with, 329; new model for, 339; sources for, 332–34; trust with, 338; as valid and substantiated, 330–31
Drug Safety Oversight Board, 336
drug supply chain safety, 40–42, *41*, 44, 45
Drug Supply Chain Security Act (DSCSA), 45–46
DTC. *See* direct-to-consumer genetic testing
DTCA. *See* direct-to-consumer advertising
DTC HIV tests, 476, 477, 479

earwax, 471, 477
Eastern Research Group (ERG), 137
economical attractiveness, of misconduct, 164
Edwards, Charles C., 20–21, 28, 312, 323n9

effectiveness: with DESI, 318, 324*n*18; of drugs, 3, 11, 309, 316, 318; of genomic information, 480, 481; postmarketing period, 97–99, 100

efficacy, 78; cost compared to, 83, 84; failures of, 98, 102; of products, 162; and safety, of new drugs, 79, 80, 83, 185–86; studies of, 100–102. *See also* drug efficacy; Drug Efficacy Studies; Drug Efficacy Study Initiative

EFPIA. *See* European Federation of Pharmaceutical Industries and Associations

Eighth Amendment, 239

electronic health records, 37, 436

elements to ensure safe use, 96–97

EMA. *See* European Medicines Agency

embryonic stem cells (hESCs), 456–57, 458

employees, 20–21, 59

empowered patient, rise of, 5, 36–37, 72*n*2; beyond AIDS, 67–70; AIDS advocacy, 65–67; cultural foundation for, 60–62; drug approval and, 59; food and drug activism rise, 62–65; for future challenges, 70–72; patient advocacy groups, 35, 36, 59; risk-benefit analysis and, 59; as special government employees, 59; unapproved drug access and, 60

end-stage pricing, 78, 82

enforcement, 101; in biomedical industry, 162–75; cooperative, 146; court enforcement actions decrease, 25; of FDA policies, 24; of FDCA, 393, 458; of individuals, 167–70, 171, 174*n*5; against pharmaceutical companies, 109–13, 146; power of, 7–8; publicity with, 24; Recall Requests, 24; Warning Letters, 24, 473, 479. *See also* federal enforcement regime

English, Andrew, 14–15, 486–500

enhancements: of FDASIA, 45; for public health, 69; for technologies, 14

"Enhancing Benefit-Risk Assessment in Regulatory Decision-Making," 70

environment: of changing health information, 61–62; for data sharing, 117; regulatory, 5. *See also* global and innovative regulatory environment

epidemic, HIV, 2, 3

epoetin alfa (Procrit), 421, 427

e-prescription: conclusion to, 497; forms for, 489; introduction to, 486–87; McGill University system with, 488; for physician and patient records, 488; potential costs of, 494–97; regularity function for, 486, 487

e-prescription, operational overview of: compliance, 489–91; monitoring, 486, 488, 491–92

e-prescription benefits: with compliance, 492, 494; with database, 493, 498*n*4; with monitoring, 492; with off-label medications, 493; with physicians, 494, 498*n*5

Equal Protection clause, 180, 182

ERG. *See* Eastern Research Group

erythropoiesis-stimulating agents (ESAs), 421

estrogen, 253

Index 529

ethnic factors: extrinsic, 506; intrinsic, 506, 513
"Ethnic Factors in the Acceptability of Foreign Clinical Data," 505, 506
ethnicity. *See* racial and ethnic categories, in biomedicine
EU. *See* European Union
European Commission, 258
European Federation of Pharmaceutical Industries and Associations (EFPIA), 124, 337, 505
European Medicines Agency (EMA), 9, 84, 117, 118, 122, 279, 408; biosimilar approval of, 419; proposed policies of, 123–24, 258, 274, 337, 340*n*6
European Union (EU), 51–52, 88, 118, 262, 408, 409; EMA and, 84, 122, 123, 124, 258
evaluation: of biosimilar interchangeability designations, 420–21; of exclusivities, 406–12; of statutory exclusivities and patent dances, 406–12
Evans, Barbara J., 6, 37, 92–103
evidence: CED, 280; of drug effectiveness, 316
evidence-based medicine, 253
evidence generation, postmarket, 278–79
evolution, of DTC genetic testing, 471–72
exclusions: authority of, 150; as criminal punishment, 154–56; as harsh remedy, 153–54; of individual, 155–56; mandatory, 150; misuse of, 154–56; under *Park* doctrine, 165, 166, 172; permissive, 150, 171; purpose of, 155; of responsible corporate officers, 149–52; rules of, 110–12
exclusive relationships, 138, 141
exclusive ties, multiple financial ties vs, 141–42
exclusivities: of biosimilar interchangeability designations, 420; of BPCIA, 403–6, 407, 409, 410–11, 415–16; with Hatch-Waxman Act, 402–3, 404–6; orphan drug, 372, 374–75; period of, 374–75; regularity, 352, 354. *See also* statutory exclusivities, patent dances and
executive culpability, 7; background of, 148–49; collateral consequences and, 146–61; conclusion to, 159; introduction to, 146–48; responsible corporate officers exclusion, 149–52; responsible corporate officers moral blameworthiness, 152–59
expanded scope, of DTC genetic testing, 472
expedited review programs, 274–75
expenditures, on drugs, 343–44
expense: of drugs, 344–45; of public health, 165, 173*n*2
extrinsic ethnic factors, 506

facilitation, of active postmarket analysis, 292–95
Facing Our Risk of Cancer Empowered (FORCE), 68
FAERS. *See* FDA Adverse Event Reporting System
failures: efficacy, 98, 102; regulatory, 76
failure-to-warn claims, 382, 383
false or misleading speech, 192–93, 229–30
FamilyTreeCNA, 474

fast track approval, 251–52, 272
FCC. *See* Federal Communications Commission
FDA. *See* Food and Drug Administration
FDAAA. *See* FDA Amendments Act of 2007
FDA Adverse Event Reporting System (FAERS), 288–89
FDA Amendments Act of 2007 (FDAAA), 6, 278–79; definitions by, 97–98; drug safety approaches of, 96; passage of, 121–22; postmarket surveillance system under, 288; regulatory powers under, 95–97; REMS and, 276; safety under, 96–97
FDA Center for Devices and Radiological Health (CDRH), 444
FDA Center for Drug Evaluation and Research, 7, 141
FDA century, 78–81
FDA draft guidance: fingerprinting in, 417; foreign product studies in, 418; totality of evidence in, 417
FDA Foreign Posts, *49*
FDAMA. *See* FDA Modernization Act
FDA Modernization Act (FDAMA), 44, 56*n*1, 503, 505
"FDA Patient Network," 70
FDA-regulated products, 39–40, 80, 93
FDA Safety and Innovation Act of 2012 (FDASIA), 5; authority of, 44; breakthrough therapy of, 273; drug supply chain safety with, 40–42, *41*, 44, 45; enhancements of, 45; patient-centered ethos with, 60, 69–70;
Title VII of, 44, 45; user fees with, 44
FDA Science Board, 27
FDASIA. *See* FDA Safety and Innovation Act of 2012
FDASIA Health IT Report, 451, 452
FDCA. *See* Federal Food, Drug, and Cosmetic Act
Federal Communications Commission (FCC), 14, 452, 453
federal enforcement regime, 170
Federal Food, Drug, and Cosmetic Act (FDCA), 17, 21, 147, 148–49, 184, 294, 330; devices under, 439; enforcement of, 393, 458; as incentive for knowledge production, 210–11; as promotional claims truth test, 204–17; requirements of, 205, 307; as test of truth, 214–16; *Wyeth v. Levine* and, 384
Federal Food and Drugs Act, 2
federal funding, for e-prescription system, 494
federal mandates, for racial and ethnic categories, 502–4
Federal Register, 23, 304, 312, 313, 316, 324*n*11
Federal Trade Commission (FTC), 13, 87, 408, 452, 453
felonies: under *Park* doctrine, 166; of pharmaceutical companies, 146–47
fenfluramine/phentermine (Fen-Phen), 187
Field Force, 18, 19
Fifth Amendment Due Process, 180, 181
filgrastim (Neupogen), 421, 427
financial ties: to FDA votes, 109–10;

multiple, 134–44; voting behavior and types of, 143
fingerprinting, 417
First Amendment, 94, 182, 187–88, 191; burden in litigation of, 213; commercial speech and, 207–9, 213; free speech and, 248; off-label promotion relating to, 8–9, 179, 180, 184, 185, 192, 215; protection of, 233, 235; Supreme Court and, 9, 180, 185, 207, 214, 216, 233, 236, 238. *See also* off-label promotion
Flynn v. Holder, 181, 182
FOIA. *See* Freedom of Information Act
FOI Act, 23
follow-on biologics, 402–13
food, 3; health threat of, 82; poisoning of, 82; products of, 81; safety of, 77, 82, 89; in schools, 88–89; subsidization of, 88; supply of, 77
food, and drug activism rise: for AIDS, 62; for breast cancer, 62, 67; campaigns relating to, 62–63; Congressional intervention relating to, 63; for empowered patients, 62–65; impact of, 69; for other diseases, 67–69; prescription drugs, 63; vitamin and mineral supplements, 63, 64
Food and Drug Administration (FDA), 443; agencies relating to, 18, 39; authorizing statute of, 1; challenges of, 1, 2, 4, 15–16, 36, 70; changes with, 35, 38, 39; concerns of, 3; conference relating to, 4; data transparency and role of, 122–23; FDAAA relating to, 99–103; financial ties to votes of, 109–10; going forward of, 465–67; individual misdemeanor prosecutions by, 147; international presence of, 48–50; jurisdiction lack of, 87; mass protests against, 63; mission of, 162, 361; oversight of, 39, 40, 77; pharmaceutical company felonies not pursued by, 146–47; in postmodern world, 76–78; prestige of, 77; previous names for, 18; program alignment of, 47; purpose of, 286; race and, 15; regulators and, 52–53; standards of, 3; technical and scientific focus of, 81; themes and development of, 4–5, 17–29
Food and Drug Administration, advisory committees of, 7, 135; as research focus, 136–37; studies about, 137–38
Food and Drug Administration, drugs and: generic, 385–86, 397n2; orphan, 376–78
Food and Drug Administration, public health imperative of, 286; conclusion to, 297–99; postmarket analysis, 287–97
Food and Drug Administration, regulations of, 3, 4, 7, 47–48; for regenerative medicine, 459–60; waning, 81–86
Food and Drug Administration Modernization Act, 2, 26, 120
food and drug policy problems, 81–86
Food Inspections Decisions, 22
food-related illnesses, 43, 77–78, 82
Food Safety and Modernization Act (FSMA), 82; food related illnesses relating to, 43; prevention-oriented standards with, 43–44

FORCE. *See* Facing Our Risk of Cancer Empowered
foreign clinical trials, 84
foreign product studies, 418
forms, for e-prescription, 489
44 Liquormart v. Rhode Island, 225, 229
framework: doctrinal, 223–25; for postmarket surveillance, 288–90, 292, 294; regulatory, 78, 94, 95, 221–23
Freedom of Information Act (FOIA), 123
Free Exercise Clause, 239
free expression, protection of, 221
free expression, theory postulates of, 233; conduct and speech regulations, 237–39; constitutional protection relating to, 236–37; government extortion, 239–40; government manipulation relating to, 234–36
free speech, 248
Friedman, Michael, 151
Friedman v. Sebelius, 166, 169
FSMA. *See* Food Safety and Modernization Act
FTC. *See* Federal Trade Commission
FTC v. Actavis, 410
funding decisions, 279–81
Future of Drug Safety, 99

gains, for negotiated rule making, 392–93
ganciclovir, 66
GAO. *See* Government and Accountability Office, U.S.
gatekeepers, 71, 77, 79, 80, 85, 86
gathering, of data, 3, 7, 37, 38
Gay Men's Health Crisis, 65
G-CSFs. *See* granulocyte colony stimulating factors

Genaissance Pharmaceuticals, 509–10
Gene By Gene, 474
General Controls, 475
generations: of biosimilars, 427–28. *See also* postmarket evidence generation; second generation applications
generic chemical drugs, 420
generic drugs, 12, 250, 317, 324n15; deaths from, 383; failure-to-warn claims against, 382, 383; FDA's regulation of, 385–86, 397n3; injuries from, 382; labeling of, 382–83, 397n1; regulation of, 414–30; tort claims with, 382. *See also* negotiated rule making, generics and
genetic biomarkers, 271
genetic groupings, surrogates for, 501
genetic testing, 1, 14, 93. *See also* direct-to-consumer genetic testing
genetic variability, 509
genetic variations, with racial and ethnic categories, 502
genome services, 435
genomic data, 14
genomic information, 14; disclosure of, 480–81; as medical device, 470, 475; regulation pathway for, 480–81, 482n3; risk classification of, 475–77; risk mitigation strategies for, 478–79; safety and effectiveness of, 480, 481; validity of, 480
genomic sequencing, 180, 470, 472, 482n1
GHTF. *See* Global Harmonization Task Force on Medical Devices
Gibson, Shannon, 9–10, 268–85
glaucoma, 440

GlaxoSmithKline (GSK), 112, 126–28, 163, 338
global and innovative regulatory environment, 39–41; conclusion to, 56; legislative authorities, 42–46; positioning for, 46–55
global collaborations and partnerships, 50; harmonization and convergence, 52–53; information sharing, 51; international coalition, 53; mutual reliance/recognition, 51–52
global drug-manufacturing supply chain, 40–42, *41*
Global Harmonization Task Force on Medical Devices (GHTF), 52, 57*n*2
globalization, 1, 3, 35, 36, 118
global marketing approval system, 36
Global Regulatory Operations and Policy Directorate (GO), 46–47
global trial recruitment, 511–12
GO. *See* Global Regulatory Operations and Policy Directorate
goals: of drug safety communication, 339; regulatory, 77; of REMS, 289
Goddard, James L., 309, 310
Goldenheim, Paul D., 151
gold standard, 252, 253
Good Guidance Practices, 23
Good Manufacturing Processes (cGMP), 347, 348
governance, with drug safety communication, 329
government: Central Hudson test interest of, 231; employees of, 59; extortion of, 239–40; manipulation of, 234–36

Government and Accountability Office, U.S. (GAO), 472
Grabowski, Henry, 13, 414–30
grandfathered drugs, 308–9, 318
granulocyte colony stimulating factors (G-CSFs), 421
Great Depression, 2
Greene, Jeremy, 10, 306–27
Greenwood, Kate, 12, 366–81
Grossman, Lewis A., 5, 36, 59–75
GSK. *See* GlaxoSmithKline
guidance: documents relating to, 23, 99–102, 228, 252, 348, 417–18, 444, 451; FDA and, 417, 418, 446–47. *See also* Draft Guidance for Industry on the Collection of Race and Ethnicity Data for Clinical Trials for FDA Regulated Products
guidance, use of: Administrative Procedure Act, 23; Compliance Policy Guides, 23; FOI Act, 23; Food Inspections Decisions, 22; Good Guidance Practices, 23; Trade Correspondence, 22–23
guidelines, for racial and ethnic categories, 504–7

Hamburg, Margaret, 165
harmonization: convergence and, 52–53; of racial and ethnic categories, 505, 513
Harter, Philip, 386–87
Harvard Law School conference, 4
Hatch-Waxman Act, 2, 409, 410, 414; accomplishments of, 362–64; BPCIA compared to, 404–6, 416, 420, 424, 428; exclusivities with, 402–3, 404–6; provisions of, 12–13, 350
HCT/Ps. *See* human cellular and tissue based products

Headquarters, 18, 20
health: food threat, 82; mobile, 1, 13; system reorganization for, 85. *See also* public health
health, care of, 15; access expanded for, 441; consumer-directed, 5; costs of, 35, 440; information for, 3, 207; programs for, 3, 7
health, information about: books relating to, 61–62; changing environment of, 61–62; Internet for, 62; from manufacturers, 62; regulatory authority and technology for, 292–95
health agency communication, 334, 335, 336, 338, 340*n*2
healthcare provider costs, of e-prescription system, 495
health care-specific causes of action, 170–71, 174*n*6
hearings, by Congress, 443
heart attacks, 486
HELP, 439
hematopoietic stem cells (HSCs), 456
hESCs. *See* embryonic stem cells
heterogeneity, of treatment effects, 99
HHS. *See* Department of Health and Human Services
hip replacements, 487
Hispanic, 508, 509, 510, 511
history and catalysts, of data transparency, 119–24
HIV: epidemic of, 2, 3; tests for, 476, 477, 479
HIV/AIDS, 35, 66
homologous use standards, 166
hormone replacement therapies, 253
House Committee on Science and Technology, 442, 444
HSCs. *See* hematopoietic stem cells

human cellular and tissue based products (HCT/Ps), 459–60, 465
Human Genome Project, 471, 501
Huntington's disease, 479, 480
Hutt, Peter Barton, 4–5, 17–31

ICH. *See* International Conference on Harmonization
ICMJE. *See* International Committee of Medical Journal Editors
ICMRA. *See* International Coalition of Medicines Regulatory Authorities
ICTRP. *See* International Clinical Trials Registry Platform
identity standards, 81
IFPTI. *See* International Food Protection Training Institute
ignorance: of corporate officers, 111; of courts, 211–14
illnesses, food-related, 43, 77–78, 82
IMDRF. *See* International Medical Device Regulators Forum
Immune Tolerance Networks' (ITN) TrialShare portal, 129
immunity, for off-label promotion, 211, 290
immunogenicity issues, 457
impacted interests, for negotiated rule making, 391–92
incentives: for biosimilars, 415–16; for innovation, 344; intellectual-property, 349; for knowledge production, 210–11; for R&D, 254, 262
inconsistent definitions: Black or African American, 504, 508; Hispanic or Latino, 508, 509, 510, 511

IND. *See* investigational new drug
individuals: enforcement of, 167–70, 171, 174n5; exclusions of, 155–56; liability of, 168; misconduct of, 168; misdemeanor prosecutions of, 147
industries: biomedical, 7, 94, 112; medical device, 18, 29n3; regulatory-naïve, 450; sponsors of, 126–27, 128; transgressions of, 109; voluntary, 125–29
ineffective drugs, 311, 312, 313, 317, 318, 319
Infectious Diseases Society, 263
information: access to, 94; changing of, 61–62; comparative, 37; about drug safety, 11; patent, 405; restriction of, 94; sharing of, 51, 249. *See also* genomic information; health
information-only product, 471, 474–75, 477, 478, 480–81
infrastructure overhaul, of e-prescription system, 495
injuries, from generic drugs, 382
innovation, 11, 162, 251, 274, 316, 324n14; incentives and exclusivity provisions, for biosimilars, 415–16; incentives for, 344; lack of, 343; regulation vs, 447–48; retardation of, 321, 325n23. *See also* global and innovative regulatory environment
innovation policy failures, 343; conclusion to, 355; pharmaceutical manufacturing today, 344–46; procedural barriers, 348–49; regulatory barriers, 347–49; technological standards, 348
innovation policy failures, change proposals: intellectual property changes, 353–55; regulatory changes, 352–53
innovation policy failures, inadequate intellectual-property incentives, 349; patents, 350; regularity exclusivity, 352, 354; trade secrecy, 351, 354
Innovative Medicines Initiative, 262
in rem proceedings, 79, 89n1
Institute of Medicine (IOM) study, 54, 55, 99, 269, 288, 336
institutional payers, 87
insurance, 85
integrity, corporate, 1, 7–8
intellectual property changes, 353–55
intellectual-property incentives, inadequate, 349
intellectual property protection, 347, 355n1
interagency coordination, 170–72
interchangeability designations, of biosimilars: acceptance of, 422–23; costs of, 423–24; evaluation of, 420–21; exclusivity of, 420; license for, 421; names issues with, 422; regulatory issue of, 420; substitution with, 421–22; value of, 421
interest: in Central Hudson test, 231; definition of, 391; negotiated rulemaking's impact on, 391–92. *See also* conflicts of interest; negotiated rule making, generics and
InterMune, 124
International Agreement on Trade-Related Aspects of Intellectual Property Rights (TRIPS), 124
International Clinical Trials Registry Platform (ICTRP), 120
international coalition, 53

International Coalition of Medicines Regulatory Authorities (ICMRA), 53
International Committee of Medical Journal Editors (ICMJE), 120–21, 125
International Conference on Harmonization (ICH), 52–53, 437, 505, 506, 508, 511
International Food Protection Training Institute (IFPTI), 55
International Medical Device Regulators Forum (IMDRF), 52–53
international presence, of FDA, 48–50
International Society for Stem Cell Research (ISSCR), 458, 466
Internet, 62
intrinsic ethnic factors, 506, 513
investigational new drug (IND), 65, 66, 67, 69, 120, 309
investment, of biosimilar firms, 418, 424–25
IOM. *See* Institute of Medicine study
ISSCR. *See* International Society for Stem Cell Research
ITN. *See* Immune Tolerance Networks' TrialShare portal

Japan, clinical studies in, 511–12
Johnson & Johnson, 110, 129, 163, 338
jurisprudence, of commercial speech, 261–62

Kadish, Stanford, 157
Kahn, Jonathan, 15, 501–15
Katzenbach v. McClung, 240
Kefauver, Estes, 80
Kefauver-Harris Drug Amendments, 2, 10, 305, 307, 309

Kelsey, Frances O., 81, 252
Kesselbaum, Aaron, 8, 184–203
Kessler, David, 87
Ketek. *See* telithromycin
Klasmeier, Coleen, 9, 219–43
knowledge, medical, 450–51
knowledge production, FDCA for, 210–11
K-V Pharmaceutical, 367, 373

labeling, 478; changes with, 97, 98; cross-labeling, 97; of generic drugs, 382–83, 397n1; issues of, for biosimilar firms, 425–26; safety-related, 96, 332; Supreme Court and issues of, 205, 334. *See also* off-label promotion; off-label promotion, regulation of
Laboratory of Neuro Imaging (LONI), 129
Laetrile. *See* amygdalin
Latino, 508, 509, 510, 511
laws: by Congress, 443; about FDA decisions, 28–29
legacy product programs, 27, 27–28
legislative action, for postmarket surveillance, 248
legislative authorities, 42; DQSA, 45–46; FDASIA, 5, 44–45; FSMA, 43–44
Lemmens, Trudo, 9–10, 268–85
Leukemia & Lymphoma Society, 72
Levitt, Geoffrey, 11, 328–42
Li, Rebecca, 6, 109, 115–33
liability: of individuals, 168; *Park* doctrine theory of, 166, 168; white-collar criminal, 7
license: for biosimilar interchangeability designations, 421. *See also* biologic license applications

Lietzan, Erika, 13, 414–30
life-cycle, of products, 1, 10
life-cycle approach, to drug approval, 269; funding decisions, 279–81; for market access, 275–78; new therapies access, 272–75; postmarket evidence generation, 278–79
life-cycle approach, to regulation, 286
life sciences, 93
Limited Population Antibacterial Drug (LPAD) Pathway legislation, 258–59
literature: about conflicts of interest, 137–38, 142; about negotiated rule making, 388; scientific, 480
litigation, First Amendment and, 213
LONI. *See* Laboratory of Neuro Imaging
Lotronex, 68
LPAD. *See* Limited Population Antibacterial Drug Pathway legislation
Lymphoma Foundation of America, 72

macromolecules, 407
Madden, Debra, 68
Makena, 12, 366–67, 372, 373, 374, 378–79, 379n1
management, of FDA: commissioners for, 20–21, 26; of employees, 20–21; Field Force, 18, 19; Headquarters, 18, 20; medical device industry, 18, 29n3; resources of, 19, *19*, 20
mandatory disclosure, 117, 123
mandatory exclusion, 150
manipulation: of autologous stem cells, 462–63; by government, 234–36

manufacturers: of biosimilars, 418–20; health information from, 62; spoken intent of, 204–7
manufacturing plants, biosimilar costs of, 419
marketing limitations, 257–59
markets, 1; approval and coverage of, 85–86; incremental approach to access of, 275–78; niche, 269, 270–71; old drugs, 317–18; unmaking of, 313–17; withdrawals from, 313–17, *314–15*, 321. *See also* postmarket; premarket approval
mass protest, against FDA, 63
mastectomies, 477, 479
McGill University system, 488
MedGuide, 333
Medicaid, 88, 146, 371
medical advances, 2
Medical Device Amendments, 22, 451
medical devices: categories of, 475–76; Class I, 445, 475, 477; Class II, 445, 447, 475, 476, 477, 478; Class III, *27*, 435, 473, 475–76, 477, 478, 479, 480; computerized, 433, 434; genomic information as, 470, 475; industry for, 18, 29n3
Medical Electronic Data Technology Enhancement for Consumers' Health (MEDTECH), 443
medical errors reduced, 440
medical knowledge regulation, 450–51
medical malpractice, 496–97
medical products, 84, 186, 487–88, 498n2
medical record costs, of e-prescription system, 495, 498n6

Medicare, 146, 371
medications, off-label, 493
medicine: access to, 85;
    computerized, 339–42,
    439–42; evidence-based, 253;
    personalized, 1, 2, 3, 94, 97,
    99, 434; practice of, 463–64;
    predictive, 99; preventive, 94–95,
    99; regenerative, 14, 434.
    See also prospective medicine;
    regenerative medicine
Medimmune v. Genentech, 411
MedSun, 493
MEDTECH. See Medical Electronic
    Data Technology Enhancement
    for Consumers' Health
Medtronic, 128–29
MedWatch, 289, 330, 492, 494,
    495
Mello, Michelle, 8, 184–203
Merrill, Richard, 93
mesenchymal stromal cells (MSCs),
    456, 457, 460, 461
Miller, Frances H., 433–37
minerals. See vitamin
Mini-Sentinel, 96
minority guidelines, of NIH, 502
misbranding, 149, 159nn1–2, 187,
    206, 207–8, 212
misconduct, 173n3; of corporate
    officers, 111; deterrence of, 163,
    164; economical attractiveness of,
    164; of individuals, 168
misdemeanor prosecutions, 147,
    166, 167, 171
misleading speech, 192–93, 229–30
misuse: of exclusions, 154–56; of
    participant-level clinical trials
    data, 116, 118
mobile health, 1, 13
models, 115, 117, 274, 339
Moffitt, Susan, 10, 306–27

molecules, 293; biosimilars and,
    414; DNA, 2; small, 1, 13, 345,
    402, 403, 404, 405, 406–9. See
    also orphan drug, from recycled
    molecule
Momenta v. Amphastar, 350
monetary penalties, 163–65, 168
monitoring: of e-prescription
    system, 486, 488, 491–92; for
    safety, 36
monoclonal antibodies, 421
moral blameworthiness, 152–59
moral responsibility, 156–57
MoreMarrowDonors.org, 181
Morris, Lewis, 165
MSCs. See mesenchymal stromal cells
multinational clinical trials, 38
multiple financial ties, 134–44;
    exclusive ties vs, 141–42
multiple physician-industry ties,
    140; theoretical accounts of,
    138–39
multiplicitous and multi-modal
    regulation, 87
Multi-Regional Clinical Trials
    Center at Harvard University, 6
Mutual Reliance Initiative, 51–52
mutual reliance/recognition, 51–52

name issues, with biosimilar
    interchangeability designations,
    422
NAS. See National Academy of
    Sciences
NASA. See National Aeronautics and
    Space Administration
NAS-NRC, 309–13, 316, 317, 318–
    19, 320, 323nn4–5, 323nn7–8
National Academy of Sciences
    (NAS), 11, 320
National Aeronautics and Space
    Administration (NASA), 81

National Cancer Institute, 503
National Health Federation (NHF), 63
National Heart Blood and Lung Institute (NHBLI), 129
National Institute for Health and Clinical Excellence (NICE), 84
National Institutes of Health (NIH), 66, 71, 120, 297, 368, 376, 439–40; minority guidelines of, 502; Statement on Sharing Research Data of, 121
National Library of Medicine, 439
National Research Council (NRC), 11
Native Hawaiian, 504
Nature Biotechnology, 502
NBIB v. Sebelius, 182
NDA. *See* New Drug Application
negative clinical trials, 336–37
negotiated rule making, generics and: better rules with, 394; committee for, 387, 389–96, 397*n*4; conclusion to, 396; cost of, 395–96, 398*n*9; faster results with, 394; introduction to, 382–83; literature about, 388; need for, 395; preemption and drug regulation, 383–85; in public interest, 387–88, 393
negotiated rule making, generics and, case for: concrete issues, 390–91, 397*nn*5–7; countervailing power, 393, 398*n*8; impacted interests, 391–92; need for rule, 389–90; potential benefits, 393–94; potential gains, 392–93; response to criticism, 395–96
Negotiated Rulemaking Act of 1990 (NRA), 387, 396
Neupogen. *See* filgrastim

new direction, in regulation, 86–89
New Drug Application (NDA), 65–67, 68, 186, 214; categories of, 313, 323*n*10; process of, 312, 318, 347, 353
new drugs: comparison studies of, 80; safety and efficacy of, 79, 80, 83, 185–86; user fees and approval system for, 24–26. *See also* abbreviated new drug application; investigational new drug; New Drug Application
new regulatory framework, 451, 453; clinical software definition, 452; Office of Software Devices created, 452; software device differences, 452
new therapies access, 272–75
NHBLI. *See* National Heart Blood and Lung Institute
NHF. *See* National Health Federation
NICE. *See* National Institute for Health and Clinical Excellence
niche markets, 269, 270–71
NIH. *See* National Institutes of Health
NIH Revitalization Act of 1993, 502, 503, 505
nonexpressive conduct regulation, 226–28
nonresponse, 98
nonunanimous votes, 142
notice of proposed rulemaking (NPRM), 383
NPRM. *See* notice of proposed rulemaking
NRA. *See* Negotiated Rulemaking Act of 1990
NRC. *See* National Research Council

OECD. *See* Organization for Economic Co-operation and Development
Office of Inspector General (OIG), 150, 151, 153, 154, 155–56; *Park* doctrine and, 169–70; public health protection of, 65, 169–70, 171; REMS and, 277; report of, 369–70
Office of International Programs (OIP), 46
Office of Management and Budget (OMB), 15; race and ethnicity standards of, 503, 504–12
Office of Pharmaceutical Quality, 353
Office of Public Health and Trade (OPHT), 48
Office of Regulatory Affairs (ORA), 46, 47
Office of Science and Engineering Laboratories, 444–45
Office of Software Devices, 452
Office of Special Populations, 503
Office of the National Coordinator for Health Information Technology (ONC), 13, 452, 453
off-label discussion, 237
off-label medications, e-prescription system and, 493
off-label promotion, 1, 146; as false or misleading speech, 192–93, 229–30; First Amendment and, 8–9, 179, 180, 184, 185, 192, 215; rationale for suppression of, 222–23; safe harbors for, 197–99, 215; *United States v. Caronia* relating to, 179–80, 184, 189–91, 227, 248, 261, 262, 294–95
off-label promotion, First Amendment and ban of, 9, 242; commercial speech doctrine and, 223–32; conclusion to, 240–41; dilemma of, 219–21; free expression and, 233–40; introduction to, 219–21; regulatory framework, 221–23; unconstitutionality of, 221
off-label promotion, regulation of, 8, 201–3; *Caronia* case, 184, 189–91; commercial speech and the courts, 187–89, 194–95, 206, 213; conclusion to, 199–200; future of, 191–99; immunity relating to, 211, 290; off-label use and, 8, 185–87
off-label use, 8, 185–87, 205–6; accepted medical practice and, 221–22; ban of, 222; value of, 222
OIG. *See* Office of Inspector General
OIP. *See* Office of International Programs
old drugs market, 317–18
oleaginous substance conviction, 206
O'Leary, Patrick, 7–8, 109–10, 162–75
OMB. *See* Office of Management and Budget
ONC. *See* Office of the National Coordinator for Health Information Technology
Oncologic Drugs Advisory Committee, 68
openSNP, 474
operational overview, of e-prescription system, 486, 489–92
OPHT. *See* Office of Public Health and Trade
ORA. *See* Office of Regulatory Affairs

Orange Book, 409, 410
Organization for Economic Co-
    operation and Development
    (OECD), 92
orphan drug, from recycled
    molecule: access to, 369–71,
    374–78; approach to, 371–73;
    conclusion to, 378–79; cost
    of, 369–71; designation of,
    372–73; drug price cap, 375–76;
    exclusivity period, 372, 374–75;
    FDA and, 376–78; introduction
    to, 366–68; limited competition,
    375; R&D for, 367, 368–69,
    370, 375, 378; 17P, 371–73
Orphan Drug Act, 12, 270, 366,
    367–69, 374, 375, 376–78
Orphan Medical, 189–90
OTC Drug Review, 22, 26
Other Pacific Islander, 504
ovarian cancer, 480
override feature, with e-prescription
    system, 490
oversight, of FDA, 39, 40, 77
OxyContin, 151

Pan American Health Organization
    (PAHO), 51, 54, 55
paper prescriptions, 486, 497*n*1
Parasidis, Efthimios, 10, 286–99
Park, John R., 149
Park doctrine, 174*n*4; abuse of,
    169; case selection under,
    170–71; concerns about, 166,
    167–70; consistency of, 172;
    exclusion under, 165, 166, 172;
    felony under, 166; liability theory
    of, 166, 168; misdemeanor
    under, 166, 167, 171; OIG and,
    169–70; prosecution under, 165,
    166, 167, 170, 172
Parkinson's disease, 435

Parkinson's Progression Marker
    Initiative (PPMI), 127
paroxetine (Paxil), 187
participant-level clinical trials data,
    119–21, 125–30; access to, 118;
    misuse of, 116, 118; privacy and
    confidentiality with, 116, 123;
    reidentification relating to, 116–
    17; sharing of, 115–16, 122–24;
    trends towards, 122–24
partnerships, 54–55. *See also* global
    collaborations and partnerships
Patent and Trademark Office
    (PTO), 409, 410, 412
patent dances. *See* statutory
    exclusivities, patent dances and
patent information, 405
Patent Office, 17–18
patent provisions, 416–17
patents, 1, 12–13, 210, 350, 354
pathway, of BPCIA, 403–4
patient-centered ethos, 60, 69–70
patient-centered initiative, 70
"Patient-Focused Drug
    Development," 70
Patient Package Insert, 332
"Patient Participation in Medical
    Product Discussion," 69
Patient Protection and Affordable
    Care Act of 2010. *See* Affordable
    Care Act of 2010
Patient-Reported Outcome (PRO)
    Consortium, 68
patients: advocacy for, 35, 36, 59;
    involvement of, 94; protection
    of, 1, 37; records for, 488, 490.
    See also empowered patient, rise
    of
patient-to-patient dialogue, 2–3
Paxil. *See* paroxetine
payer sophistication, 85
penalties, 163–65, 168

permissive exclusions, 150, 171
personalized medicine, 1, 2, 3, 94, 97, 99, 434
PFDA. *See* Pure Food and Drugs Act
Pfizer, 110, 164
Pham-Kanter, Genevieve, 7, 112–13, 134–45
pharmaceutical companies: enforcement against, 109–13, 146; felonies of, 146–47; as profitable, 296; settlements by, 110, 163–64
pharmaceutical manufacturing today, 344–46
pharmaceutical R&D, 15, 362–63, 364
pharmaceutical regulation, of FDA, 87
Pharmaceutical Research and Manufacturers of America (PhRMA), 109, 337, 507, 512
pharmaceuticals, 39, 78, 83, 85–86
Pharmacia, 507, 508, 509, 511, 512
pharmacogenetic discovery, 102
pharmacogenetic testing, 93, 97
pharmacogenomic drug products, 276–77
pharmacological action, 97–98
PhRMA. *See* Pharmaceutical Research and Manufacturers of America
PHSA. *See* Public Health Service Act
physicians: e-prescription and, 494, 498*n*5; prescribing, 495–96; records of, 488. *See also* multiple physician-industry ties
Physicians' Desk Reference, 62
Pike, Elizabeth R., 14, 470–85
The Pill Book, 62
placebos, 80, 84

platforms: academic-private partnership, 128–29; company-specific, 126–27; disease-specific, 127–28; public organization-specific, 129
PLIVA, Inc. v. Mensing, 382, 384–85, 386, 389–90, 391, 395
pluripotent cells, 457, 459
PMA. *See* premarket approval
policies: of EMA, 123–24, 258, 274, 337, 340*n*6; makers of, 119; Medicaid reimbursement, 88. *See also* management, of FDA
policies, of FDA, 119; conclusion to, 29; enforcement, 24; introduction to, 17–18; legacy product programs, *27*, 27–28; rulemaking, 21–22; science, law, and, 28–29; use of guidance, 22–23; user fee impact, 24–27
Policy Advisory Committee, 11, 310
policy problems, of food and drug, 81–86
pooled results, for data, 116
pooling, with racial and ethnic categories, 502
"Population Pharmacokinetics," 503
positioning, for global and innovative regulatory environment: FDA's international presence, 48–50; FDA's program alignment, 47; global collaborations and partnerships deepening, 50; GO, 46–47; proactive approach on trade development, 47–48; regulatory systems strengthened, 54–55
postmarket activities costs, 424–26
postmarket analysis: drug review process and, 287–91; FDA's public health imperative for, 287–97

postmarket evidence generation, 278–79
postmarketing efficacy studies, 100–102
postmarketing period effectiveness, 97–99, 100
Postmarketing Studies and Clinical Trials (Guidance), 99–101
postmarket organization infrastructure, 424–26
postmarket risk, 290
postmarket safety measures, 261–62
postmarket studies, 293
postmarket surveillance, 9–10; legislative action for, 248; passive framework for, 288–90, 292, 294; regulation, 248–50; system under, 288
postmodern world, 76–77; conclusion to, 89; FDA century, 78–81; food and drug policy problems, 81–86; new direction in, 86–89
power: of enforcement, 7–8; for negotiated rule making, 393, 398n8
PPMI. See Parkinson's Progression Marker Initiative
practice, of medicine, 463–64. See also accepted medical practice, off-label use and
PRAIS. See Regional Platform on Access and Innovation for Health Technologies
predictive medicine, 99
predictive technologies, 93
preemption, drug regulation and, 383–85
pregnancy testing, 97
premarket approval (PMA), 9, 95, 247, 435, 444, 476

premarket clinical trials, 95, 247–48, 254–55
premarket investments, for biosimilars, 418
premarket regulatory authority, 361–62
premarket syndrome, 9–10; addressing of, 268–71; conclusion to, 281; life-cycle approach, to drug approval, 269–81; in niche markets, 269, 270–71; trends relating to, 269
prescribing physicians, 495–96
Prescription Drug User Fee Act, 26
Prescription Drug User Fee Amendments of 2012, 70
prescriptions: for drugs, 63, 88; paper, 486, 497n1. See also e-prescription
Preventing Regulatory Overreach To Enhance Care Technology (PROTECT), 443
prevention-oriented standards, with FSMA, 43–44
preventive medicine, 94–95, 99
preventive technologies, 93
Price, W. Nicholson, 11, 247–50, 343–58
pricing, 78, 82, 86
Priority Review designation, 272
privacy: confidentiality and, 116, 123; e-prescription system costs of, 496, 498n7
PRO. See Patient-Reported Outcome Consortium
probation, of corporate officers, 155
procedural barriers, 348–49
Procrit. See epoetin alfa
product information, from biosimilar firms, 425

products: adulterated, 42, 79, 82, 361; animal, 3; benefits of, 36, 37; biological regulation of, 3; cost of, 76; efficacy of, 162; FDA-regulated, 39–40, 80, 93; food, 81; information-only, 471, 474–75, 477, 478, 480–81; life cycle of, 1; medical, 84, 186, 487–88, 498n2; pharmacogenomic drug, 276–77; risks of, 36, 37; safety of, 76, 162; therapeutic, 77, 78. *See also* legacy product programs
product-specific focus, 82
progesterone, 253
prognosis, of diseases, 93
Program Alignment Group, 47
Project Data Share, 128
Project Inform, 65
Promethease, 474
PROMIS, 439
promotion: of unapproved drugs, 7, 8. *See also* off-label promotion; off-label promotion, regulation of
promotional claims, for dietary supplements, 208–9
promotional claims, truth test for: court ignorance, 211–14; FDCA as truth test, 214–16; FDCA relating to, 210–11; manufacturer's spoken intent, 204–7; presumptions of truth, 207–9, 212
propoxyphene, 487
prosecutions: criminal, 171; misdemeanor, 147, 166, 167, 171; under *Park* doctrine, 165, 166, 167, 170, 172
prospective medicine, 6, 35, 37; attributes of, 93; challenges of, 93–95; conclusion to, 103; FDAAA regulatory powers, 95–97; FDA powers, under FDAAA, 99–103; introduction of, 92–93; postmarketing period effectiveness of, 97–99, 100
PROTECT. *See* Preventing Regulatory Overreach To Enhance Care Technology
PROTECT Act, 451
protected expression, 233
protection: constitutional, 236–37; of consumers, 228–29; in First Amendment, 233, 235; of free expression, 221, 233, 236–37; of patients, 1, 37; of public health, 7–8, 65, 69, 162–74
protection safeguards, for data, 117
PTO. *See* Patent and Trademark Office
PTO Patent Trial and Appeals Board, 412
Public Citizen, 163, 173n1
public disclosure, 293
public health: advancement of, 162; enhancing of, 69; expense of, 165, 173n2; prioritizing of, 170–72; protection of, 7–8, 65, 69, 162–74
Public Health Service Act (PHSA), 434, 458, 460, 461–62, 465
public interest, of negotiated rule making, 387–88, 393
publicity, with enforcement, 24
public organization-specific platforms, 129
public trust, 6
Purdue Pharma case, 150–52, 155, 156
Pure Food and Drugs Act (PFDA), 78, 79

QSR. *See* Quality Systems Regulations
quality, of biosimilars, 420

Quality-by-Design, 353
Quality Systems Regulations (QSR), 445
questionable accuracy, 509–11, 512

race, FDA and, 15, 501–3
racial and ethnic categories, in biomedicine, 437; accuracy of, 509–11, 512; American Indian or Alaska Native, 504, 510; appropriate use of, 501; Asian, 504, 510; Black or African American, 504, 508; concerns with, 502; data for, 504; federal mandates relating to, 502–4; genetic variations with, 502; guidelines for, 504–7; harmonization of, 505, 513; inconsistent definitions for, 507–8; Native Hawaiian or Other Pacific Islander, 504; OMB standards for, 503, 504–12; pooling with, 502; surrogates relating to, 501; White, 504
Raggi, Reena, 190
Rai, Arti, 12–13, 402–13
randomized, controlled clinical trials (RCTs), 101–3, 330
RAPS. *See* Regulatory Affairs Professional Society
RCTs. *See* randomized, controlled clinical trials
R&D. *See* research and development
recall: of drugs, 343; of software, 442
Recall Requests, 24
recruitment. *See* global trial recruitment
recycled molecule. *See* orphan drug, from recycled molecule
Redish, Martin H., 9, 219–43
reference listed drug (RLD), 385, 397*n*2
reform proposal, of biomedical industry, 167–72
regenerative medicine, 14, 433, 434, 455; cell sources, 456–57; cell therapy as drug, 462–63; Commerce Clause, 464–65; conclusion to, 467; contamination with, 461–62; definition of, 456; FDA going forward, 465–67; future issues of, 465; natural human substances used for, 456; practice of, 463–64; safety issues with, 457; tissue products relating to, 459
regenerative medicine, regulation of, 458; challenges to, 460–65; FDA with, 459–60
Regenerative Sciences, 460–61, 463, 464, 465
Regenerative Sciences stem cell procedure, 435–36
Regenexx, 461, 465
Regional Platform on Access and Innovation for Health Technologies (PRAIS), 51
registration requirements, 120–21
regularity exclusivity, 352, 354
regularity function, for e-prescription, 486, 487
regularity lock-in, 349
regularity multiplicity, 78
regularity statutes, 78
regulating drug efficacy, in retrospect, 308–13
regulation pathway, for genomic information, 480–81, 482*n*3
regulations: of biological product, 3; CBE, 384, 385–86; of digital world, 445; of FDA, 3, 4, 7; innovation vs, 447–48; to life-cycle approach, 286;

regulations (*continued*)
multiplicitous and multi-modal, 87; new direction in, 86–89; of off-label promotion, 8, 184–203; by payment, 87; of postmarket surveillance, 248–50; of regenerative medicine, 458–65; of speech, 212, 237–39; of stem-cell therapies, 1. *See also* Food and Drug Administration
regulations, for biosimilars, 364–65; costs of, 418–20, 424–26
regulations, of drugs, 1–2, 10, 80, 383–85; generic, 414–30
regulations, of genetic testing, 1, 14; DTC, 472–73
regulators, 162, 164, 165, 212; approval of, 127, 135; of communication, 329, 335, 340*n*4; data relating to, 119, 122; FDA and, 52–53
Regulatory Affairs Professional Society (RAPS), 55
regulatory agencies, 46–47, 52–53
regulatory authority: active postmarket analysis, 292–95; health information technology and, 292–95; premarket, 361–62; truth with, 303–4
regulatory barriers, 347–49
regulatory changes, 352–53
regulatory competencies, 78, 82
regulatory compliance, of biosimilar firms, 424–26
regulatory environment, 5. *See also* global and innovative regulatory environment
regulatory failures, 76
regulatory framework, 78, 94, 95, 221–23. *See also* new regulatory framework
regulatory goals, 77

regulatory issue, with biosimilar interchangeability designations, 420
Regulatory Letter, 22
regulatory-naïve industry, 450
regulatory powers, under FDAAA, 95–97
regulatory priorities, 5–6
regulatory standards, of FDA draft guidance, 417–18
regulatory systems, 54–55
reidentification, 116–17
REMS. *See* Risk Evaluation and Mitigation Strategy
requirements: of FDCA, 205, 307; registration, 120–21
research: data on, 330, 340*n*1; DESI as tool for, 319–22; FDA's focus on, 136–37
research and development (R&D), 86; costs of, 210, 263, 280; incentives for, 254, 262; increase in, 346; for orphan drug, 367, 368–69, 370, 375, 378; pharmaceutical, 15, 362–63, 364; restructure of, 309
resources, 19, *19*, 20
responsibility, moral, 156–57
responsibility, of corporate officers, 149–51; moral blameworthiness of, 152–59; subordinates of, 156–59
restriction, of information, 94
reverse payment settlements, 410
rights revolution, 61
Riley, Margaret Foster, 14
risk-benefit analysis, 59, 98
risk classification, of genomic information, 475–77
Risk Evaluation and Mitigation Strategy (REMS), 71, 81, 249, 261; goals of, 289; safety relating

to, 96, 97, 98–99, 276–77;
surveillance and, 255, 256, 293, 294
risk mitigation strategies, for genomic information, 478–79
risks: analysis system and identification of, 95–96; of data transparency, benefits, issues, and, 116–19; of diseases, 476; postmarket, 290; of products, 36, 37; serious, 100
RLD. *See* reference listed drug
Robertson, Christopher, 8, 109–14, 204–17
rofecoxib (Vioxx), 35, 70, 77, 186, 254, 275, 335–36
Roin, Benjamin N., 361–65
Rosenberg, David, 14–15, 486–500
rosiglitazone (Avandia), 187, 254, 486
Ruger, Theodore W., 5–6, 37, 76–91
rule making. *See* negotiated rule making, generics and
rules: better, 394; of exclusions, 110–12

saccharin, 64
SACGHS. *See* Secretary's Advisory Committee on Genetics, Health, and Society
safe harbors, for off-label promotion, 197–99, 215
safety, 78; communicating about, 11; cost and, 80; of drug supply chain, 40–42, *41*, 44, 45; and effectiveness, of drugs, 3, 11, 309, 318; under FDAAA, 96–97; of food, 77, 82, 89; of genomic information, 480, 481; monitoring for, 36; of new drugs, 79, 80, 83, 185–86; of products, 76, 162; with regenerative medicine, 457; speed and, 9, 35, 36
SAFETY Act, 451
safety-related labeling, 96, 332
Salmonella, 77
Sandoz, Inc. v. Amgen, Inc. et al, 406, 410
Schmidt, Eric, 303
schools food, 88–89
science, policy, and law, about FDA decisions, 28–29
scientific focus, of FDA, 81
scientific literature, 480
second generation applications, 416
Secretary's Advisory Committee on Genetics, Health, and Society (SACGHS), 472
Sentinel System, 96, 99, 288–89, 298, 331, 493, 494
serious adverse drug experience, 100
settlements: by pharmaceutical companies, 110, 163–64; reverse payment, 410
17-alpha hydroxyprogesterone caproate (17P) (Delalutin), 371–73
sharing: of information, 51, 249. *See also* data
shortages, of drugs, 343
Shulman, Robert, 318–19, 324n19
Sidelines National Support Network, 372, 373
Sklamberg, Howard, 5, 36, 39–58
small molecules, 13, 345, 402, 403, 404, 405, 406–9
smoking, 87
software, 448–49; clinical, 452; recalls of, 442
software, devices of: differences in, 452; errors with, 442
Sorrell v. IMS Health, 188–89, 261

sources, for drug safety communication, 332–34

Special Controls, 476, 478

Spector-Bagdady, Kayte, 14, 470–85

speech: as false or misleading, 192–93, 229–30; regulations of, 212, 237–39. *See also* commercial speech

speed, safety and: accelerated approval relating to, 255–56; emphasis on, 252–56; problems with, 254–55

speed, safety vs, 9, 35, 36, 264–67; accelerated approval, 252; advertising, 259–61; commercial-speech jurisprudence, 261–62; conditional approval, 262–63; fast track approval for, 251–52, 272; final comments, 263; marketing limitations, 257–59; postmarket safety measures, 261–62; speed and safety, 252–56

spoken intent, of manufacturers, 204–7

sponsors, 69, 138, 140–41, 142, 143; active postmarket analysis and, 10, 287, 288–89, 290, 291, 292, 293, 294, 295, 296, 297–98; data sharing with, 116, 117, 118, 119; drug, 334–38, 339; industry, 126–27, 128; responsibility of, 194, 249–50, 258, 259, 280; trials and, 120, 124, 135

spontaneous adverse event reporting, 330

staggered approval model, 274

standardization, 306–8, 323n1

standards: for biosimilar regulation, 417–20; categorical, 224–25, 229; of FDA, 3; gold, 252, 253; homologous use, 166; of identity, 81; prevention-oriented, 43–44; for race and ethnicity, 503, 504–12; technological, 348. *See also* regulatory standards

Stanford University Medical Experimental Computer-Artificial Intelligence in Medicine (SUMEX-AIM), 440

Statement on Sharing Research Date, 121

statutory exclusivities, patent dances and: BPCIA exclusivities, 403–6, 407, 409, 410–11; evaluation of, 406–12; introduction to, 402–3

stem cells: adult, 456, 460; clinic treatments of, 460–61; hESCs, 456–57, 458; HSCs, 456; Regenerative Sciences procedure for, 435–36. *See also* autologous stem cells

stem cells, therapies of: delays of, 466; regulation of, 1

stem cell tourism, 465

strategic options, for biosimilars, 426–28

strokes, 486

studies. *See specific studies*

subordinates, 156–59

Subpart E regulation, 66

subsidization, of food, 88

substantiated drug safety communication, 330–31

substitution, of biosimilar interchangeability designations, 421–22

sulfanilamide, 35

SUMEX-AIM. *See* Stanford University Medical Experimental Computer-Artificial Intelligence in Medicine

supplements, vitamin and mineral, 63, 64

supply, of food, 77
supply chain, drug, 40–42, *41*, 44, 45
suppression, rationale for, 222–23
Supreme Court, 8, 79, 112, 169; commercial speech and, 187–89, 194–95, 206, 213, 224–26; decisions of, 12, 87, 148, 149, 166, 316, 407, 410, 411, 426; First Amendment and, 9, 180, 185, 207, 214, 216, 233, 236, 238; labeling issues and, 205, 334; truthfulness and, 207, 208, 209
surrogates, for genetic groupings, 501
surveillance: REMS and, 255, 256, 293, 294. *See also* postmarket surveillance
Symms, Steven D., 64–65
system of disease, 92

targeting, of corporate officers, 165–67
Task force on Computers and Software as Medical Devices, 444
tbo-filgrastim, 427
technical focus, of FDA, 81
technological standards, 348
technologies: advances for, 434; complicated regulations of, 433; emerging, 13; enhancing, 14; for health information, 292–95; predictive, 93; preventive, 93
telithromycin (Ketek), 187
teratogenic drugs, 97
testing: of genetics, 1, 14, 93; for HIV, 476, 477, 479; pharmacogenetic, 93, 97; pregnancy, 97. *See also* direct-to-consumer genetic testing
thalidomide, 19, 35, 252–53

themes and development, of FDA. *See* policies, of FDA
Therac-25, 444
therapeutic products, 77, 78
therapies. *See specific therapies*
Thompson v. W. States Med. Ctr., 225
Title II, of DQSA, 45–46
Title VII, of FDASIA, 44, 45
tobacco, 3, 18, 19, 45, 47, 82, 87
tort claims: with generic drugs, 382; limitations of, 290–91
tort liability, of biosimilar firms, 426
totality of evidence, with biosimilarity, 417
totipotent cells, 457
Trade Correspondence, 22–23
trade development, 47–48
trade secrecy, 351, 354
transparency, 279, 337, 338, 354. *See also* data
treatment effects, of heterogeneity, 99
Treatment IND rule, 65
treatments: autologous MSC, 465; off-shoring of, 465; response of, 102; at stem cell clinics, 460–61
trends, towards participant-level data sharing, 122–24
trials: sponsors and, 120, 124, 135. *See also* clinical trials
triazolam, 487
trimetrexate, 66
TRIPS. *See* International Agreement on Trade-Related Aspects of Intellectual Property Rights
trust: decline of, 60–61; with drug safety communication, 338; public, 6
truth: with regulatory authority, 303–4. *See also* promotional claims
truthfulness concept, 207, 208, 209
tumorigenicity, 457

twelve-year exclusivity period, 428
23andMe, 435, 470, 471, 473, 474

Udell, Howard, 151
unanimous votes, 142
unapproved drugs, 7, 8, 60
unconstitutionality, of off-label promotion, 221
uncurated data, 117
United States Adopted Names (USAN) Council, 422
United States Department of Agriculture (USDA), 18, 88
United States v. Caronia: decision in, 191–92, 193, 196–97, 199, 208, 240; off-label promotion relating to, 179–80, 184, 189–91, 227, 248, 261, 262, 294–95
United States v. Dotterweich, 148–49, 166
United States v. Park, 166
United States v. Regenerative Sciences, 434
United States v. The Purdue Frederick Co., Inc., 151
United States v. Two Barrels Desiccated Eggs, 79
unmediated data, 117
Untitled Letters, 470, 473, 474
Upjohn Company, 316
USAN. *See* United States Adopted Names Council
USDA. *See* United States Department of Agriculture
user fees: authority for, 336, 337; for biosimilars, 26, 44; drug application, 331
user fees, impact of: on coal-tar colors, 24; on new drug approval system, 24–26; on programs, 26; repair of, 27
U.S. Sentencing Commission, 45

vaccines, 2
validity, of genomic information, 480
valid safety communication, 330–31
Vioxx. *See* rofecoxib
Virginia State Board of Pharmacy v. Virginia Citizens Consumer Council, Inc., 223
vitamin: deregulation of, 63; and mineral supplements, 63, 64
Vitamin-Mineral Amendments of 1976, 63
voluntary industry, 125–29
votes: of FDA, 109–10; nonunanimous, 142; unanimous, 142
voting behavior and types of financial ties, 143

warfarin, 471
Warning Letters, 24, 473, 479
Washington Legal Foundation, 179–80, 190
WebMD, 2, 62, 293
Weinberger v. Bentex Pharmaceuticals, Inc, 316
Weinberger v. Hynson, 316–17
Westcott & Dunning; USV Pharmaceutical Corp v. Weinberger, 316
Western States Medical Center, 209
WHI. *See* Women's Health Initiative
White, 504
white-collar criminal liability, 7
WHO. *See* World Health Organization
withdrawals, from markets, 313–17, *314–15*, 321
Women's Health Initiative (WHI), 253
world: changing, 5. *See also* postmodern world
World Bank, 54

World Health Assembly, 54
World Health Organization (WHO), 54, 55, 120, 314, 422
World Wide Web, 2, 62
Wyeth v. Levine, 382, 383, 384

Yale University Open Data Access (YODA), 128–29
Yan, Huaou, 14–15, 486–500
YODA. *See* Yale University Open Data Access